About the Author

Danuta Kruk received her Ph.D. from Jagiellonian University in Cracow. She spent several years at Physical Chemistry Arrhenius Laboratory in Stockholm and at Solid State Physic Department of Technical University in Darmstadt working on theoretical aspects of Nuclear Magnetic Resonance and Electron Spin Resonance. In Stockholm she was involved in a development of nuclear and electron spin relaxation theories for molecular liquid systems containing transition ion complexes. During her stay in Darmstadt her interest was focused on solid state systems containing quadrupolar nuclei, in particular on theory of field dependent relaxation processes and polarization transfer phenomena. Now she is back at the Jagiellonian University and would like to share her experience with experimentalists concerned with problems of molecular dynamics and wishing to become conversant with the pertinent body of mathematics, and theorists interested in quantum mechanics combined with classical motion.

Theory of Evolution and Relaxation of Multi-spin Systems

Application to Nuclear Magnetic Resonance and Electron Spin Resonance

Danuta Kruk

Institute of Physics, Jagiellonian University, Krakow, Poland

Institut für Festkörperphysik, Technische Universität Darmstadt, Germany

Published 2007 by arima publishing

www.arimapublishing.com

ISBN 978-1-84549-176-5

© Danuta Kruk 2007

All rights reserved

This book is copyright. Subject to statutory exception and to provisions of relevant collective licensing agreements, no part of this publication may be reproduced, stored in a retrieval system, or transmitted in any form or by any means, without the prior written permission of the author.

Printed and bound in the United Kingdom

Typeset in Times New Roman

This book is sold subject to the conditions that it shall not, by way of trade or otherwise, be lent, re-sold, hired out, or otherwise circulated without the publisher's prior consent in any form of binding or cover other than that which it is published and without a similar condition including this condition being imposed on the subsequent purchaser.

abramis is an imprint of arima publishing

arima publishing
ASK House, Northgate Avenue
Bury St Edmunds, Suffolk IP32 6BB
t: (+44) 01284 700321

www.arimapublishing.com

*To my family:
Robert,
Sabina,
Przemek,
Karolina*

PREFACE

Modern Nuclear Magnetic Resonance (NMR) and Electron Spin Resonance (ESR) experimental techniques provide a great opportunity of studying details of molecular dynamics in condensed matter. To profit fully from the advanced experimental methodology, it is of primary importance to apply appropriate theoretical models, linking quantities observed in NMR and ESR experiments to dynamical and structural properties of molecular systems. The theoretical difficulties are connected with the complexity of investigated systems in which a quantum-mechanical component is in interaction with a complex environment. The purpose of this book is to present complete theories of spin evolution (in particular relaxation processes and polarization transfer effects) and apply them to multi-spin systems containing mutually coupled dipolar, quadrupole and electron spins. The formalism adopted here is based on the Liouville - von Neumann equation, the perturbation theory, and the Wigner-Eckart theorem. The text is intended to appeal to experimentalists concerned with problems of molecular dynamics who wish to become conversant with the pertinent body of mathematics, and to theorists interested in quantum mechanics combined with classical motion.

I feel that this book should benefit the readers because of some aspects which I would like to highlight.

In this book I take full advantage of the common Hamiltonian formalisms for quadrupole and zero field splitting interactions. It shows the readers how to apply the same methodology to essentially different problems.

The first few chapters deal with basic theories of relaxation processes, and provide in this way the background of further, more advanced treatments. I have assumed that a detailed treatment of a number of classical topics (such as the derivation of the Solomon – Bloembergen -Morgan theory) would be very instructive. Rather than choosing a formal and abstract method of presentation I have opted for a more applied method using examples of problems and systems of increasing complexity. In my view, such a method is the most appropriate way to introduce the reader into the subject. Relevant steps of derivations presented in this book are discussed in great detail. Thus, an acquaintance with this text should prepare the reader for a detailed study of more complicated problems. At the same time, the book presents a number of topics at an advanced level, far beyond textbook considerations.

The applicability of 'classical' relaxation theories presented in literature is strongly limited; in particular they break down at low magnetic fields and may not be applied to spin systems which exhibit relatively strong quadrupolar coupling or zero field splitting. This leads to misinterpretations of field dependent NMR experiments. Therefore, in this book, attention is given to providing theoretical approaches valid for an arbitrary magnetic field. Much effort is put into the explanations on why for the proper interpretation of experiments, it is absolutely necessary to apply the much more advanced and complex treatments.

I demonstrate in several examples that, within the framework of the second order perturbation theory, one can solve complicated problems resulting from interplay between various spin interactions. I have attempted to discuss in detail validity regimes of the presented perturbation treatments, that the reader can really understand their physical limitations. I regret to say that this subject is too often lacking in the literature.

I discuss a general theory of the evolution of spin systems based on a full solution of the Liouville - von Neuman equation by using a multipole representation of tensor operators and the Wigner-Eckart theorem. This approach is valid for arbitrary motional conditions and interaction strengths. Properties of this general theory make it possible to adapt it to situations when perturbation treatments break down. Such an approach is particularly important for studying slow and anisotropic motion. Unfortunately, because of computational and conceptual complexity this approach is presented relatively seldom in literature.

In presenting theoretical models I refer to appropriate experimental examples. I consider this aspect of my book as very beneficial for experimentalists. Working with experimental groups as a theorist I learned to appreciate the great opportunity to test theoretical models against experiments. Writing this book, I have tried to build a bridge between the experimental community and theorists describing 'the world' by abstract formalisms.

The theory is developed using a uniform formalism that stresses the unity of Nuclear Magnetic Resonance and Electron Spin Resonance. The Liouville operator formalism is employed throughout, and it constitutes the unifying feature of this book.

I believe that I have succeeded in finding an acceptably simple version of the mathematics needed for a rigorous treatment of the presented problems. Since the book is the outcome of research work, including recent research results, it could be used as a reference book for researchers working in the field of molecular dynamics.

January 2007, Kraków

Acknowledgments

My great debt of thanks goes to Prof. Robert Sharp and Prof. Rainer Kimmich for very careful reading of this manuscript and very useful comments and advises.

I am very grateful to Prof. Jozef Kowalewski for the very fruitful time in Stockholm when I learned a lot on relaxation processes in paramagnetic systems, for his support in the process of writing this book and his comments on the manuscript. Without him this book would not appear.

I would like to express my deep gratitude to Prof. Fujara for his unstinting support, encouragement and help.

My special thanks go to Prof. Alexander Gutsche for his support.

My thanks are due to Prof. Józef Mościcki for motivating me to complete this book.

I thank Prof. Jerzy Blicharski for getting me interested in NMR and relaxation processes, in particular.

I would very much like to thank the group in Darmstadt, who I have had the pleasure to work with for almost four years, in particular to Alexei Privalov and Patryk Gumann. Performing many beautiful experiments they gave me the opportunity to face theory with the complicated, real world.

I am also grateful to Prof. Hartmut Benner for careful reading of some parts of the manuscript.

My family, my husband Robert and my children: Sabina, Przemek and Karolina, made it possible to focus on the book to the exclusion of nearly everything else. They stood by me all this time. To them goes my gratitude. Thanks to Sabina for helping in the index preparation.

German Science Foundation (DFG) is gratefully acknowledged for financing this publication: Grant number FU309/9.

CONTENTS

Outline	*1*
I. Introduction	***8***
I.1. Nuclear Magnetic Resonance and Electron Spin Resonance (NMR and ESR)	*8*
I.2. Introductory aspects of relaxation	*10*
I.3. The concept of the density matrix	*13*
I.4. Time correlation functions and spectral densities	*14*
II. Perturbation theory of relaxation processes	***17***
II.1. The Liouville -von Neumann equation. Hilbert and Liouville space formalisms	*18*
II.2. Perturbation treatment of the Liouville – von Neumann equation	*22*
II.3. Matrix formulation of the Wangsness-Bloch-Redfield (WBR) relaxation theory	*31*
II.4. General representation of spin-lattice interactions and spectral densities	*35*
II.5. Relaxation theory within the Liouville operator formalism	*37*
II.6. Multipole representation of relaxation superoperators	*40*

III. Interaction Hamiltonians and motional models — 53

III.1. Interaction Hamiltonians — 54
III.2. Examples of spin systems and motional models — 60

IV. Examples of relaxation processes treated within the perturbation approach — 72

IV.1. Two equivalent spins 1/2 relaxing via mutual dipole-dipole coupling — 72
IV.2. Dipolar relaxation in systems of two non-equivalent spins ½ — 77
IV.3. Electron spin relaxation caused by fluctuations of zero field splitting (ZFS) tensor — 81
IV.4. Quantum-mechanical correlation function and the Solomon-Bloembergen- Morgan (SBM) theory — 84
IV.5. Relaxation processes in nuclear spin – electron spin system at low magnetic field. Effects of the zero field splitting — 97
IV.6. Effects of non-zero averaged quadrupolar coupling on relaxation of dipolar spin - quadrupolar spin systems — 108
IV.7. Validity conditions of the presented treatments — 116

V. Electron and quadrupolar spin relaxation: unified treatment — 122

V.1. Field dependent electron spin relaxation for axially symmetric ZFS and the spin quantum number S=1: the simplest example — 125
V.2. Theory of electron spin relaxation for slowly rotating systems of low symmetry — 128
V.3. Field dependent electron (quadrupolar) spin relaxation processes for spin quantum number 3/2 — 138
V.4. Multiexponential relaxation processes for high spin quantum numbers at low and high magnetic fields — 149

VI. Nuclear spin relaxation in the presence of neighboring electron spins for slowly rotating systems — 158

VI.1. Closed form description of nuclear spin-lattice relaxation at low and high magnetic fields for electron spin quantum numbers 3/2 and 7/2 — 159

VI.2. Theory of field dependent spin-lattice relaxation of nuclear spins 1/2 coupled to an electron spin: the general formulation — 173

VI.3. Nuclear spin relaxation in intermediate magnetic field range – examples of the semi-analytical treatment — 181

VII. Dipolar spin relaxation in solid state systems containing quadrupolar spins — 186

VII.1. Spin-lattice relaxation of spins 1/2 due to dipole-dipole couplings to quadrupolar spins in crystals — 186

VII.2. Examples of dipolar spin relaxation in the presence of neighboring quadrupolar spins — 197

VII.3. Comments on dipolar contribution to the quadrupolar spin relaxation — 202

VIII. Evolution of systems of mutually coupled spins under time independent interactions — 210

VIII.1. Time evolution of spin observables — 211

VIII.2. Polarization transfer processes within a two spin system — 212

VIII.3. Multi-quantum polarization transfer effects in systems containing several dipolar and quadrupolar spins — 222

VIII.4. Motional conditions of polarization transfer effects and relaxation enhancement — 227

IX. Motional conditions and relaxation processes in multi-spin systems: examples and warnings — 233

IX.1. Effects of fast modulations of ZFS tensor on electron as well as nuclear spin dynamics — 234
IX.2. Relaxation in spin systems of high symmetry at low magnetic field — 245
IX.3. Validity regimes of relaxation theories based on perturbation treatment — 253

X. Dynamics of spin systems beyond validity regimes of the second order perturbation theory — 263

X.1. General description of ESR and quadrupolar spectra — 265
X.2. Dipolar spin relaxation treated within the general formalism — 279

XI. Effects of neighboring dipolar spins on dynamics of quadrupolar spin — 290

XI.1. Quadrupolar spin relaxation due to dipole-dipole couplings to dipolar spins — 291
XI.2. Influence of dipole-dipole relaxation mechanism on quadrupolar spectra — 296
XI.3. Time evolution of spin observables in the presence of relaxation processes — 303

XII. Relaxation processes caused by translational diffusion of interacting spins — 314

XII.1. Analytical description of dipolar spin relaxation in the presence of translational diffusion — 315
XII.2. General treatment of relaxation mediated by translational diffusion — 329
XII.3. Effects of intermolecular forces — 338

XIII. Quantum vibrations as an origin of electron spin relaxation — 345

XIII.1. Quantum description of vibrational dynamics — 347
XIII.2. Electron spin spectral densities resulting from molecular vibrations — 350

Appendix — 362

List of selected symbols — 366

Index — 372

OUTLINE

Before we start, I would like to make some comments on the individual chapters of this book.

I. Introduction

The material presented in this book assumes knowledge of the fundamentals of quantum mechanics, NMR (Nuclear Magnetic Resonance) and ESR (Electron Spin Resonance) phenomena. Nevertheless, for completeness of the presentation I introduce the reader to some fundamental aspects of NMR and EPR. I discuss, in particular, the physical origin of relaxation processes. I outline also the concept of density operator in conjunction with those specific problems in which it is used in the following chapters.

II. Perturbation theory of relaxation processes

In this chapter I present two equivalent formalisms originating from early works on relaxation and being commonly used. The first one is referred to as WBR (Wangness – Bloch- Redfield) relaxation theory and is based on a matrix representation of density operators, while the second one is more closely related to the operator formalism proposed by Abragam, and is called multipole representation treatment. My goal is to provide a compact and clear description of the two formalisms for performing evaluations of practical usefulness. I do not restrict the considerations to systems where it is a good approximation to treat the lattice classically. On the contrary, I emphasize that especially the multipole representation formalism is very suitable for dealing with spin systems, which experience classical as well as quantum mechanical degrees of freedom of the environment.

III. Interaction Hamiltonians and motional models

This chapter contains a concise compilation of spin interactions and their Hamiltonians, knowledge of which is a necessary prerequisite for an understanding of the following chapters. The emphasis is on a uniform representation of zero field splitting (ZFS) and quadrupole interactions. Further topics treated in this chapter are some examples of systems under interest such as paramagnetic complexes or selected crystal lattices and some models of motion. In this way I explain definitions of static (averaged) and transient (fluctuating) parts of quadrupole and ZFS interactions, intensively used in the following chapters. Clear presentation of motional models leading to the definitions is crucial for understanding the role of these interactions (typically the static parts contribute to energy level structure, while the transient ones act as a perturbation). This chapter includes also a discussion of correlation functions (spectral densities) resulting from these motional models.

IV. Examples of relaxation processes treated within the perturbation approach

In this chapter the main emphasis is on the use of the theoretical tools presented in Chapter II for evaluating characteristic relaxation time (rate) constants. I demonstrate that the Redfield relaxation theory employing the density matrix formalism is appropriate for a subsequent treatment of mutually interacting spin systems of increasing complexity. I begin with 'classical' examples, such as dipolar relaxation of two equivalent spins 1/2 or the Solomon-Bloembergen-Morgan theory. Using the second example I demonstrate the equivalence of the formalisms discussed in Chapter II and introduce, in a practical manner, a concept of 'composite lattice'. The basis idea of the 'composite lattice' is to treat one of the participating spins as a part of the lattice and limit the spin part of the problem to the second spin. Next, I discuss how to adjust the perturbation treatment (usually applied to cases of high magnetic field) to describe properly relaxation processes at low magnetic field, when other interactions (like ZFS) dominate the energy level structure of, at least, one of the participating spins. This chapter introduces the reader into very important aspects of relaxation theories and provides a background for further more advanced considerations.

V. Electron and quadrupolar spin relaxation: unified treatment

In this chapter I focus attention on field dependent relaxation processes of quadrupole and electron spins of an arbitrary spin quantum number (≥ 1). I deal with a situation when the spins are placed in a molecular surrounding of rather low symmetry, and therefore they exhibit a non-zero averaged quadrupole coupling or zero field splitting, respectively. I demonstrate how to apply the Redfield relaxation theory to describe properly multiexponential relaxation dynamics of such spin systems for an arbitrary strength of the applied magnetic field. The reader has opportunity to get experience with non-trivial applications of the Redfield relaxation theory far beyond cases usually presented in text books.

VI. Nuclear spin relaxation in the presence of neighboring electron spins. Paramagnetic relaxation enhancement effects (PRE) for slowly rotating systems

This chapter contains an in-depth look at relaxation processes in a wide range of applied magnetic fields. Here I am concerned with spin systems containing nuclear spins coupled by dipole-dipole interactions to electron spins, which are characterized by a non-Zeeman energy level structure and possess their own (independent of the nuclear spin) relaxation mechanism, considered in detail in the previous chapter. The presence of the electron spin enhances the nuclear spin relaxation. The paramagnetic relaxation enhancement (PRE) is caused by random variations of the electron spin – nuclear spin coupling. Recently one can observe a significant progress in theories of these effects, stimulated to a certain extent, by fast development of field cycling techniques. Relaxation processes of the electron spin contribute to the fluctuations of the dipole-dipole coupling in a manner similar to other stochastic processes, such as rotational motion or exchange dynamics. Such systems are very demanding from the point of view of a proper theoretical treatment, however very attractive due to quite unique effects of the field dependent electron spin dynamics on the nuclear spin relaxation. The calculations are performed within the Liouville superoperator formalism, employing the concept of the composed lattice (Chapter IV), which includes the electron spin subsystem.

VII. Dipolar spin relaxation in solid state systems containing quadrupolar spins

In this chapter I deal with a theory of field dependent relaxation processes in solid state systems containing mutually coupled dipolar and quadrupole spins. The theory is formulated following the analogies between quadrupole interaction and zero field splitting. The approach is valid for an arbitrary magnetic field. It can be treated as a 'quadrupole' counterpart of the paramagnetic relaxation enhancement theory presented in Chapter VI.

VIII. Evolution of systems of mutually coupled spins under time independent interactions

This chapter concerns complex systems of dipolar and quadrupolar spins mutually coupled by time independent interactions. It is assumed that the evolution of the spin system is not affected by any other dynamic processes; in particular the spins do not relax. I present theory of polarization transfer by single- as well as multi-quantum pathways. I emphasize, that the mechanisms of the polarization transfer effects and the relaxation processes are essentially different. The polarization transfer originates from an efficient, time-independent coupling between participating spins, while the relaxation processes are caused by interactions fluctuating in time.

IX. Motional conditions and relaxation processes in multi-spin systems: examples and warnings

In this chapter the reader is faced with a situation where the electron (quadrupole) spin system does not fulfill the requirements of the second order perturbation theory and therefore one cannot describe its evolution in terms of well defined, time-independent relaxation rates. I discuss how motional conditions combined with relative strengths of relevant spin interactions alter relaxation dynamics of quadrupole (electron) as well as dipolar spins. This chapter is meant to be a warning that in the case of several interactions affected by different motional processes the validity conditions of the perturbation theory must be checked on with special caution at every stage of the calculations; otherwise the whole description can very easily break down. The discussion and the presented examples underline the necessity of a much more general approach, presented in Chapter X.

X. Dynamics of spin systems beyond validity regimes of the second order perturbation theory

In this chapter I present a general theoretical tool appropriate for describing evolution of spin systems under arbitrary motional conditions. The presented approach is based on solution of the Liouville equation employing the multipole representation formalism presented in Chapter II. The conceptual problem of the theory is to set up an appropriate matrix representation of the lattice Liouvillians in a basis (in principle infinite) including classical as well as quantum mechanical degrees of freedom of the lattice. Attention is given to how to adapt the general treatment to spin systems characterized by various interactions and various types of motional processes. I illustrate this subject by applying the general treatment to evaluation of ESR (quadrupole) spectra and PRE effects. Thus the reader becomes acquainted with practical applications of this general formalism and should be able to adjust it to other spin systems under interest and different models of motion. An important example of flexibility of this approach is presented in Chapter XII.

XI. Effects of neighboring dipolar spins on dynamics of quadrupolar spin

This chapter is concerned mainly with the way in which neighboring dipolar spins contribute to the relaxation processes of quadrupole spins. I discuss the problem of a proper evaluation of quadrupole spectra taking into account that there are two relaxation pathways for the quadrupole spin: it relaxes due to its own relaxation mechanism created by fluctuations of the electric field gradient tensor and due to the dipole-dipole coupling to neighboring dipolar spins. The second part of this chapter I dedicate to time evolution of spin observables in the presence of relaxation processes. The treatment presented here is an extension of the theory of Chapter VII by including relaxation effects.

XII. Relaxation processes caused by translational diffusion of interacting spins

This chapter is meant to discuss complex problems when one has to deal simultaneously with translation degrees of freedom, effects of intermolecular forces upon them, electron spin dynamics affecting nuclear spin relaxation and molecular tumbling. Further topic treated in this chapter is how to extend the general treatment, discussed in Chapter X, to make it suitable for the translational motion of the participating spins.

XIII. Quantum vibrations as an origin of electron spin relaxation

This chapter deals with a theory of electron spin relaxation resulting from molecular vibrations. I critically discuss models of the ZFS coupling, linking electron spin variables to the vibrational degrees of freedom of the lattice. A special emphasis is on an appropriate, quantum-mechanical treatment of the vibrational motion. I point out the hierarchy of events: electron spin senses lattice vibrations by ZFS interaction, and next, the resulting electron spin dynamics is sensed via electron spin - nuclear spin dipole-dipole coupling by nuclear spin. I formulate general analogies between electron spin relaxation caused by quantum vibrations and nuclear spin relaxation caused by quantum dynamics of electron spin. This chapter is meant to provide an instructive example of relaxation theories dealing with a quantum-mechanical lattice.

CHAPTER I

Introduction

There are a host of excellent books and reviews that cover the general aspects of quantum mechanics and statistical physics as they apply towards Nuclear Magnetic Resonance (NMR) and Electron Spin Resonance (ESR) (for example [1-11]). In this book the physics of NMR and EPR are investigated by understanding the physical meaning of correlation functions and density operators representing an averaged behavior of many independent quantum systems. Therefore, I outline here these relatively standard topics.

I.1. Nuclear Magnetic Resonance and Electron Spin Resonance (NMR and EPR)

In Nuclear Magnetic Resonance and Electron Spin Resonance relies on measuring the magnetic moments of nuclei and unpaired electrons. These magnetic moments arise from the ability of the particles to possess an angular momentum as if they were spinning, *i.e.* the spin. Magnetic moments of nuclei and electrons, μ_I and μ_S, respectively, are determined by their spins:

$$\mu_I = g_I \frac{e\hbar}{2m_p} \sqrt{I(I+1)} \qquad (1.1a)$$

and, in analogy

$$\mu_S = g_e \frac{e\hbar}{2m_e}\sqrt{S(S+1)} \qquad (1.1b)$$

where I and S denote the corresponding spin quantum numbers (for a single nucleon or for a single electron $I = S = \frac{1}{2}$), e is the elementary charge of the proton, m_p and m_e are the proton and electron masses and $g_e = 2.0023$ is called free electron factor, while g_I depends on the considered nucleus. The ratio of μ_S and μ_I for the electron and the proton results from the ratio between their masses and is 658.2. The quantities $\frac{e\hbar}{2m_p}$ and $\frac{e\hbar}{2m_e}$ are referred to as the nuclear and electron Bohr magneton, μ_N and μ_B, respectively. The I and S manifolds split in an external magnetic field $\vec{B}_0$ into $2I+1$ ($2S+1$) allowed components of the nuclear (electron) spin along the magnetic field; they are characterized by the magnetic spin quantum number m_I (m_S) ranging between $\pm I$ ($\pm S$), differing by $|\Delta m_I| = 1$ ($|\Delta m_S| = 1$). Therefore, in an external magnetic field the available states are described by $2I+1$ ($2S+1$) wavefunctions which form a complete basis $\{|I,m_I\rangle\}$ ($\{|S,m_S\rangle\}$). Generally, in the presence of other spin interactions (examples of which are given in Chapter III; such as nuclear electric quadrupole coupling, zero field splitting, dipole-dipole coupling), the basis functions do not correspond to eigenfunctions of the considered nuclear or electron spin. However, because of the very introductory character of this chapter, at this stage I do not wish to comment more on this subject. If we restrict ourselves only to the interaction of a spin with an external magnetic field, the functions $|I,m_I\rangle$ ($|S,m_S\rangle$) describe the eigenstates of the spin (they are the eigenfunctions). The interaction energies (eigenvalues) depend on the spin state; for a nuclear spin one has:

$$E_I = \frac{g_I \mu_N}{\hbar} B_0 m_I = \gamma_I B_0 m_I = \omega_I m_I \qquad (1.2a)$$

while for an electron spin

$$E_S = \frac{g_e \mu_B}{\hbar} B_0 m_S = \gamma_S B_0 m_S = \omega_S m_S \qquad (1.2b)$$

The coefficients γ_I and γ_S are referred to as gyromagnetic ratios (factors), while the frequencies ω_I and ω_S are called Larmor frequencies. This nomenclature originates from the classical picture of NMR and ESR, where ω_I (ω_S) is the precession frequency of the nuclear (electric) magnetic moment around the external magnetic field. For transitions between two 'neighboring' energy levels (for which $|\Delta m_{I(S)}| = 1$) ω_I and ω_S are simply the transition frequencies.

An ensemble of spins in the presence of a static external magnetic field $\vec{B}_0$ leads to a macroscopic magnetization aligned with $\vec{B}_0$. Magnetic resonance and electron spin resonance experiments generally rely on having an external magnetic field $\vec{B}_0$ applied across a sample as well as an oscillating magnetic field $\vec{B}_1$ (perpendicular to $\vec{B}_0$), that induces transitions between the energy levels of the spin. The same role is played by time dependent spin interactions (the dipole-dipole coupling, the quadrupolar interaction, *etc...*); they cause transitions between the energy levels leading, in this way, to relaxation processes.

I.2. Introductory aspects of relaxation

Relaxation is a general term for phenomena that bring spin systems back towards their equilibrium state. In Nuclear Magnetic Resonance or Electron Spin Resonance it is a reinstatement of nuclear or electron magnetization, respectively, to its equilibrium configuration after it has been perturbed. The longitudinal component of the magnetization (parallel to the applied static magnetic field) recovers to the equilibrium magnetization in a longitudinal relaxation time T_1, called a spin-lattice relaxation time, while the transverse magnetization (perpendicular to magnetic field $\vec{B}_0$) disappears with a spin-spin relaxation time T_2. Due to the transverse relaxation mechanism (the spin-spin relaxation) the spin system establishes thermal equilibrium within itself, while the longitudinal (spin-lattice) relaxation brings the spin system to thermal equilibrium with the lattice.

The term 'lattice' is used to depict all thermal degrees of freedom available for the considered molecular system as a whole. The consideration of only two relaxation times is a simplistic view. Nevertheless, this simplified picture is very useful in understanding the basis of spin relaxation. There are many other subtle effects, especially in the case of multispin systems, and in this book I shall deal with many of them in later chapters.

There is only one way to perturb a spin system from its equilibrium configuration or to take it back to its equilibrium state: one must induce transitions. The longitudinal relaxation involves energy exchange events between the spin states and the lattice. If the population of the high energy state of a particular spin transition is larger than its equilibrium value, it will relax back towards equilibrium through events in which individual spins flip from a high energy to a low energy state while the lattice accepts the energy released (for instance by accelerating the molecular tumbling). Similarly, if the population of the high energy state is smaller than its equilibrium value, then corresponding decelerations of molecular motions would provide the energy needed to promote spins to the high energy state. However, the relaxation transitions can only occur if there is an interaction that couples the motions of the molecule to the spin states, *i.e.* if there is a spin – lattice coupling. Depending on the molecular system, relaxation processes are caused by the various interactions to which nuclear and electron spins are subjected. All elementary magnetic moments in a macroscopic sample sense local magnetic fields arising from various effects. One of the most common relaxation mechanisms is that due to dipole-dipole interactions between the magnetic moments. Beside this relaxation mechanism I shall consider in this book relaxation processes caused by nuclear electric quadrupole and zero field splitting interactions. As anticipated above, to cause relaxation the interactions must fluctuate in time. This statement is absolutely crucial for understanding the physical meaning of the relaxation processes. The spin interactions are mediated by various motional processes, like molecular tumbling, translational diffusion, exchange motion, molecular distortions, lattice vibrations, *etc*, depending on the considered molecular system. Thus, the local magnetic fields created by these interactions fluctuate in time and can therefore induce transitions. The fluctuations have to occur on an appropriate time scale to provide an efficient mechanism for spin transitions. Generally the efficiency of the relaxation processes depends upon the magnitude and the fluctuation rate of the local fields. In particular, in the simplest case of two spins 1/2 when the magnetic fields change at a rate equal to the Larmor frequency ω_0, they act on the spins in the same way as does the $\vec{B}_1$ field. However motional frequencies at $2\omega_0$ also contribute to relaxation transitions. This particular issue shall be clarified in Chapter II.

All the mechanisms which contribute to spin-lattice relaxation also contribute to the spin-spin relaxation, because the restitution of the equilibrium populations brings zero magnetization in the plane perpendicular to the $\vec{B}_0$ direction. There are, however, processes which influence the transverse relaxation but do not affect the longitudinal relaxation. The two processes are significantly different. Only energy exchanges with the lattice contribute to the longitudinal process. Spin-spin flip-flop transitions do not involve energy exchange between the spin and the lattice and therefore they do not contribute to the spin-lattice relaxation, but they do contribute to the spin-spin relaxation. For this reason the transverse relaxation is called spin-spin relaxation. These two types of relaxation are sometimes linked, in the sense that one influences the other, and sometimes they are not linked. It should also be emphasized that relaxation processes are single exponential only in some limiting cases that exclude complex multispin systems. The single exponential relaxation processes can be described in a phenomenological, very simplified way by the set of Bloch equations:

$$\frac{dM_z}{dt} = \frac{M_0 - M_z}{T_1} \tag{1.3a}$$

$$\frac{dM_x}{dt} = M_y \omega_{off} - \frac{M_x}{T_2} \tag{1.3b}$$

$$\frac{dM_y}{dt} = -M_y \omega_{off} - \frac{M_y}{T_2} \tag{1.3c}$$

where M_x, M_y, M_z are the magnetization components along the x, y, z axes ($\vec{B}_0 \| z$), M_0 is the equilibrium magnetization of the sample, while ω_{off} is the deviation from the Larmor frequency. However, nonexponential relaxation may take place even if the system consists of identical nuclear (electron) spins. I shall develop this issue further in the forthcoming chapters of this book.

This very short introduction to the problems of relaxation brings us to the conclusion that the characteristic relaxation time constants contain structural information, through the interactions themselves, as well as dynamic information through the time fluctuations of these interactions.

I.3. The concept of the density matrix

All our observations concern an average behavior of an ensemble of quantum spin systems. Let us choose a complete set of basis functions $\{|u_n\rangle\}$. The basis functions can be the eigenfunctions of the Hamiltonian characterizing the considered spin system, or they can be any complete set of functions. A quantum state of any individual spin can be characterized by a wavefunction being a superposition of these basis functions:

$$\psi(t) = \sum_n c_n(t) |u_n\rangle \quad (1.4)$$

where the expansion coefficients $c_n(t)$ generally vary in time. The density matrix is defined as:

$$\sigma_{mn}(t) = \langle c_m(t) c_n^*(t) \rangle \quad (1.5)$$

where we average over all spins; the star denotes complex conjugate. The matrix elements depend upon the representation used, *i.e.* upon the complete set of basis functions $\{u_n\}$. The matrix transforms in the same way as any quantum operator when one changes the representation. Therefore, the density matrix is commonly referred to as a density operator σ. In particular it appears in this form in the Liouville von-Neuman equation, which is a master equation describing the evolution of an arbitrary spin system under an arbitrary Hamiltonian. In fact, all theories of spin evolution and relaxation start from this equation, as we shall soon see. Any physical observables, in particular the nuclear or electron spin magnetization, are nothing more than expectation values of the corresponding quantities characterizing individual quantum spin systems, weighted by the density operator.

I.4. Time correlation functions and spectral densities

A fundamental quantity characterizing stochastic processes is a time correlation function. Let us consider two physical quantities $A(x)$ and $B(x')$ taken when the molecular system is in the states generally denoted as x and x_0, respectively. Since the states of the system depend on time $x \equiv x(\tau)$ ($x_0 \equiv x_0(0)$) the two quantities are in fact time dependent. The time correlation function is defined as:

$$\langle A^*(\tau)B(0)\rangle = \int\int A(x)B(x_0)P(x,x_0,\tau)P_{eq}(x_0)dx_0 dx \qquad (1.6)$$

The function $P(x,x_0,\tau)$ describes the probability that our system is in the state x at time τ if it has been in the state x' at time zero, while P_{eq} is the equilibrium distribution of states; in particular $P_{eq}(x_0)$ describes the probability of finding the state x_0 in equilibrium. This definition is quite general. Since we are interested in describing molecular motion we can associate the states x and x' with positions of particles carrying the spins under interest, described by the position vectors $\vec{r}$ and $\vec{r}'$, respectively. The conditional probability density $P(x,x_0,\tau)$, that a particle is found at a position $\vec{r}$ at the time τ if it was placed at a position $\vec{r}_0$ in the initial time point can be obtained by solving the diffusion equation. Let us restrict the derivations to the most popular types of molecular motion, namely rotational and translational diffusion. Then the function $P(x,x_0,\tau)$ is determined by solving the equation:

$$\frac{\partial P(\vec{r},\vec{r}_0,\tau)}{\partial t} = -\hat{S}(\vec{r},\tau)P(\vec{r},\vec{r}_0,\tau) \qquad (1.7)$$

where the operator $\hat{S}(\vec{r},\tau)$ includes the rotational and translational terms:

$$\hat{S}(\vec{r},\tau) = \vec{K}\cdot\hat{D}_{tr}\cdot[\vec{K}+\vec{K}V(\vec{r})] + \vec{J}\cdot\hat{D}_{rot}\cdot[\vec{J}+\vec{J}V(\vec{r})] \qquad (1.8)$$

where $\vec{K} = -i\nabla$, $\vec{J} = -i\vec{r}\times\nabla$, while $\hat{D}_{tr}$ and $\hat{D}_{rot}$ are translational and rotational diffusion tensors, respectively. This equation includes the effects of intermolecular forces by the interaction potential $U(\vec{r})$ scaled in thermal energy units,

$V(\vec{r}) = \dfrac{U(\vec{r})}{k_B T}$. In the simplest case of isotropic and free ($U(\vec{r}) = 0$) rotational motion the equation takes the form:

$$\frac{\partial P(\vec{r}, \vec{r}_0, \tau)}{\partial t} = -D_{rot} \nabla^2_{rot} P(\vec{r}, \vec{r}_0, \tau) \tag{1.8a}$$

where the Laplace operator ∇^2_{rot} depends now only on the orientation of the interspin vector $\nabla^2_{rot} \equiv \nabla^2_{rot}(\vartheta, \varphi, r = 1)$; the diffusion tensor $\hat{D}_{rot}$ has been replaced by the diffusion coefficient D_{rot}.

The time correlation function of Eq.1.6 is called the auto-correlation function as it expresses the correlation of the same physical quantities at different times, *i.e.* $A \equiv B$; otherwise it is called the cross-correlation function ($A \neq B$). The Fourier transform of the correlation function give us a spectrum of frequencies characterizing the considered motional process. The spectrum is described by the spectral density function $J(\omega)$:

$$J(\omega) = \int_0^\infty \langle A^*(\tau) B(0) \rangle \exp(-i\omega\tau) d\tau \tag{1.9}$$

We shall soon learn that spectral densities determine probabilities of transitions between spin states and in consequence the efficiency of relaxation processes. They are of primary importance for relaxation theories.

References:

1. A. Abragam, The principles of nuclear magnetism, *Oxford University Press, Oxford* (1961)
2. J. McConnel, The theory of nuclear magnetic relaxation in liquids, *Cambridge University Press, Cambridge* (1987)
3. Ch. P. Poole, Jr, H.A. Farach, Theory of magnetic resonance, A Wiley – Interscience publication, *John Wiley & Sons* (1987)
4. C.P. Slichter, Principles of magnetic resonance, *Springer - Verlag, Berlin* (1990)
5. J. W. Hennel, J. Klimowski, Fundamentals of nuclear magnetic resonance, *Harlow, Longman* (1993)
6. R.R. Ernst, G. Bodenhausen, A, Wokaun, Principles of nuclear magnetic resonance in one and two dimensions, *Clarendon Press, Oxford* (1994)
7. R. Kimmich, NMR - Tomography, Diffusometry, Relaxometry, *Springer – Verlag Berlin* (1997)
8. M. H. Levitt, Spin Dynamics, *Chichester, Wiley* (2001)
9. J. Kowalewski, L. Mäler, Nuclear spin relaxation in liquids: theory, experiments and applications, Series in Chemical Physics, *Taylor & Francis Group* (2006)
10. A. Abragam, B. Bleaney, Electron paramagnetic resonance of transition ions, *Clarendon Press, Oxford* (1970)
11. J. E. Wertz, J. R. Bolton, Electron Spin Resonance: Elementary theory and practical applications, *McGraw-Hill, New York* (1972)

CHAPTER II

Perturbation theory of relaxation processes

Relaxation is a consequence of interactions between the spin and the rest of the system, referred to as the lattice. From the spin point of view, the lattice is a thermal reservoir with many degrees of freedom, and the interaction between the spin system and the lattice serves as the path for spin relaxation. The objective of relaxation theory is to take into account the random time-dependent effects on the spin system caused by the environment. The theoretical framework needed to evaluate spin relaxation is conveniently constructed in terms of the time evolution of the density operator under the influence of a Hamiltonian as expressed by the Liouville – von Neumann equation [1-12]. A perturbation treatment of this equation leads to well defined quantities describing relaxation dynamics. The idea and the main formalism of the relaxation theory were developed by Purcell and Pound [13], Solomon, Bloembergen and Morgan [14-17], Wangsness and Bloch [18-19], Abragam [1], Redfield [20,21] and Kubo and Tomita [10, 22,23].

In this chapter I shall present two equivalent formalisms, resulting from the early works and being in common use. The first one is referred to as WBR (Wangsness – Bloch- Redfield) relaxation theory and is based on a matrix representation of the spin density operator [1-12], while the second one is closer to the operator concept proposed by Abragam [1] and is called the multipole representation treatment. I do not intend to discuss the deep theoretical background of the formalisms, for which complete treatments can be found in the literature [1-31]. My goal is to provide a compact and clear description of the two formulations of the relaxation theory for performing relaxation calculations of practical usefulness. I do not restrict the considerations to systems where it is a good approximation to treat the lattice classically. The formalisms presented are suitable for a full and general treatment of spin systems experiencing classical as well as quantum mechanical degrees of freedom of the environment. I shall apply it in the next chapters to very different molecular systems and, in consequence, various types of lattice dynamics.

II.1. The Liouville - von Neumann equation. Hilbert and Liouville space formalisms

To begin the theoretical derivations we decompose the total Hamiltonian H which describes the selected spin I interacting with the lattice L, into three parts: $H = H_I + H_L + H_{IL}$. The first two terms represent the pure spin system and pure lattice contributions, respectively, while the last one describes the coupling between them and contains parameters of both the spin system and the lattice. The term lattice refers to all degrees of freedom not directly associated with the spin in which we are interested. For example, rotational, translational or vibrational degrees of freedom of the molecule bearing the spin of interest as well as states and dynamics of any other spin in the surrounding can contribute to the complex lattice. A schematic illustration of this idea is presented in Fig.2.1.

The entire system containing the spin and the lattice can be described by a density operator $\rho(t)$. For practical purposes to perform calculations utilizing the concept of the density operator it is necessary to set up its proper representation. The density operator can be represented, for example, in the basis constructed from eigenvectors (eigenfunctions) of the Hamiltonian H_0 containing the pure spin and pure lattice interactions $H_0 = H_I + H_L$.

Chapter II – Perturbation theory of relaxation processes

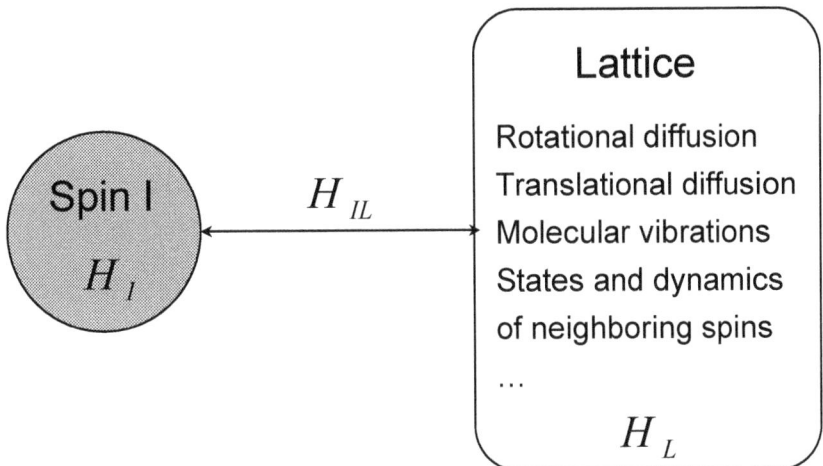

Fig.2.1
A pictorial view of an entire molecular system divided into the spin subsystem and the lattice including all degrees of freedom relevant for the spin under interest. The Hamiltonians H_I and H_L characterize the spin subsystem and the lattice, respectively, while the Hamiltonian H_{IL} represent a coupling between them and is responsible for relaxation processes of the I spin.

Actually, for the majority of cases discussed in this book, it is possible to separate the spin and lattice contributions to the entire density operator and in consequence consider separately their representations within the spin and lattice eigenbasis. On the whole, the basis used to represent the density operator does not need to contain eigenstates of any part of the system, as it has been anticipated in Section I.2. This can be an arbitrarily chosen complete and orthonormal set of functions (kets) $\{|\mu\rangle\}$ ($|\mu\rangle$ denotes the individual functions) covering all degrees of freedom of this part of the system we want to describe within the density operator formalism. One can create a matrix representation of the density operator $\rho(t)$ calculating the elements $\rho_{\mu\nu}(t) = \langle\mu|\rho(t)|\nu\rangle$. This approach is called in the literature the Hilbert space representation [3-7, 27, 29, 30] (the chosen basis is referred to as the Hilbert space). There is an alternative approach, named the Liouville space (or operator) representation [3-7, 27, 29, 30]. The Hilbert space functions $\{|\mu\rangle\}$ can be used to create a set of 'bracket' operators: $\{|\mu\rangle\langle\nu|\}$, which

19

defines the Liouville space of dimension N^2 when the corresponding Hilbert space is of N dimension. In fact, other complete and orthonormal sets of operators obtained as appropriate linear combinations of the $|\mu\rangle\langle v|$ vectors create Liouville spaces equivalent to the 'original' one. Obviously, every operator can be expanded into the Liouville basis; particularly the density operator $\rho(t)$. In this book I shall denote operators O represented in Liouville space as $|O\rangle$. In particular, the density operator treated within the Liouville space formalism is indicated as $|\rho(t)\rangle$. This concerns also the basis vectors $|\mu\rangle\langle v|$ which can be written as : $|\mu\rangle\langle v| = |\mu, v\rangle$. Operators O acting on the Liouville space (superspace) elements are called superoperators and often denoted by $\hat{O}$. The superoperators most frequently invoked in this book, are the Liouville superoperator $\hat{L}$ generated by the corresponding Hamiltonian H: $\hat{L} = [H,...]$ and the relaxation superoperator $\hat{R}$, so far not defined. The discussed relations between the Hilbert and Liouville spaces are presented graphically in Fig.2.2a,b.

a)

	$|\mu_1\rangle$	...	$|\mu_i\rangle$	...	$|\mu_N\rangle$
$\langle\mu_1|$					
...					
$\langle\mu_j|$			$O_{\mu_j\mu_i} = \langle\mu_j|O|\mu_i\rangle$ $\rho_{\mu_j\mu_i} = \langle\mu_j|\rho|\mu_i\rangle$		
...					
$\langle\mu_N|$					

b)

	$\|\mu_1,\mu_1)$	...	$\|\mu_N,\mu_N)$	$\|\mu_1,\mu_2)$	...	$\|\mu_i,\mu_j)$ $i \neq j$	...	$\|\mu_{N-1},\mu_N)$
$(\mu_1,\mu_1\|$								
...		$(\mu_i,\mu_i\|O\|\mu_i,\mu_i)$ $(\mu_i,\mu_i\|\rho\|\mu_i,\mu_i)$						
$(\mu_N,\mu_N\|$								
$(\mu_1,\mu_2\|$								
...								
$(\mu_i,\mu_j\|$ $i \neq j$						$(\mu_i,\mu_j\|O\|\mu_i,\mu_j)$ $(\mu_i,\mu_j\|\rho\|\mu_i,\mu_j)$		
...								
$(\mu_{N-1},\mu_N\|$								

Fig.2.2a,b

Matrix representation of an arbitrary operator O (in particular the density operator ρ) within a) the Hilbert space $\{|\mu_i\rangle\}$ of the dimension N $(i, j = 1,..., N)$ b) the Liouville space $\{|\mu_i\rangle\langle\mu_j| \equiv |\mu_i,\mu_j)\}$ associated with the Hilbert space $\{|\mu_i\rangle\}$ The operator matrix elements within the Liouville basis, $(\mu_i,\mu_j|O|\mu_i,\mu_j)$, can be evaluated as follows:

$$(\mu_i,\mu_j|O|\mu_i,\mu_j) = |\mu_i\rangle\langle\mu_j|O|\mu_i\rangle\langle\mu_j| = \langle\mu_j|O|\mu_i\rangle|\mu_i\rangle\langle\mu_j| =$$
$$O_{\mu_j\mu_i}|\mu_i\rangle\langle\mu_j| = O_{\mu_j\mu_i}|\mu_i,\mu_j)$$

Solid 'fat' line in Fig.2.2b divides the entire matrix (of the dimension N^2) into the block corresponding to these basis vectors of the Liouville space, which are formed by the same Hilbert space vectors: $|\mu_i,\mu_i) \equiv |\mu_i\rangle\langle\mu_i|$ (the dimension of this block is N) and the remaining part corresponding to the vectors $|\mu_i,\mu_j)$, where $i \neq j$; the dimension of this part is $N(N-1)$. We shall see in Section II.2 that this ordering of the Liouville space vectors reflects the two types of relaxation processes (spin-lattice and spin-spin relaxation) mentioned in Chapter I.

where the Liouville superoperator $\hat{\hat{L}} = \hat{\hat{L}}_I + \hat{\hat{L}}_L + \hat{\hat{L}}_{IL}$ contains the spin, lattice and spin-lattice terms, generated by the corresponding parts of the Hamiltonian H. In the next section I shall discuss a perturbation solution of this fundamental equation. The solution is the background of relaxation theories, as we shall see in the next sections.

Time evolution of the entire density operator $\rho(t)$ under the Hamiltonian H is governed by the Liouville - von Neumann equation:

$$\frac{d}{dt}\rho(t) = -i[H, \rho(t)] \quad (2.1a)$$

The equation can be written equivalently in the operator formalism as:

$$\frac{d}{dt}|\rho(t))) = -i\hat{\hat{L}}|\rho(t))) \quad (2.1b)$$

II.2. Perturbation treatment of the Liouville - von Neumann equation

Solution of the Liouville - von Neumann equation may be obtained by a perturbation treatment. However, the perturbation theory is based on a series of assumptions, significantly limiting its applicability. The first assumption, crucial for the further considerations, is that the total Hamiltonian H can be divided into a main part $H_0 = H_I + H_L$, ($\hat{\hat{L}}_0 = \hat{\hat{L}}_I + \hat{\hat{L}}_L$) which determines the energy level structure for the considered system, and a small perturbing term H_{IL}, ($\hat{\hat{L}}_{IL}$), causing transitions between the energy levels. So far, this statement is rather general and does not specify exact conditions which must be fulfilled by an interaction to be included into the main or into the perturbing part of the Hamiltonian H. A rigorous, mathematical formulation of this fundamental requirement of the perturbation theory is essential for a proper description of time evolution of any spin system. However, since such a formulation requires introducing the concept of a correlation time, a quantity not defined yet, I shall postpone the detailed discussion to a later part of this chapter. At this moment we just assume that the decoupling of the Hamiltonian H into the main and perturbing parts is allowed and continue our derivations.

To remove the pure lattice (H_L) and pure spin (H_I) contributions and to single out the perturbation spin-lattice coupling (H_{IL}) one transforms the density operator $\rho(t)$ and the perturbing Hamiltonian H_{IL} into the interaction representation [3-7, 26-30], generated by the main Hamiltonian H_0:

$$\rho'(t) = \exp(-iH_0 t)\rho(t)\exp(iH_0 t) \tag{2.2}$$

$$H'_{IL}(t) = \exp(-iH_0 t)H_{IL}(t)\exp(iH_0 t) \tag{2.3}$$

The transformation leads to the new time dependent quantities $\rho'(t)$ and $H'_{IL}(t)$ referred to as interaction representations of the operators $\rho(t)$ and $H_{IL}(t)$, respectively. This procedure is a part of an ordinary and common treatment of the Liouville - von Neumann equation. The corresponding transformation performed in the Liouville space can be written in an analogous manner:

$$|\rho'(t)\rangle\rangle = \exp\left(i\hat{L}_0 t\right)|\rho(t)\rangle\rangle \tag{2.4}$$

$$\hat{L}'_{IL}(t) = \exp\left(i\hat{L}_0 t\right)\hat{L}_{IL}(t)\exp\left(-i\hat{L}_0 t\right) \tag{2.5}$$

where $|\rho'(t)\rangle\rangle$ and $\hat{L}'_{IL}(t)$ are the Liouville space counterparts of $\rho'(t)$ and $H'_{IL}(t)$. It is important to note that there are two sources of time dependence of the perturbing spin-lattice coupling $H'_{IL}(t)$: that due to some motional degrees of freedom of the lattice and that due to the transformation to the interaction representation. The transformation simplifies the Liouville - von Neumann equation to the form:

$$\frac{d}{dt}\rho'(t) = -i\left[H'_{IL}(t), \rho'(t)\right] \tag{2.6a}$$

involving explicitly only the spin-lattice interaction $H'_{IL}(t)$. Correspondingly, the operator form of the interaction representation of the Liouville – von Neumann equation can be written as:

$$\frac{d}{dt}|\rho'(t))=-i\hat{L}_{IL}(t)|\rho'(t))\tag{2.6b}$$

The perturbation approach to this differential equation is rewriting it in an integration form and then substituting it into itself iteratively. By a formal integration of Eq. 2.6a one obtains:

$$\rho'(t)=\rho'(0)-i\int_0^t[H'_{IL}(t_1),\rho'(0)]dt_1-\int_0^t dt_1\int_0^{t_1}dt_2[H'_{IL}(t_1),[H'_{IL}(t_2),\rho'(0)]]+\ldots\tag{2.7a}$$

where $\rho'(0)$ is the interaction representation of the initial density operator $\rho(0)$: $\rho'(0)=\exp(-iH_0 t)\rho(0)\exp(iH_0 t)$. Following the concept of the Hilbert and Liouville space treatments, presented in parallel, I write down the perturbation expansion also in the operator formalism:

$$|\rho'(t))=|\rho'(0))-i\int_0^t \hat{L}_{IL}(t_1)|\rho'(0))dt_1-\int_0^t dt_1\int_0^{t_1}dt_2\hat{L}_{IL}(t_1)\hat{L}_{IL}(t_2)|\rho'(0))+\ldots\tag{2.7b}$$

where $|\rho'(0))=\exp(i\hat{L}_0 t)|\rho(0))$. The expressions of Eq. 2.7a and 2.7b are rigorous. However, for the purpose of explicit evaluations of the density matrix evolution and a clean theory of relaxation processes the perturbation expansion must be restricted to a finite number of terms. At this point some words of caution are in order. The assumption that the second order perturbation theory describes well the system (*i.e.* the higher order terms in this expansion are negligible) is called 'the Redfield limit' [1-7, 21, 22, 26, 27, 29, 30] and is the second fundamental assumption of the Redfield relaxation theory. A strict mathematical form of this assumption is on the same primary level of importance as the first one concerning the decomposition of the Hamiltonian H and it also involves the concept of a correlation time. I shall turn to a detailed discussion of both the fundamental requirements soon. At present, being aware of the Redfield condition, we just restrict the perturbation expansion to the second order terms and proceed further.

To link the time derivative of the density operator $\frac{d\rho'(t)}{dt}$ to its current value $\rho'(t)$ we take the derivative of both sides of the equation, yielding:

$$\frac{d\rho'(t)}{dt} = -i[H'_{IL}(t), \rho'(0)] - \int_0^t dt_1 [H'_{IL}(t), [H'_{IL}(t_1), \rho'(0)]] \qquad (2.8a)$$

or in the operator formulation:

$$\frac{d|\rho'(t))}{dt} = -i\hat{L}'_{IL}(t)|\rho'(0)) - \int_0^t dt_1 \hat{L}'_{IL}(t)\hat{L}'_{IL}(t_1)|\rho'(0)) \qquad (2.8b)$$

From this point, I follow several steps where certain 'minor' approximations must be invoked in order to reach the goal to describe the dynamics of the spin I in terms of specific relaxation rates associated with individual density matrix elements.

From the point of view of the relaxation theory it is desirable to get rid of the first term of the right-hand side of Eqs.2.8a and 2.8b. The first step to achieve this is to bring to attention that the macroscopic sample can be considered as an ensemble of isolated systems each consisting of a single spin system of interest, immersed in a lattice. Each of the members of the ensemble experiences the coupling to the lattice which fluctuates in time in a stochastic way. The time evolution pathway of a given spin system is determined by one realization of the lattice dynamics within the available motional degrees of freedom. It leads to different density operators $\rho'(t)$ for the particular spin systems, resulting from their different evolution history. We assume that the stochastic fluctuations do not keep in memory the initial state of the lattice encoded into the density operator $\rho'(0)$. This assumption is equivalent to the statement that $H'_{IL}(t)$ and $\rho'(0)$ are not correlated and can be separately averaged over the spin ensemble: $\langle H'_{IL}(t)\rho'(0)\rangle = \langle H'_{IL}(t)\rangle\langle\rho'(0)\rangle$ ($\langle\ \rangle$ denotes the ensemble averaging). Next, to eliminate the first order term we have to assume that the average effect of the interaction Hamiltonian at any point of time is zero: $\langle H'_{IL}(t)\rangle = 0$. For stationary perturbations the ensemble average is equivalent to the time average. This is a direct conclusion from the definition of stationary processes as those for which the probability of the realization of the $A(t+\tau)$ configuration is determined by the configuration at the time t, $A(t)$, and the time interval τ. This equivalence of time and ensemble averaging implies that we can omit the linear terms in Eq.2.8a and Eq.2.8b if the stochastic fluctuations of the spin-lattice coupling, $H'_{IL}(t)$, are on a rapid timescale compared with the evolution of the density operator through relaxation. To better envision the physical meaning of this condition one can say that the spin-lattice coupling fluctuates so fast with respect to the interaction frame that from the perspective of the spin I the spin-lattice Hamiltonian averaged over a time period relevant for its relaxation dynamics has zero value. Even though we have not yet defined explicitly the relaxation times reflecting directly the time scale of relaxation processes, it is constructive for the considerations to invoke at this point the common formulation of this requirement that the spin relaxation must not be faster than the motion causing this relaxation. This formulation is obvious in the context of the derivations presented so far. Further discussions will bring us to the conclusion that if the perturbing Hamiltonian satisfies the Redfield condition, this requirement is fulfilled as well.

If the spin-lattice interaction fluctuates on the same timescale as the spin dynamics, the perturbation approach breaks down and it is not possible to define explicitly spin relaxation rates. A proper treatment of the evolution of a spin system under such conditions requires a general solution of the Liouville - von Neumann equation. Chapter X is devoted to this problem.

It has been pointed out already that for describing the time evolution of stationary interactions it is enough to consider time intervals (absolute time values are irrelevant). This implies that the integrals in Eq.2.8a,b really depend upon the difference $\tau = t - t_1$ so the variable of integration can be changed in both formulations as follows:

$$\frac{d\rho'(t)}{dt} = -\int_0^t d\tau \left\langle \left[H'_{IL}(t), \left[H'_{IL}(t-\tau), \rho'(0)\right]\right]\right\rangle \qquad (2.9a)$$

$$\frac{d|\rho'(t)\rangle}{dt} = -\int_0^t d\tau \left\langle \hat{\hat{L}}_{IL}(t)\hat{\hat{L}}_{IL}(t-\tau)|\rho'(0)\rangle\right\rangle \qquad (2.9b)$$

neglecting at the same time the linear terms. At present it is worthwhile to once again call attention to the fact that we describe an averaged behavior of a statistical ensemble of systems. The above form of the evolution equation results from the ensemble average (represented by the brackets) of both sides of Eqs.

2.8a,b. For simplicity I omit the explicit averaging symbol for the density operator. The present formulations provide a link between the changes of the density operator at time t, represented by the derivative $\frac{d\rho'(t)}{dt}$, and its initial value $\rho'(0)$. The relationship results from the development of the spin-lattice coupling experienced by the spin system over the time period t, and averaged over all available dynamic pathways of the lattice.

It is fundamental for a clear description of relaxation processes to be able to relate the derivative of the density operator to its current value $\rho'(t)$ instead to referring to the initial conditions described by the operator $\rho'(0)$. For this purpose one needs to assume also that the density operator does not change rapidly over the range of the integral, so one can replace $\rho'(0)$ by $\rho'(t)$ not affecting significantly the description. Let us state at this moment that this operation is allowed; it will soon be justified on the basis of the two fundamental assumptions of the perturbation treatment invoked above. Nevertheless, performing the integration in the rhs of Eqs.2.9a,b one would get a quantity dependent on the upper limit of integration, t. This effect is quite easy to visualize in a simplified, however quite illustrative way. The instantaneous time evolution of the system is reflected by the changes of the density operator. The development starts at the time zero and over an initial time period does not occur in a 'linear' manner, which leads to a time dependent derivative of the density operator. However, the evolution of the system over long time scales of the system becomes more 'linear' so that after a long time t the subsequent relevant changes of the density operator occurring in short time intervals $t, t+\tau$; $t+\tau, t+2\tau$, ... become independent of t. It brings us to the concept of well defined spin relaxation rates as quantities independent of time. To formulate such a description it must be permissible to extend the integration limit to infinity, which means that t is long enough that the integral will have decayed to zero and no error will be introduced by this extension. The statement is nothing else than a mathematical formulation of this simplified picture of the time evolution of the system. To understand its physical background, one should realize that the memory between the spin-lattice

Hamiltonian taken at different time-points $H'_{IL}(t)$ and $H'_{IL}(t-\tau)$ only lasts for a short time interval $\tau \leq \tau_c$. The characteristic time constant τ_c is referred to as the correlation time [1-7, 26-34]. The correlation time obviously reflects the timescale of the stochastic fluctuations of the spin-lattice coupling; however one should not treat this fact as a definition of the correlation time. Although the timescale τ_c over which the quantity $\langle H'_{IL}(t)H'_{IL}(t-\tau_c)\rangle$ decays by a substantial amount is determined first of all by the time scale of the random fluctuations of the spin-lattice interaction, it depends also on the physical mechanism of the lattice motion. I shall illustrate this statement by some examples in Chapter XII devoted to translation diffusion. Independently of the nature of the stochastic degrees of freedom of the lattice, if we consider times $t \gg \tau_c$, we may extend the upper limit of integration in Eqs. 2.9a,b to infinity, that is a crucial step in the way to define time independent relaxation coefficients. Nevertheless, I still did not give a convincing explanation why the density operator does not change significantly over the t range, so that one can assume: $\rho'(0) \cong \rho'(t)$, which seems to be especially problematic in the context of the fact that we want to extend the integration limit to infinity. Now is an appropriate time to profit from one of the two main assumptions of the perturbation theory, namely the Redfield condition stating that in the perturbation solution of the Liouville - von Neumann equation terms of higher than second order can be neglected. Making use of the correlation time concept, one can establish the mathematical formulation of this requirement as: $|\omega_{IL}\tau_c| \ll 1$, where ω_{IL} is the amplitude of the spin-lattice interaction represented by the Hamiltonian H_{IL}, expressed in angular frequency units. The relative change of the density operator $\dfrac{\rho(t)-\rho(0)}{\rho(0)}$ over the time range from zero to t is determined by the factor $t|\omega_{IL}|^2\tau_c$. The requirement that it is permissible to set $\rho(t) \cong \rho(0)$ means that $t|\omega_{IL}|^2\tau_c \ll 1$. This condition cannot contradict the relation $t > \tau_c$ needed for the extension of the integration limit. However, we have already concluded that only values of t being of the order of a few times τ_c

contribute to the integral (for longer time delays the memory between $H'_{IL}(t)$ and $H'_{IL}(t-\tau)$ is lost). Thus, if the condition $|\omega_{IL}\tau_c| \ll 1$ is fulfilled, the statement that $t|\omega_{IL}|^2 \tau_c \ll 1$ is also fulfilled. In this way, we can set up the requested differential equation linking the derivative of the density operator $\dfrac{d\rho'(t)}{dt}$ to its current value $\rho'(t)$ through quantities independent of the integration limit:

$$\frac{d\rho'(t)}{dt} = -\int_0^\infty \langle [H'_{IL}(t),[H'_{IL}(t-\tau),\rho'(t)]]\rangle d\tau \qquad (2.10a)$$

$$\frac{d|\rho'(t)\rangle}{dt} = -\int_0^\infty \langle \hat{L}_{IL}(t)\hat{L}_{IL}(t-\tau)|\rho'(t)\rangle\rangle d\tau \qquad (2.10b)$$

In further derivations I shall call Eq.2.10a and Eq.2.10b Redfield relaxation equation in the Hamiltonian and Liouville operator forms, respectively.

Finishing this quite general part of the derivations I wish to stress clearly that the physical significance of this extension of the integration limit, leading finally to Eqs.2.10a,b, is that we cannot expect to obtain from the present approach the evolution of the spin system over time intervals $t \leq \tau_c$. The assumptions and limitations of the perturbation approach are schematically illustrated in Fig.2.3.

The time scale separation between the fluctuations of the spin-lattice coupling and the resulting evolution of the spin system has further, very important consequences. Until now we discuss a treatment of the whole spin-lattice system represented by the entire density operator $\rho(t)$.

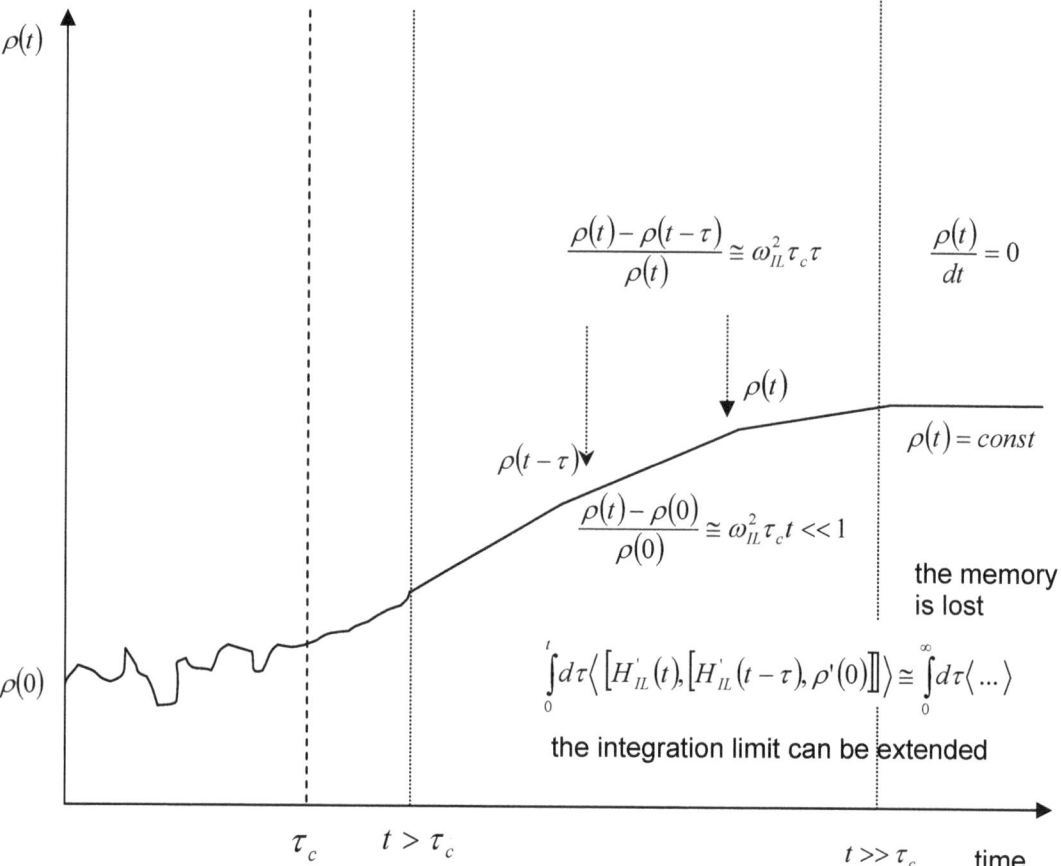

Fig.2.3
Graphical illustration of the assumptions invoked to obtain Eqs.2.10a,b. For $t > \tau_c$ one can extend the integration limit to infinity, as discussed in the text. The relative change of the density operator $\frac{\rho(t)-\rho(0)}{\rho(0)}$ is, in a good approximation, proportional to the time t. This increasing departure of the density operator $\rho(t)$ from $\rho(0)$ can be danger since we need that $\rho(t) \cong \rho(0)$. However, this effect saturates because the memory between $H'_{IL}(t)$ and $H'_{IL}(t-\tau)$ disappears with time. When the memory is lost (for $t \gg \tau_c$) the density operator does not change any more.

The spin and lattice contributions can be separated under the fundamental assumption, *i.e.* the Redfield condition that the stochastic motion of the lattice under the Hamiltonian H_L is on a rapid timescale in comparison with the I spin relaxation resulting from the H_{IL} coupling. It implies that the spin density operator $\rho_I(t)$ and the lattice density operator $\rho_L(t)$ are statistically uncorrelated and one can consider the total density operator as an outer product of the two quantities: $\rho(t) = \rho_I(t) \otimes \rho_L(t)$. Despite the lack of correlation between the lattice and spin density operators, the rapid motion in the lattice keeps it, from the perspective of the spin system, at equilibrium. In other words, any disturbances of the lattice caused by the presence of the spin (for example energy transferred from the spin system to the lattice due to spin relaxation processes) dissipate out over the lattice degrees of freedom very rapidly. The fast dissipation processes leave the lattice density operator at any time equal to the equilibrium value $\rho_L(t) \cong \rho_L^{eq}$. Thus, the entire density operator can be considered as: $\rho(t) = \rho_I(t) \otimes \rho_L^{eq}$. Since the lattice and spin Hamiltonians constituting the main Hamiltonian H_0 commute, this stochastic independence of the spin and lattice density operators is obeyed also in the interaction representation.

II.3. Matrix formulation of the Wangsness-Bloch-Redfield (WBR) relaxation theory

In this chapter we proceed further with the perturbation relaxation theory treating the Hilbert representation of the relaxation equation, Eq.2.10, as a starting point. At this moment, it is essential to depart from the treatment of the entire system of spins and lattice and focus only of the smaller system of interest, namely, the spin system. The further treatment is directly based on the decomposition of the density operator $\rho(t) = \rho_I(t) \otimes \rho_L^{eq}$ and, consequently, correct only if the separation is permissible.

Representing the spin density operator $\rho_I'(t)$ in the basis constructed from the eigenstates $|\alpha\rangle$ of the spin Hamiltonian H_I, one can transform Eq.2.10a to a set of coupled equations for particular matrix elements of the density operator $(\rho'_I(t))_{\alpha\alpha'} = \langle\alpha|\rho'_I(t)|\alpha'\rangle$:

$$\frac{d(\rho'_I(t))_{\alpha\alpha'}}{dt} = \sum_{\beta\beta'} \Gamma_{\alpha\alpha'\beta\beta'} \exp[i(\omega_{\alpha\alpha'} - \omega_{\beta\beta'})t](\rho'_I(t))_{\beta\beta'} \qquad (2.11)$$

For simplicity I shall omit from now on the index 'I' referring explicitly to the spin system and use just the labeling $\rho'_{\alpha\alpha'}$. If the density operator includes the lattice, I shall point out this fact explicitly. The frequencies $\omega_{\alpha\alpha'}$ are transition frequencies between the eigenstates $|\alpha\rangle$ and $|\alpha'\rangle$ of the spin system: $\omega_{\alpha\alpha'} = \omega_\alpha - \omega_{\alpha'}$. One can see from Eq.2.11 that the time dependencies of the density matrix elements $\rho'_{\alpha\alpha'}(t)$ are given by products of the time independent coefficients $\Gamma_{\alpha\alpha'\beta\beta'}$ and the terms oscillating in time with the corresponding frequency $\omega_{\alpha\alpha'} - \omega_{\beta\beta'}$. In order to eliminate the effects of oscillations and to obtain a system of differential equations with constant coefficients which couple the time evolution of the element $\rho_{\alpha\alpha'}$ to the elements $\rho_{\beta\beta'}$, all the expressions $\omega_{\alpha\alpha'} - \omega_{\beta\beta'}$, unless equal to zero, must be effectively averaged to zero on the time scale relevant for changes of the density matrix; this occurs if $(\omega_{\alpha\alpha'} - \omega_{\beta\beta'}) \gg \Gamma_{\alpha\alpha'\beta\beta'}$. One can formulate this condition in an equivalent manner, saying that the Hamiltonian H_I chosen as the main interaction for the spin system and therefore determining the oscillation frequencies, must provide a stationary basis set for the considered spin. Taking into account that the timescale of the relaxation processes is determined by the factor $|\omega_{IL}^2 \tau_c|$ one can express mathematically this requirement as: $|\omega_I| \gg |\omega_{IL}|^2 \tau_c$, (the frequencies ω_I and ω_{IL} describe the amplitudes of the Hamiltonians H_I and H_{IL}, expressed here in angular frequency units). This relation is known as the secular approximation condition [1, 2, 26, 27, 29, 30]. It can be also treated as the mathematical relation between the interaction Hamiltonians H_I and H_{IL} which specifies their roles as the interaction generating the energy level structure and the interaction causing the transitions between these levels, respectively. Under this condition only terms for which $\omega_{\alpha\alpha'} = \omega_{\beta\beta'}$ are retained, and the Redfield relaxation equation is transformed back into the laboratory frame. It now takes [3, 4, 7, 27, 29, 30]:

$$\frac{d\rho_{\alpha\alpha'}(t)}{dt} = -i\omega_{\alpha\alpha'}\rho_{\alpha\alpha'}(t) + \sum_{\substack{\beta\beta' \\ \omega_{\alpha\alpha'}=\omega_{\beta\beta'}}} \Gamma_{\alpha\alpha'\beta\beta'}\rho_{\beta\beta'}(t) \qquad (2.12)$$

The matrix Γ with the elements $\Gamma_{\alpha\alpha'\beta\beta'}$ is called Redfield dynamic matrix, while the matrix elements $\rho_{\alpha\alpha'}$ are referred to as coherences. The number of coherences, including populations (*i.e.* the density matrix elements $\rho_{\alpha\alpha}$), is N^2, (where N denotes the number of eigenstates of the spin system) and is equal to the dimension of the Liouville space. Now, one can see the connection between the Liouville space vectors $|\alpha\rangle\langle\alpha'|$, being so far nothing more than mathematical objects, and the density matrix elements $\rho_{\alpha\alpha'}$. The two objects are equivalent; it is only the matter of representation. Actually, one arrives at the concept of the N^2-dimensional Liouville space covering the set of mutually coupled coherences, by just analyzing the evolution of the spin system. In fact the N^2-dimensional relaxation matrix containing coefficients $\Gamma_{\alpha\alpha'\beta\beta'}$, which link the time evolution of the element $\rho_{\alpha\alpha'}$ to the elements $\rho_{\beta\beta'}$, has been set up in the Liouville space. The secular approximation decouples the time evolution of the diagonal element of the density operator $\rho_{\alpha\alpha}$ (the zero-quantum coherences or populations) from the off-diagonal elements $\rho_{\alpha\alpha'}(\alpha \neq \alpha')$. In addition, the time evolution of the off-diagonal elements is also decoupled into various multiquantum orders. The term 'multiquantum order' is especially easy to explain if the Zeeman coupling is the main interaction for the spin. In this case the frequencies of various coherences are given by a multiplication of the transition frequency ω_0: $\omega_{\alpha\alpha'} = n\omega_0$, where n describes the 'multiquantum order'. In the context of the secular approximation coherences with different transition frequencies (*i.e.* of different orders) are decoupled. On the other hand the time evolution of all the diagonal elements is mutually coupled, that leads to a significantly more complicated description of their relaxation dynamics. Fig.2.4 shows the structure of the Redfield relaxation matrix under the secular approximation condition.

The matrix elements $\Gamma_{\alpha\alpha'\beta\beta'}$ are expressed in terms of quantities $\Im_{\mu\mu'\nu\nu'}(\omega)$ including corresponding matrix elements of the Hamiltonian H_{IL} taken at t and $t-\tau$:

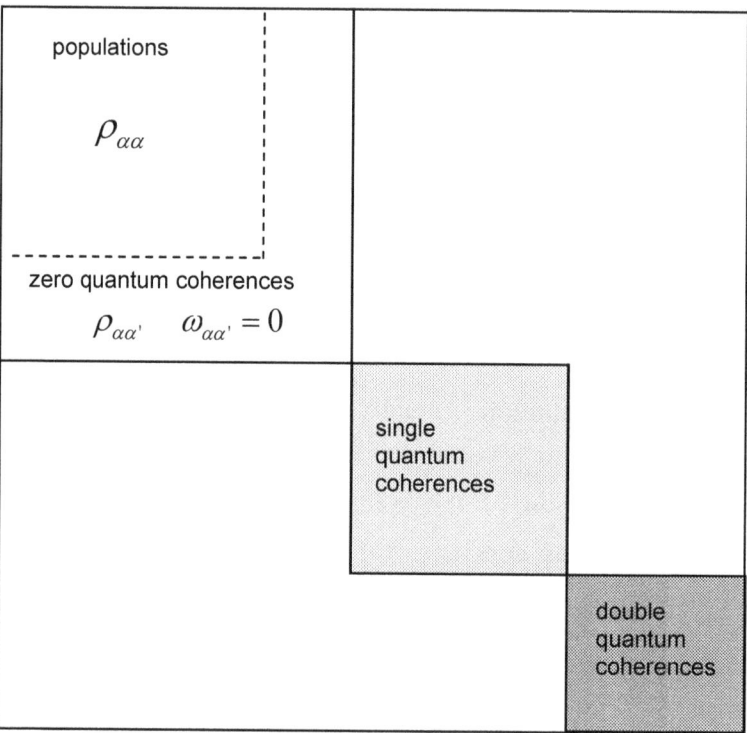

Fig.2.4

The structure of the Redfield relaxation matrix. Density matrix elements associated with transitions of different frequencies are not coupled.

$$\Im_{\mu\mu'\nu\nu'}(\omega) = \int_0^\infty \langle\langle\mu|H_{IL}(t)|\mu'\rangle\langle\nu|H_{IL}(t-\tau)|\nu'\rangle\rangle \exp(-i\omega\tau)d\tau \qquad (2.13)$$

The frequency ω corresponds to one of the transition frequencies between the participating eigenstates of the spin system: $|\mu\rangle$, $|\mu'\rangle$, $|\nu\rangle$ and $|\nu'\rangle$. Since the quantities $\Im_{\mu\mu'\nu\nu'}$ are merely Fourier transforms of the ensemble averaged terms $\langle\langle\mu|H_{IL}(t)|\mu'\rangle\langle\nu|H_{IL}(t-\tau)|\nu'\rangle\rangle$, representing correlation between the spin-lattice Hamiltonian matrix elements, they are called spectral densities (see Section I.4). The particular coefficients $\Gamma_{\alpha\alpha'\beta\beta'}$ are given as:

$$\Gamma_{\alpha\alpha'\beta\beta'} = R_{\alpha\alpha'\beta\beta'} - iL_{\alpha\alpha'\beta\beta'}$$
$$= \Im_{\alpha\beta\alpha'\beta'}(\omega_{\alpha\beta}) + \Im_{\alpha\beta\alpha'\beta'}(\omega_{\beta'\alpha'}) - \delta_{\alpha'\beta'}\sum_{\gamma}\Im_{\alpha\gamma\beta\gamma}(\omega_{\gamma\beta}) - \delta_{\alpha\beta}\sum_{\gamma}\Im_{\beta'\gamma\alpha'\gamma}(\omega_{\beta'\gamma}) \quad (2.14)$$

Being Fourier transforms over positive time only, the spectral densities are complex. This implies that the dynamic matrix Γ is composed of real and imaginary parts, referred to as the relaxation matrix and the dynamic frequency shift matrix, respectively. The relaxation matrix elements $R_{\alpha\alpha'\beta\beta'}$ are formed by the real parts of the corresponding coefficients $\Gamma_{\alpha\alpha'\beta\beta'}$ and in consequence by the real parts of the spectral densities involved in Eq.2.14. The imaginary parts $L_{\alpha\alpha'\beta\beta'}$ of the coefficients $\Gamma_{\alpha\alpha'\beta\beta'}$ result from the imaginary parts of the spectral densities and produce shifts of the resonance frequencies for particular coherences. The dynamic frequency shift becomes important only under special conditions, illustrated by some examples in Chapter X. Furthermore, there is no contribution to the dynamic frequency shift from the zero-frequency spectral densities. In the next section I shall discuss a general form of the spectral densities, resulting from a common representation of the spin-lattice coupling.

II.4. General representation of spin - lattice interactions and spectral densities

The usual way to describe spin-lattice interactions is to partition them into a symmetrized product of lattice functions $F_{-m}^{l}(t)$, which are random functions of time and spin tensor operators I_m^l:

$$H_{IL} = \xi_{IL}\sum_{m=-l}^{l}(-1)^m I_m^l F_{-m}^l(t) \quad (2.15)$$

The constant ξ_{IL} describes the interaction strengths (amplitudes). Employing the general form of the interaction Hamiltonian H_{IL} the spectral densities $\mathfrak{I}_{\mu\mu'\nu\nu'}$ can be written as:

$$\mathfrak{I}_{\mu\mu'\nu\nu'}(\omega) = \xi_{IL}^2 \sum_{m,m'=-l}^{l} (-1)^{m+m'} \langle \mu|I_m^l|\mu'\rangle \langle \nu|I_{-m'}^l|\nu'\rangle \int_0^\infty \langle F_{-m}^{l*}(t) F_{-m'}^l(t-\tau)\rangle \exp(-i\omega\tau) d\tau \quad (2.16)$$

It is important to point out that the decoupling between the average over the spin part and that over the lattice part performed in Eq.2.16 is allowed because of the timescale separation of the relaxation processes and the lattice dynamics being much faster than the former process. The ensemble average in Eq.2.16 includes only the lattice degrees of freedom and is called the lattice correlation function (see the general definition of Eq.1.6):

$$C_{mm'}^l(\tau) = \langle F_{-m}^{l*}(t) F_{-m'}^l(t-\tau)\rangle = \langle F_{-m}^{l*}(\tau) F_{-m'}^l(0)\rangle \quad (2.17)$$

The last equality reflects the stationary character of the stochastically fluctuating lattice functions, leading to a correlation function which only depends on the time interval τ [32-34], as already explained in Section I.3 The Fourier transform of the lattice correlation gives, according to Eq.1.9, the lattice density function:

$$J_{mm'}^l(\omega) = \int_0^\infty \langle F_{-m}^{l*}(\tau) F_{-m'}^l(0)\rangle \exp(-i\omega\tau) d\tau \quad (2.18)$$

Although the formalism presented is especially appropriate for systems where it is a good approximation to treat the lattice classically, it can also be adapted to systems requiring a quantum mechanical treatment of spin and lattice variables. The density operator describing the state of the whole system, including both spin and lattice variables must be evaluated in a space constructed as an outer product of the spin and lattice bases. The treatment becomes much more cumbersome if the lattice contains quantum mechanical as well as classical degrees of freedom. The latter require an infinite basis set, representing the classical continuum in the common (classical) description of the lattice. It is relatively easier to perform such calculations within the Liouville operator formalism which we left at the stage of Eq.2.10b. Of course, by changing the formalism one cannot escape from the problem of the infinite basis set. The advantage of the Liouville representation is, however, the more straightforward mathematical treatment which is just based on the definition of the Liouville operator replacing the 'commutator' Hamiltonian formalism. Since the book is devoted mainly to spin systems coupled to a lattice containing classical as well as quantum mechanical degrees of freedom, where the Liouville formalism is very profitable, I shall develop it further in the next section.

II.5. Relaxation theory within the Liouville operator formalism

In this section, first of all, I aim at deriving within the Liouville formalism, a general expression for the dynamic superoperator $\hat{\hat{\Gamma}} = \hat{\hat{R}} - i\hat{\hat{L}}$, instead of giving immediately a recipe for determining its matrix representation in the basis generated by the eigenstates of the system, as was done in the previous section. The general formulation of the dynamic superoperator results directly from the representation of Eq.2.10b in the Liouville space. It will turn out to be very useful when applied to spin systems analyzed in the forthcoming chapters of this book. Under the two main conditions: $|\omega_{IL}|\tau_c \ll 1$ and $|\omega_I| \gg |\omega_{IL}|^2 \tau_c$ (allowing for a clean definition of time independent quantities describing the spin relaxation processes as already discussed) the dynamic superoperator $\hat{\hat{\Gamma}}$ links just the time derivative of the spin density operator $\frac{d|\rho'(t))}{dt}$ to its current value $|\rho'(t))$:

$$\frac{d|\rho'(t))}{dt} = \hat{\hat{\Gamma}}|\rho'(t)) \qquad (2.19)$$

This equation is the Liouville space counterpart of Eq.2.11 with the oscillating part being averaged out due to the secular approximation. In this context Eq.2.10b gives us straightforwardly the general expression for the operator $\hat{\hat{\Gamma}}$:

$$\hat{\hat{\Gamma}} = -\int_0^\infty Tr_L\left\{\hat{\hat{L}}_{IL}(t)\hat{\hat{L}}_{IL}(t-\tau)|\rho_L^{eq})\right\}d\tau \qquad (2.20)$$

Although the formulation is quite obvious, some words of caution are appropriate. The timescale separation of spin and lattice dynamics, which leads to the decomposition of the total density operator into the spin part and the lattice equilibrium density operator, is crucial for the current formulation in the same manner as it has been for Eq.2.11. The equilibrium lattice density operator is explicitly present in Eq.2.20. Since we shall apply the Liouville formalism to spin systems coupled to a lattice with various degrees of freedom, classical as well as quantum ones, it is important to define in a general sense the lattice density operator at thermal equilibrium:

$$|\rho_L^{eq}) = \frac{\exp(-H_L/k_B T)}{Tr_L\{\exp(-H_L/k_B T)\}} \qquad (2.21)$$

Since classical degrees of freedom are characterized by a continuum of eigenstates, $\rho_L^{eq} = 1$, and therefore it does not explicitly occur in Eq.2.11. The next comment concerns the mathematical formalism: we have replaced the brackets $\langle \rangle$ denoting the ensemble average over the lattice degrees of freedom by the trace operation (Tr_L). It has been done for 'compatibility' reasons; the further chapters of this book refer to the literature, where such a notation is in use.

The relation of Eq.2.19 can be transformed back to the laboratory frame. This transformation leads to the Liouville space counterpart of the Redfield relaxation equation expressed by Eq.2.12:

$$\frac{d|\rho(t))}{dt} = \left[-i\hat{\hat{L}}_I + \hat{\hat{\Gamma}}\right]|\rho(t)) \qquad (2.22)$$

where the dynamic superoperator $\hat{\hat{\Gamma}}$ is also defined in terms of quantities expressed in the laboratory frame:

$$\hat{\hat{\Gamma}} = -\int_0^\infty Tr_L\left\{\hat{\hat{L}}_{IL}\left[\exp\left(-i\left(\hat{\hat{L}}_I + \hat{\hat{L}}_L\right)\tau\right)\hat{\hat{L}}_{IL}\right]\rho_L^{eq}\right\}d\tau \qquad (2.23)$$

The formulation provides a complete and elegant tool to describe relaxation processes for an arbitrary spin system obeying the requirements of the perturbation treatment of the Liouville -von Neumann equation. Nevertheless, it is necessary for practical applications to link it to the common form of the spin-lattice interaction Hamiltonian, given by Eq.2.15. To point out that we set up a generalized theoretical framework appropriate for a treatment of quantum mechanical as well as classical lattice functions, we use instead of the symbols F_{-m}^l associated in our mind with the classical lattice, the new ones: T_{-m}^l. This replacement is not only motivated by the generalization, it has also a practical reason to keep consistence with literature. If the H_{IL} Hamiltonian is inserted into Eq.2.21, the following expression for the quantity $\hat{\Gamma}|\rho(t))$ is obtained:

$$\hat{\Gamma}|\rho(t)) = -(\xi_{IL})^2 \sum_{m,m'} (-1)^{(m+m')} \left(\begin{array}{c} \int_0^\infty Tr_L\left\{T_{-m'}^l \exp(-i\hat{L}_L\tau)T_{-m}^{l+}|\rho_L^{eq}\right\}\left[I_{m'}^l, \exp(-i\hat{L}_I\tau)I_m^l|\rho(t))\right]d\tau + \\ \int_0^\infty Tr_L\left\{\exp(-i\hat{L}_L\tau)T_{-m}^l T_{-m'}^{l+}|\rho_L^{eq}\right\}\left[|\rho(t))\exp(-i\hat{L}_I\tau)I_m^l, I_{m'}^l\right]d\tau \end{array} \right) \quad (2.24)$$

where index '+' denotes the adjoint tensor components. The spin part is separated from the lattice correlation function and written as a commutator. It is also assumed that the spin-lattice coupling is of one type only, so that Eq.2.24 does not contain any interference terms. Actually, I am not going to discuss in this book interference effects at all. Nevertheless, I wish to stress that the Hamiltonian as well as operator formalisms can be easily adapted for the purpose of taking them into account, if necessary.

The lattice correlation functions which appear in Eq.2.24 are called in the literature [35-39] the quantum mechanical correlation functions, to point out that they include quantum degrees of freedom of the lattice. The quantum mechanical correlation functions are defined as:

$$G_{m,m'}(\tau) = (-1)^{(m+m')} Tr_L\left\{\left[\exp(-i\hat{L}_L\tau)T_{-m}^l T_{-m'}^{l+}\right]\rho_L^{eq}\right\} = Tr_L\left\{\left[\exp(-i\hat{L}_L\tau)T_m^{l+} T_{m'}^l\right]\rho_L^{eq}\right\} \quad (2.25a)$$

$$G_{m',m}(-\tau) = (-1)^{(m+m')} Tr_L\left\{T_{-m'}^l\left[\exp(-i\hat{L}_L\tau)T_{-m}^{l+}\right]\rho_L^{eq}\right\} = Tr_L\left\{\left[T_{m'}^{l+}\exp(-i\hat{L}_L\tau)T_m^l\right]\rho_L^{eq}\right\} \quad (2.25b)$$

In the last equalities we have utilized the symmetry properties of the lattice tensor functions: $T_m^{l+} = (-1)^m T_{-m}^l$. Since the Tr operation is invariant under cyclic permutations of the operators one can conclude that $G_{m,m'}^*(\tau) = G_{m',m}(-\tau)$ ('*' denotes the complex conjugation). Actually, employing the orthogonality and symmetry properties of the tensor operators forming the spin-lattice Hamiltonian [40-45], one can set up several relations between the quantum mechanical correlation functions that are very important for practical calculations. It is of particular interest to mention two of them: the cross-correlation functions within the irreducible tensor operator representation vanish: $G_{m,m'}(\tau) = \delta_{mm'} G_{m,m}(\tau)$, and the time reversal symmetry: $G_{m,m}(\tau) = G_{m,m}(-\tau)$.

The Fourier transform of the quantum mechanical correlation function, which is relevant in the Redfield relaxation theory, is referred to as the quantum mechanical spectral density [35-39]:

$$K_{-m,-m}(\omega) = \int_0^\infty G_{-m,-m}(-\tau)\exp(-i\omega\tau)d\tau \qquad (2.26)$$

The exponential term containing an appropriate transition frequency ω originates from the spin part of the formulation of Eq.2.22. Although we introduced and precisely defined within the Liouville operator formalism the relaxation superoperator $\hat{\hat{R}} = \text{Re}(\hat{\hat{\Gamma}})$, to obtain individual elements of this operator determining, for example, the spin-lattice relaxation rate, we need to set up its representation in an appropriate Liouville basis. Many observables in the relaxation experiment do not identify with individual elements of the relaxation matrix. Various linear combinations of diagonal elements of the density operator correspond to magnetization modes, whereas combinations of various off-diagonal elements are identified as coherence modes; several examples are given in Chapter IV. These magnetization and coherences modes are normal modes or eigenvectors of the Redfield relaxation matrix and can be obtained by its diagonalization. In the next section I discuss an alternative representation which does not require this procedure.

II.6. Multipole representation of relaxation superoperators

For an arbitrary spin quantum number I one can construct the standard Liouville basis $\{|\alpha_Z\rangle\langle\beta_Z|\}$ formed by the Zeeman eigenstates of the spin $|\alpha_Z\rangle = |I, m_I\rangle$, where m_I is the magnetic quantum number. This representation can of course be used to obtain explicit forms of required relaxation rates (describing, for example, the spin-lattice relaxation processes). However, because of symmetry properties of the tensor operators it is very convenient to use the complete and orthonormal set of basis operators $\{|\Sigma,\sigma\rangle\}$ defined as [46-48]:

$$|\Sigma,\sigma\rangle = \sum_{m_I}(-1)^{I-m_I-\sigma}\sqrt{2\Sigma+1}\begin{pmatrix} I & I & \Sigma \\ m_I+\sigma & -m_I & -\sigma \end{pmatrix}|I,m_I+\sigma\rangle\langle I,m_I| \qquad (2.27)$$

where the symbols $\begin{pmatrix} I & I & \Sigma \\ m_I+\sigma & -m_I & -\sigma \end{pmatrix}$ denote the 3j symbols [49-53]. The orthonormal, irreducible tensor operators $|\Sigma,\sigma\rangle$ are related to the standard irreducible spin operators I_σ^Σ by normalization constants. The relationships for the first and second rank tensors are following:

$$|1,\sigma\rangle = \sqrt{\frac{3}{(2I+1)(I+1)I}}\, I_\sigma^1 \qquad (2.28a)$$

$$|2,\sigma\rangle = \sqrt{\frac{30}{(2I+3)(2I+1)(I+1)(2I-1)}}\, I_\sigma^2 \qquad (2.28b)$$

The tensor operators I_σ^Σ are defined in terms of the angular momentum operators I_z and $I_\pm$:

$$I_0^1 = I_z, \quad I_{\pm 1}^1 = \mp\frac{1}{\sqrt{2}}I_\pm \qquad (2.29a)$$

$$I_0^2 = \frac{1}{\sqrt{6}}\left[3I_z^2 - I(I+1)\right], \quad I_{\pm 1}^2 = \mp\frac{1}{2}(I_z I_\pm + I_\pm I_z), \quad I_{\pm 2}^2 = \frac{1}{2}I_\pm I_\pm \qquad (2.29b)$$

One can expand the spin density operator $|\rho(t)\rangle$ into the $\{|\Sigma,\sigma\rangle\}$ basis, yielding:

$$|\rho(t)\rangle = \frac{1}{2I+1} + \sum_{\Sigma=1}^{2I}\sum_{\sigma=-\Sigma}^{\Sigma} P_\sigma^\Sigma(t)|\Sigma,\sigma\rangle \qquad (2.30)$$

The expansion coefficients $P_\sigma^\Sigma(t)$ are called state multipoles. They are just the expectation values of the tensor operators $\langle|\Sigma,\sigma\rangle(t)\rangle$ obtained as the scalar product of the tensor and the spin density matrix: $P_\sigma^\Sigma(t) = \langle|\Sigma,\sigma\rangle(t)\rangle = (\Sigma,\sigma|\rho(t))$. The Redfield relaxation equation, Eq.2.20, can be written in terms of the state multipoles [23, 46, 47] as:

$$\frac{dP_\sigma^\Sigma(t)}{dt} = \sum_{\Sigma'=1}^{2I}\sum_{\sigma'=-\Sigma'}^{\Sigma'}\left[-i(\Sigma,\sigma|\hat{L}_I|\Sigma',\sigma') + (\Sigma,\sigma|\hat{\hat{\Gamma}}|\Sigma',\sigma')\right]P_{\sigma'}^{\Sigma'}(t) \qquad (2.31)$$

The multipoles with $\sigma = 0$ are associated to the population elements of the standard Liouville space, $|I,m_I\rangle\langle I,m_I|$. These multipoles are referred to as polarization; at the same time as those with $\sigma \neq 0$ describe the multiquantum coherences of the σ order. The general formulation of Eq.2.31 can be set up, for example, for the expectation value of the I_z operator, $\langle I_z\rangle(t)$, represented by the multipole $P_0^1(t)$. In this context, the Bloch equation is nothing else than a phenomenological counterpart of the multipole representation of Redfield relaxation equation set up for the quantity $\langle I_z\rangle = \langle I_0\rangle = P_1^0$. The I_z operator can always be expressed as a linear combination of the populations $|I,m_I\rangle\langle I,m_I|$, where the states with positive and negative magnetic quantum numbers m_I are taken with opposite signs. The appropriate coefficients, depending on the spin quantum number I, are actually provided by Eq.2.29 defining the basis in the multipole representation. The same coefficients can also be obtained by diagonalisation of the Redfield relaxation matrix with the elements given by Eq.2.14. The analogy is more general. The multipoles (operators) $|\Sigma,\sigma\rangle$ representing individual spin coherences are equivalent to the eigenvectors of the Redfield relaxation matrix. It leads to the conclusion that the relaxation rates obtained as eigenvalues of the Redfield relaxation matrix are available directly from the multipole treatment. The explicit expressions for the relaxation supermatrix elements $R_{\sigma,\sigma'}^{\Sigma,\Sigma'} \equiv \left(\Sigma,\sigma\left|\hat{\hat{R}}\right|\Sigma',\sigma'\right)$ depend on the form of the spin-lattice coupling, represented by Eq.2.14, in particular on the order l of the spin tensor operators I_m^l (determining automatically the order of the lattice tensors F_{-m}^l). In further considerations, employing this multipole representation treatment, we shall describe the spin-lattice interactions in terms of the first rank tensor components I_m^1. It is sufficient in this context to set up the specific expression for the relaxation supermatrix elements for the first rank spin tensor operators:

$$R_{\sigma,\sigma'}^{\Sigma,\Sigma'} = -\xi_{IL}^2 \operatorname{Re} \sum_n \sum_\Lambda \sum_\lambda K_{-n,-n}(n\omega_I) \times$$

$$(2I+1)(I+1)I(2\Lambda+1)\sqrt{(2\Sigma+1)(2\Sigma'+1)}\left[(-1)^{\Sigma+\Lambda+1}-1\right]\left[1-(-1)^{\Sigma'+\Lambda+1}\right] \times \quad (2.32)$$

$$\begin{pmatrix} \Sigma & 1 & \Lambda \\ -\sigma & n & \lambda \end{pmatrix} \begin{pmatrix} \Lambda & 1 & \Sigma' \\ \lambda & n & -\sigma' \end{pmatrix} \begin{Bmatrix} \Sigma & 1 & \Lambda \\ I & I & I \end{Bmatrix} \begin{Bmatrix} \Lambda & 1 & \Sigma' \\ I & I & I \end{Bmatrix}$$

where the symbols $\begin{Bmatrix} \Sigma & 1 & \Lambda \\ I & I & I \end{Bmatrix}$ are the 6j symbols [49-53]. The number of non-zero relaxation matrix elements is effectively reduced by symmetry properties of the 3j and 6j symbols. Although this last result is valid generally for arbitrary spin quantum number, some comments are appropriate. The multipole representation basis has been build up in terms of the 'bra-ket' elements formed by the Zeeman eigenstates: $|I, m_I\rangle$. It implies that the multipole description is the most suitable way for a discussion of relaxation effects, when the main Hamiltonian determining the energy level structure of the spin of interest is dominated by the Zeeman interaction, but it is not necessarily the most appropriate way of dealing with other systems. The spectral densities $K_{-n,-n}(n\omega_I)$ present in Eq.2.32 are explicitly taken at the frequencies corresponding to n-th quantum transitions between the Zeeman energy levels. This announces conceptual and computational difficulties which can appear if the considered spin does not exhibit the Zeeman energy level structure. This treatment is especially useful for the case of the spins of $I = \frac{1}{2}$ coupled to other spin of an arbitrary spin quantum number which forms a quantum-mechanical lattice for the first one. The explicit expressions for the spin-lattice and spin-spin relaxation rates ($R_{1I} = T_{1I}^{-1}$ and $R_{2I} = T_{2I}^{-1}$, respectively) for $I = \frac{1}{2}$ have the form:

$$T_{1I}^{-1} = R_{0,0}^{1,1} = \xi_{IL}^2 \operatorname{Re}\{K_{1,1}(-\omega_I) + K_{-1,-1}(\omega_I)\} = 2\xi_{IL}^2 \operatorname{Re}\{K_{1,1}(-\omega_I)\} \quad (2.33)$$

$$T_{2I}^{-1} = R_{1,1}^{1,1} = R_{-1,-1}^{1,1} = \xi_{IL}^2 \operatorname{Re}\{K_{1,1}(-\omega_I) - K_{0,0}(0)\} \quad (2.34)$$

Thus, in this case one can take the advantage of the simple form of the relaxation rates obtained within the multipole representation framework and profit from the generalized description of the lattice incorporated into the quantum-mechanical spectral densities [54-57]. Eq.2.33 will be invoked very often in this book as a starting point of relaxation calculations for a spin system coupled to a lattice with classical as well as quantum mechanical degrees of freedom.

The simple mathematical form of the relaxation expressions of Eqs.2.32 and 2.33 results from the fact that the components of the irreducible tensor operators I_m^1 (Eq.2.13) are eigenoperators of the spin Liouvillian $\hat{\hat{L}}_I$, thus $\hat{\hat{L}}_I I_n^1 = n\omega_I I_n^1$. If the spin Hamiltonian (Liouvilian) contains terms which do not obey this condition (like, for example, quadrupolar or zero field splitting interactions discussed in the next chapter), the multipole representation treatment becomes more cumbersome. Independently of the applied formalism, to describe properly relaxation dynamics one must identify eigenstates (energy levels) and eigenvectors (eigenfunctions) of the spin system under interest. To extend the multipole approach to systems with non-Zeeman energy level structure it is necessary to link its eigenfunctions $|\mu\rangle$ to the Zeeman basis functions $|I, m_I\rangle$:

$$|\mu\rangle = \sum_{m_I=-I}^{I} c_{\mu, m_I} |I, m_I\rangle.$$ Employing this relationship one can establish the relation:

$$|I, m_I\rangle = \sum_{\mu=-I}^{I} [c]^{-1}_{m_I \mu} |\mu\rangle$$ (where $[c]^{-1}_{m_I \mu}$ are elements of the inverted matrix c^{-1} containing the eigenvector coefficients) and express the vectors $|I, m_I + \sigma\rangle\langle I, m_I|$ in terms of the Lioville space vectors $|\mu\rangle\langle v|$, formed from the eigenstates of the system: $|I, m_I + \sigma\rangle\langle I, m_I| = \sum_{\mu, v=-I}^{I} c^*_{m_I+\sigma, \mu} c_{m_I, v} |\mu\rangle\langle v|$. Inserting this relation into Eq.2.27 leads to the multipole representation $|\Sigma, \sigma)$ expanded into the basis $|\mu\rangle\langle v|$. In the next step, one needs to express the spin tensor components I_n^1, involved in the formulation of the spin-lattice coupling, into the same basis. One can achieve this in the same manner, utilizing the representation of the operators I_n^1 in the terms of the Zeeman vectors $|I, m_I + \sigma\rangle\langle I, m_I|$ and the relation between the eigenstates of the system and the Zeeman vectors. The Redfield relaxation equation, Eq.2.31,

Chapter II – Perturbation theory of relaxation processes

operating on the eigenstates representation of the state multipoles and spin tensor operators gives in principle explicit expressions for the relaxation matrix elements $R_{\sigma\sigma'}^{\Sigma\Sigma'}$. Nevertheless, the required computational effort is much higher than in the case of spin systems characterized by the Zeeman Hamiltonian only. Actually, the relaxation matrix is, in general, not diagonal (while for the Zeeman case it is) and the resulting expressions are much more complicated. The briefly sketched procedure, leading to the relaxation matrix elements, suggests that there are cases when it could be more convenient to use the matrix formalism presented in Section II.3. The multipole treatment also requires a diagonalization of the spin Hamiltonian, if it contains other (non-Zeeman) terms. Instead of the already outlined calculation steps needed within the multipole state formalism, in the Hilbert space formulation one needs to set up the relaxation matrix with the elements given by Eq.2.14 and diagonalize it. Resolving the problem of choosing a more convenient formalism to describe relaxation processes of a given spin system one should also take into account advantages of the quantum-mechanical correlation function when the lattice requires to be treated quantum mechanically. The matrix formulation can be also adapted to such cases; nevertheless it necessitates some conceptual effort and is also computationally complex. I am aware that the last two statements are so far not sufficiently elucidated. Subsequent chapters of this book provide various examples of spin systems treated within the Hamiltonian as well as the Liouville operator (multipole states) formalisms. The examples shall explain and illustrate the two approaches. Besides the goal to provide just a theory of relaxation processes, I anticipate in this chapter to discuss the two relaxation formalisms in a way which gives the reader some flexibility to choose one of them, depending on circumstances.

To illustrate better the equivalence of the matrix and operator formalisms we shall derive Eq.2.33 and Eq.2.34 from the relaxation matrix established according to Eq.2.14. The basis for spin $I = \frac{1}{2}$ consists of two Zeemann eigenstates corresponding to the magnetic quantum numbers $m_I = \pm\frac{1}{2}$; we label them as follows: $|1\rangle = \left|\frac{1}{2}\right\rangle$ and $|2\rangle = \left|-\frac{1}{2}\right\rangle$. The expectation value $\langle I_z \rangle$ is proportional to the difference between the populations of the two states: $\langle I_z \rangle = \frac{1}{2}(\rho_{11} - \rho_{22})$. It means that only the 2×2 part of the population block of the Redfield relaxation matrix needs to be considered to get an expression for the T_1 relaxation time:

$$\frac{d}{dt}\begin{pmatrix} \rho_{11} \\ \rho_{22} \end{pmatrix} = -\begin{pmatrix} R_{1111} & R_{1122} \\ R_{2211} & R_{2222} \end{pmatrix}\begin{pmatrix} \rho_{11} \\ \rho_{22} \end{pmatrix} \qquad (2.35)$$

The relaxation matrix elements are given by Eq.2.13. According to this formulation one gets: $R_{1111} = R_{2222} = -R_{1122} = -R_{2211} = -\Im_{1212}(\omega_I) - \Im_{1212}(-\omega_I)$. The quantum mechanical correlation function incorporated into the Liouville space superoperator formalism has obviously its Hilbert space formulation:

$$\langle F_m^*(t)F_{m'}(t-\tau)\rangle = Tr_L\left\{F_m^*\left[\exp\left(i\hat{L}_L \tau\right)F_{m'}\right]\rho_L^{eq}\right\} \qquad (2.36)$$

This implies that the spectral density $J_{mm'}(\omega_I)$ (derived within the Hilbert formalism) can be treated as a counterpart of the $K_{mm'}(\omega_I)$ spectral density (derived within the Liouville formalism); i.e.: $J_{mm'}(\omega_I) \equiv K_{mm'}(\omega_I)$. It brings us to the conclusion that $\Im_{1212}(\omega_I) = \xi_{IL}^2 \frac{1}{2} K_{11}(\omega_I)$. Since the spectral density $\Im_{1212}(\omega)$ involves the spin tensor component I_1^1 (related to the angular momentum operator, $I_1^1 = \frac{1}{\sqrt{2}} I_+$) connecting the two Zeeman states of $m_I = \pm\frac{1}{2}$, it corresponds to the function $K_{11}(\omega)$. The factor $\frac{1}{2}$ comes from the Zeeman representation of the spin tensor component $\langle \frac{1}{2}|I_1^1|-\frac{1}{2}\rangle = \frac{1}{\sqrt{2}}$ incorporated into the definition of the spectral density $\Im_{1212}(\omega_I)$. Diagonalizing the relevant part of the relaxation matrix (Eq.2.35) one gets immediately Eq.2.34. Analogous derivations can be addressed to the T_2 relaxation time. The density matrix element ρ_{12} corresponding to the single quantum coherence between the states $|1\rangle = \left|\frac{1}{2}\right\rangle$ and $|2\rangle = \left|-\frac{1}{2}\right\rangle$, $\rho_{12} = |1\rangle\langle 2|$, evolves exponentially in time with the rate constant R_{1212}. The expression of

Eq.2.14 for the relaxation matrix element R_{1212} together with the relation between the quantities $\Im_{\mu\mu'\nu\nu'}(\omega)$ and $K_{mm'}(\omega)$:

$$\Im_{\mu\mu'\nu\nu'}(\omega) = \xi_{IL}^2 \sum_{m,m'=-l}^{l} (-1)^{m+m'} \langle\mu|I_m^l|\mu'\rangle\langle\nu|I_{-m'}^l|\nu'\rangle K_{mm'}(\omega)$$ lead to the formulation of Eq.2.33.

Finishing this chapter I would like to point out that the perturbation description is valid only under the two essential conditions: $|\omega_{IL}\tau_c| \ll 1$ and $|\omega_I| \gg |\omega_{IL}\tau_c|^2$. The first one leads to a very important statement that one can define the relaxation matrix elements only if the motion modulating the relevant interaction that leads to the relaxation is much faster than the relaxation itself. Taking into account that the timescale of the relaxation processes is determined by the factor $|\omega_{IL}^2\tau_c|$ one can express mathematically this requirement as: $|\omega_{IL}^2\tau_c| \ll \tau_c^{-1}$; it is obviously fulfilled for $|\omega_{IL}\tau_c| \ll 1$.

If there is no coupling between the spin system and the lattice, the spin system evolves in time just under its own Hamiltonian H_I according to the Liouville - von Neumann equation written for the spin density operator and involving the Hamiltonian H_I only. The interaction representation of spin density operator $\rho(t)$ is now generated by the simplified transformation: $\rho'(t) = \exp(iH_I t)\rho(t)\exp(-iH_I t)$. According to the Liouville - von Neumann equation $\frac{d\rho'(t)}{dt} = 0$ and the operator $\rho'(t)$ remains unchanged in time and equal to the initial value $\rho'(t) = \rho(0)$. We shall deal with this subject in Chapter VIII.

References:

12. A. Abragam, The principles of nuclear magnetism, *Oxford University Press, Oxford* (1961)
13. J. McConnel, The theory of nuclear magnetic relaxation in liquids, *Cambridge University Press, Cambridge* (1987)
14. C.P. Slichter, Principles of magnetic resonance, *Springer - Verlag, Berlin* (1990)
15. R.R. Ernst, G. Bodenhausen, A, Wokaun, Principles of nuclear magnetic resonance in one and two dimensions, *Clarendon Press, Oxford* (1994)
16. R. Kimmich, NMR - Tomography, Diffusometry, Relaxometry, *Springer – Verlag Berlin* (1997)
17. M. H. Levitt, Spin Dynamics, *Chichester, Wiley* (2001)
18. J. Kowalewski, L. Mäler, Nuclear spin relaxation in liquids: theory, experiments and applications, Series in Chemical Physics, *Taylor & Francis Group* (2006)
19. U. Fano, Description of states in quantum mechanics by density matrix and operator techniques, *Rev. Mod. Phys.* **29** (1957) 74-93
20. P.S. Hubbard, Quantum-mechanical and semiclassical forms of the density operator theory of relaxation, *Rev. Mod. Phys.* **33** (1961) 249-264
21. R. Kubo, Stochastic Liouville equations, *J. Math. Phys.* **4** (1963) 174-183
22. K. Blum, Density matrix theory and applications, *Plenum Press, New York* (1989)
23. L.T. Muss, P.W. Atkins (Eds.), Electron spin relaxation in liquids, *Plenum Press, New York* (1972)
24. N. Bloembergen, E.M. Purcell, R.V. Pound, Relaxation effects in nuclear magnetic resonance absorption, *Phys. Rev.* **73** (1948) 679-712
25. I. Solomon, Relaxation processes in a system of two spins, *Phys. Rev.* **99** (1955) 559-565
26. I. Solomon, N. Bloembergen, Nuclear magnetic interactions in the HF molecule, *J. Chem. Phys.* **25** (1956) 261-266
27. N. Bloembergen, Proton relaxation times in paramagnetic solutions, *J. Chem. Phys.* **27** (1957) 572-573

28. N. Bloembergen, L.O. Morgan, Proton relaxation times in paramagnetic solutions: Effects of electron spin relaxation, *J.Chem.Phys.* **34** (1961) 842-850
29. R.K. Wangsness, F. Bloch, The dynamic theory of nuclear induction, *Phys. Rev.* **89** (1953) 728-739
30. F. Bloch, Dynamical theory of nuclear induction. 2., *Phys. Rev.* **102** (1956) 104-135
31. F. Bloch, Generalized theory of relaxation, *Phys. Rev.* **105** (1957) 1206-1222
32. A.G. Redfield, On the theory of relaxation processes, *IBM J. Res. Dev.* **1** (1957) 19-31
33. A.G. Redfield, The theory of relaxation processes, *Adv. Magn. Reson.* **1** (1965) 1-32
34. R. Kubo, K. Tomita, A general theory of magnetic resonance absorption, *J. Phys. Soc. Jpn.* **9** (1954) 888-819
35. K. Tomita, A general theory of magnetic resonance saturation, *Progr. Theor. Phys.* **19** (1958) 541-580
36. J. Jeener, Superoperators in magnetic resonance, *Adv. Magn. Reson.* **10** (1982) 1-51
37. A.G. Redfield, Relaxation theory: density matrix formulation, in Encyclopedia of Nuclear Magnetic Resonance, D.M. Grant, R.K. Harris (Eds.) *Wiley, Chichester* (1996) 4085-4092
38. C.L. Mayne, S.A. Smith, Relaxation processes in coupled-spin systems, in Encyclopedia of Nuclear Magnetic Resonance, D.M. Grant, R.K. Harris, (Eds.) *Wiley, Chichester* (1996) 4053-4071
39. D. Canet, Relaxation mechanisms: magnetization modes, in Encyclopedia of Nuclear Magnetic Resonance, D.M. Grant, R.K. Harris, (Eds.) *Wiley, Chichester* (1996) 4046-4053
40. A. Kumar, R.C.R. Grace, P.K. Madhu, Cross-correlations in NMR, *Prog. Nucl. Magn. Reson. Spectr.* **37** (2000) 191-319
41. M. Goldman, Formal theory of spin – lattice relaxation, *J. Magn. Reson.* **149** (2001) 160-187
42. D. Canet, Introduction: General theory of nuclear relaxation, *Advances in Inorganic Chemistry* **57** 3-40, (2005)

43. N.G. van Kampen, Stochastic processes in physics and chemistry, *North-Holland, Amsterdam* (1983)
44. D.T. Gillespie, Markov Processes, *Academic Press, Boston* (1992)
45. M. Doi, S. F. Edwards, The theory of polymer dynamics, *Clarendon Press, Oxford* (1986)
46. N. Benetis, J. Kowalewski, L. Nordenskiöld, H. Wennerström, P.-O. Westlund, Nuclear spin relaxation in paramagnetic systems. The slow motion problem for electron spin relaxation, *Mol. Phys.* **48** (1983) 329-346
47. N. Benetis, J. Kowalewski, L. Nordenskiöld, H. Wennerström, P.-O. Westlund, Nuclear – spin relaxation in paramagnetic systems (S=1) in the slow-motion regime for the electron spin. 2. The dipolar T2 and the role of scalar interaction, *J. Magn. Reson.* **58** (1984) 261-281
48. N. Benetis, J. Kowalewski, L. Nordenskiöld, H. Wennerström, P.-O. Westlund, Nuclear – spin relaxation in a paramagnetic nickel (II) complex – an experimental test of new theoretical models, *J. Magn. Reson.* **58** (1984) 282-293
49. P.-O. Westlund, H. Wennerström, L. Nordenskiöld, J. Kowalewski, N. Benetis, Nuclear spin – lattice and spin – spin relaxation in paramagnetic systems in the slow motion regime for the electron-spin. 3. Dipole – dipole and scalar spin – spin interaction for S=3/2 and S=5/2, *J. Magn. Reson.* **59** (1984) 91-109
50. J. Kowalewski, L. Nordenskiöld, N. Benetis, P.-O. Westlund, Theory of nuclear spin relaxation in paramagnetic systems, *Progr. NMR. Spectr.* **17** (1985) 141 – 185
51. P.S: Hubbard, Some properties of correlation functions of irreducible tensor operators, *Phys. Rev.* **180** (1969) 319-326
52. M. Tinkham, Group theory and quantum mechanics, *McGraw-Hill, New York* (1964)
53. C.D.H. Chisholm, Group theoretical techniques in Quantum Chemistry, *Academic Press, London* (1976)
54. N.C. Pyper, Theory of symmetry in nuclear magnetic relaxation including applications to high resolution NMR line shapes, *Mol. Phys.* **21** (1971) 1-33

55. N.C. Pyper, Theory of symmetry in nuclear magnetic relaxation: Part II, *Mol. Phys.* **22** (1971) 433-458
56. S. Szymanski, A.M. Gryff-Keller, G. Binsch, A Liouville space formulation of Wangsness – Bloch – Redfield theory of nuclear spin relaxation suitable for machine computation: I. Fundamental aspects, *J. Magn. Reson.* **68** (1986) 399-432
57. B.C. Sanktuary, Multipole operators for an arbitrary number of spins, *J. Chem. Phys.* **64** (1976) 4352-4361
58. B.C. Sanktuary, Multipole NMR: III. Multiplet spin theory, *Mol. Phys.* **48** (1983) 1155 – 1176
59. G.J. Bowden, W.D. Hutchison, Tensor operator formalism for multiple-quantum NMR.1. Spin-1 nuclei, *J. Magn.Reson.* **67** (1986) 403-414
60. M. Rotenberg, R. Bivins, N. Metropolis, J.K. Wooten, The 3-J and 6-J symbols, *Technology Press, Cambridge* (1959)
61. M. E. Rose, Elementary theory of angular momentum, *Wiley, New York* (1957)
62. A. R. Edmunds, Angular momentum in quantum mechanics, *Princeton University Press, Princeton* (1974)
63. D. M. Brink, G.R. Satchler, Angular momentum, *Clarendon Press, Oxford* (1979)
64. D.A. Varshalovich, A.N. Moskalev, V.K. Khersonkii, Quantum theory of angular momentum, *Word Scientific Publishing, Singapore* (1988)
65. P.-O. Westlund, Nuclear paramagnetic spin relaxation theory: Paramagnetic spin probes in homogenous and microheterogenous solutions, in Dynamics of solutions and fluid mixtures by NMR, J. J. Delpuech (Ed.), *Wiley, Chichester* (1995) 173-229
66. R. Sharp, L. Lohr, J. Miller, Paramagnetic NMR relaxation enhancement: recent advances in theory, *Prog. Nucl. Magn. Reson. Spectr.* **38** (2001) 115-158
67. J. Kowalewski, D. Kruk, G. Parigi, NMR relaxation in solution of paramagnetic complexes: Recent theoretical progress for $S \geq 1$, *Advances in Inorganic Chemistry* **57** (2005) 41-104

68. N. Schaefle, R.Sharp, Four complementary theoretical approaches for the analysis of NMR paramagnetic relaxation, *J. Magn. Reson.* **176** (2005) 160-170

CHAPTER III

Interaction Hamiltonians and motional models

Presenting in the previous chapter the relaxation formalisms we have used the general spherical tensor form of the spin-lattice interactions of Eq.2.15 which involves the spin tensor operators I_m^l and the corresponding lattice functions $F_{-m}^l(t)$. Discussing in Section II.5 the relaxation theory within the Liouville operator formalism we have replaced the functions $F_{-m}^l(t)$ (describing a classical lattice) by the T_{-m}^l quantities which include quantum mechanical as well as classical components of the lattice (T_{-m}^l are, generally, operators on the lattice states). The treatments presented in Chapter II apply to an arbitrary interaction described by the Hamiltonian H_{IL}, but in this book we shall deal mainly with dipole-dipole couplings, quadrupolar interactions and zero field splitting interactions. Therefore, in this chapter I shall define the interactions and present detailed forms of their Hamiltonians. Before I shall provide the detailed expressions, it is very important to note that both tensor operators contributing to the spin-lattice Hamiltonian, I_m^l and T_{-m}^l (F_{-m}^l), transform like spherical harmonics under rotation, independently of the physical origin of the spin-lattice couplings.

III.1. Interaction Hamiltonians

Nuclei with spin quantum numbers S higher than 1/2 have a non-spherical charge distribution, which results in a nuclear electric quadrupolar moment. The nuclear quadrupolar moment, eQ, of a nucleus interacts with the electric field gradient at the nucleus site. This interaction is described by the Hamiltonian [1-10]:

$$H_Q(S) = \frac{1}{2}\sqrt{\frac{3}{2}} \frac{a_Q}{S(2S-1)} \sum_{m=-2}^{2}(-1)^m A_{-m}^2 T_m^2(S) = \sum_{m=-2}^{2}(-1)^m \tilde{A}_{-m}^2 T_m^2(S) \qquad (3.1)$$

At this moment the reader can have some problems to identify this expression with the general form of Eq.2.15 and some explanations may be appropriate. The amplitude ξ_{IL} is here defined as: $\xi_{IL} = \frac{1}{2}\sqrt{\frac{3}{2}} \frac{a_Q}{S(2S-1)}$. The quadrupolar coupling constant is defined as $a_Q = e^2 qQ/\hbar$, where Q is called the quadrupolar moment of the nucleus and q is the zz component of the electric field gradient tensor. The symbol $T_m^2(S)$ should not be mixed with the lattice tensor components T_{-m}^l; it is pointed out clearly that they are associated with the S spin. The $T_m^2(S)$ quantities are components of the second rank ($l=2$) spin tensor operator (this is a standard symbol and therefore I introduce it here) $T_m^2(S) \equiv S_m^2$ and are defined, according to Eq.2.29b, as: $T_0^2(S) = \frac{1}{\sqrt{6}}[3S_z^2 - S(S+1)]$, $T_{\pm 1}^2(S) = \mp\frac{1}{2}[S_z S_\pm + S_\pm S_z]$ and $T_{\pm 2}^2(S) = \frac{1}{2}S_\pm S_\pm$. I have changed here the spin labeling, instead of I used in Eq.2.15 (and 2.29b) I denote the spin as S. We shall deal in the next chapters with relaxation and evolution of multi-spin systems including various kinds of spins. This needs a consistent labeling to be introduced. I shall denote spins of the spin quantum number $1/2$ as I, while spins of higher spin quantum numbers shall be labeled as S (I wish also to be consistent with the literature). In this context, the functions A_{-m}^2 correspond obviously to the lattice functions F_{-m}^l (it would only be confusing to use the same labeling for all interactions). Looking at Eq.3.1 one can see that we have also expressed the quadrupolar Hamiltonian in terms of the $\widetilde{A}^2_{-m}$ quantities, which include the interaction constant ξ_{IL}. Such a representation is very convenient for practical calculations, and we use it also defining the next interactions. The A^2_{-m} and $\widetilde{A}^2_{-m}$ functions depend on the reference frame. In the principal axis system of the electric field gradient tensor, denoted as (P), they are given as: $A_0^{(P)} = 1$, $A_{\pm 1}^{(P)} = 0$, $A_{\pm 2}^{(P)} = \eta/\sqrt{6}$ ($\widetilde{A}_0^{(P)} = \frac{1}{2}\sqrt{\frac{3}{2}} \frac{a_Q}{S(2S-1)}$, $\widetilde{A}_{\pm 1}^{(P)} = 0$, $\widetilde{A}_{\pm 2}^{(P)} = \frac{a_Q}{4S(2S-1)}\eta$), where η is the asymmetry parameter. Thus, the quadrupolar Hamiltonian expressed in the principal axis system of the electric field gradient tensor has the form:

$$H_Q^{(P)}(S) = \frac{1}{2}\sqrt{\frac{3}{2}} \frac{a_Q}{S(2S-1)}\left[T_0^2(S) + \frac{\eta}{\sqrt{6}}\left(T_{-2}^2(S) + T_2^2(S)\right)\right] \quad (3.2)$$

If we wish to express the quadrupolar Hamiltonian in another reference frame, we need to employ the transformation rules for tensor components [11-14], as already anticipated. The relationship between tensor functions F^l_{-m} expressed in reference frames (P_1) and (P_2) is:

$$F_{-m}^{l(P_2)} = \sum_{k=-l}^{l} F_k^{l(P_1)} D_{k,-m}^l(\Omega_{P_1 P_2}) \quad (3.3)$$

where the Wigner rotation matrices $D_{k,-m}^l(\Omega_{P_1 P_2})$ with the Euler angles $\Omega_{P_1 P_2}$ describe the relative orientation of the frames. We shall use this transformation many times in this book, applying it to various spin interactions. Let us use it now to express the quadrupolar Hamiltonian of Eq.3.2 in the laboratory frame (L) (the laboratory axis system is determined by the direction of the external magnetic field). We obtain:

$$H_Q^{(L)}(S)(t) =$$
$$\frac{1}{2}\sqrt{\frac{3}{2}} \frac{a_Q}{S(2S-1)} \sum_{m=-2}^{2}(-1)^m A_{-m}^{2(L)}(t)T_m^2(S) = \sum_{m=-2}^{2}(-1)^m \widetilde{A}_{-m}^{2(L)}(t)T_m^2(S) \quad (3.4)$$

where:

$$A_{-m}^{2(L)}(t) = \sum_{k=-2}^{2} A_k^{2(P)} D_{k,-m}^2(\Omega_{PL}(t)), \quad \tilde{A}_{-m}^{2(L)}(t) = \sum_{k=-2}^{2} \tilde{A}_k^{2(P)} D_{k,-m}^2(\Omega_{PL}(t)) \qquad (3.5)$$

The Euler angle Ω_{PL} describes the orientation of the principal axis system of the electric field gradient tensor with respect to the laboratory frame; the transformation is illustrated in Fig.3.1a.

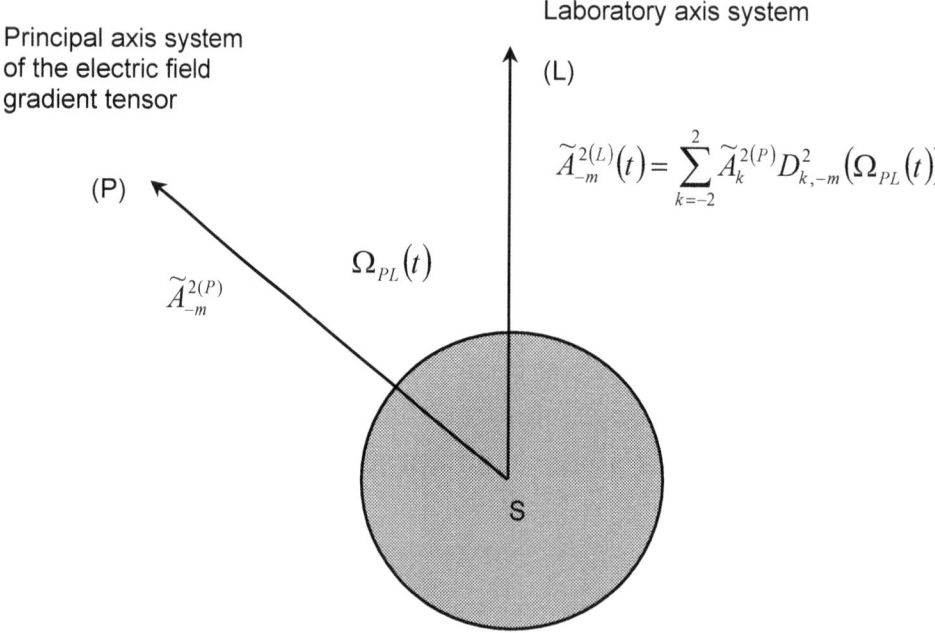

Fig.3.1a
The relationship between the spatial components of the quadrupolar Hamiltonian expressed in the principal axis system of the electric field gradient tensor and in the laboratory frame.

Zero field splitting interactions characterize electron spins of $S \geq 1$. They have different physical origins, depending on the spin system. In the case of transition metal complexes, it is due to second order effects in the spin – orbit coupling; the S manifold is split due to the indirect interaction between unpaired electrons through the spin-orbit coupling. For systems such as radicals, a direct dipole-dipole interaction between the unpaired electrons creates the zero field splitting (ZFS). Independently of the physical origin of the ZFS interaction its Hamiltonian can be written as [15-22]:

$$H_{ZFS}(S) = \sqrt{\frac{2}{3}} D \sum_{m=-2}^{2} (-1)^m V_{-m}^2 T_m^2(S) = \sum_{m=-2}^{2} (-1)^m \widetilde{V}_{-m}^2 T_m^2(S) \qquad (3.6)$$

Obviously the spatial tensor components V_{-m}^2 take the simplest form in the principal axis system of the ZFS interaction, we shall denote it also by (P) (it should not cause confusions with the quadrupolar interaction). In this frame they are defined as: $V_0^{2(P)} = 1$, $V_{\pm 1}^{2(P)} = 0$,, $V_{\pm 2}^{2(P)} = \sqrt{\frac{3}{2}} \frac{E}{D}$, $(\widetilde{V}_0^{2(P)} = \sqrt{\frac{2}{3}} D$, $\widetilde{V}_{\pm 1}^{2(P)} = 0$, $\widetilde{V}_{\pm 2}^{2(P)} = E$) where the parameters D and E describe axial and rhombic components of the ZFS, respectively. The permanent (static) ZFS has axial symmetry if the ligand field is tetragonal or trigonal. If the symmetry of the ligand field is reduced further, then the rhombic terms appear in the ZFS tensor. The ZFS Hamiltonian includes only the second order terms; in this book I shall not deal with higher order contributions to the ZFS interaction.

Relaxation by dipolar interactions is the most commonly treated relaxation mechanism in NMR. In this book we also focus our attention on the dipolar interaction and the dipole-dipole relaxation mechanism. The dipole-dipole Hamiltonian $H_{DD}(I,S)$ representing the $I-S$ dipole-dipole coupling is usually expressed in the laboratory frame as [1-9]:

$$H_{DD}(I,S) = a_{DD}^{IS} \sum_{m=-2}^{2} (-1)^m F_{-m}^2 T_m^2(I,S) = \sum_{m=-2}^{2} (-1)^m \widetilde{F}_{-m}^2 T_m^2(I,S) \qquad (3.7)$$

The components $T_m^2(I,S)$ of the two-spin tensor operator have the form:

$$T_0^2(I,S) = \frac{1}{\sqrt{6}} \left[2I_z S_z - \frac{1}{2}(I_+ S_- + I_- S_+) \right], \qquad T_{\pm 1}^2(I,S) = \mp \frac{1}{2} [I_z S_\pm + I_\pm S_z]$$

and $T_{\pm 2}^2(I,S) = \frac{1}{2} I_\pm S_\pm$. The dipole-dipole coupling constant is defined as:

$a_{DD}^{IS} = \sqrt{6} \frac{\mu_0}{4\pi} \frac{\gamma_I \gamma_S \hbar^2}{r_{IS}^3}$ where r_{IS} is the inter-spin distance, γ_I and γ_S are

57

gyromagnetic factors for the spins I and S, respectively. The fact that the spin part of the dipole-dipole Hamiltonian includes the two spin tensor operators does not contradict the general expression of Eq.2.15. This equation states only that the spin-lattice interactions can be expressed in terms of tensor operators (generally denoted as I_m^l) corresponding to the spin (spins) under interest and the corresponding lattice functions. If we are interested only in one of the spins, we can decompose the two-spin tensor operators into first rank tensors related to the individual spins [23-28] and treat one of them as a part of the lattice. To keep the notation consistent we should denote the lattice functions as T_{-m}^l in this case. We shall proceed in this way in the next chapter. The dipole-dipole interaction is axially symmetric. In the dipole-dipole frame the functions F_{-m}^2 have a very simple form: $F_0^{2(DD)} = 1$, $F_{\pm 1, \pm 2}^{2(DD)} = 0$ ($\widetilde{F}_0^{2(DD)} = 1$, $\widetilde{F}_{\pm 1, \pm 2}^{2(DD)} = 0$). This implies that when we transform the dipole-dipole Hamiltonian to another reference frame, for example to the laboratory one, according to the transformation of Eq.3.3, we obtain an expression which involves only two Euler angles, namely:

$$F_{-m}^{2(L)}(t) = \sum_{k=-2}^{2} F_k^{2(DD)} D_{k,-m}^2(\Omega_{DDL}(t)) = D_{0,-m}^2(0, \beta_{DDL}(t), \gamma_{DDL}(t)), \text{ while}$$

$$\widetilde{F}_{-m}^{2(L)}(t) = a_{DD}^{IS} D_{0,-m}^2(0, \beta_{DDL}(t), \gamma_{DDL}(t)).$$

The transformation has been illustrated in Fig.3.1b. For completeness of this short overview let us remind that the Zeeman Hamiltonian takes in the laboratory frame the simple form:

$$H_Z^{(L)}(P) = \gamma_P B_0 P_z \qquad P = I, S \qquad (3.8)$$

where γ_P is an appropriate gyromagnetic factor. Employing the transformation rules we can obviously transform the Zeeman Hamiltonian to a different reference frame, but the expression will be more complicated. The Zeeman interaction should be treated in most cases as a spin interaction (generally represented by the Hamiltonian $H_{I(S)}$), especially when expressed in the laboratory frame, not a spin-lattice coupling. Nevertheless, there are situations when the Zeeman interaction includes some degrees of freedom of the lattice. It is easy to understand this if one realizes that, for example, the Zeeman Hamiltonian expressed in a reference frame which rotates with respect to the laboratory axis, contains the rotational degrees of freedom, but let us leave this subject at this moment.

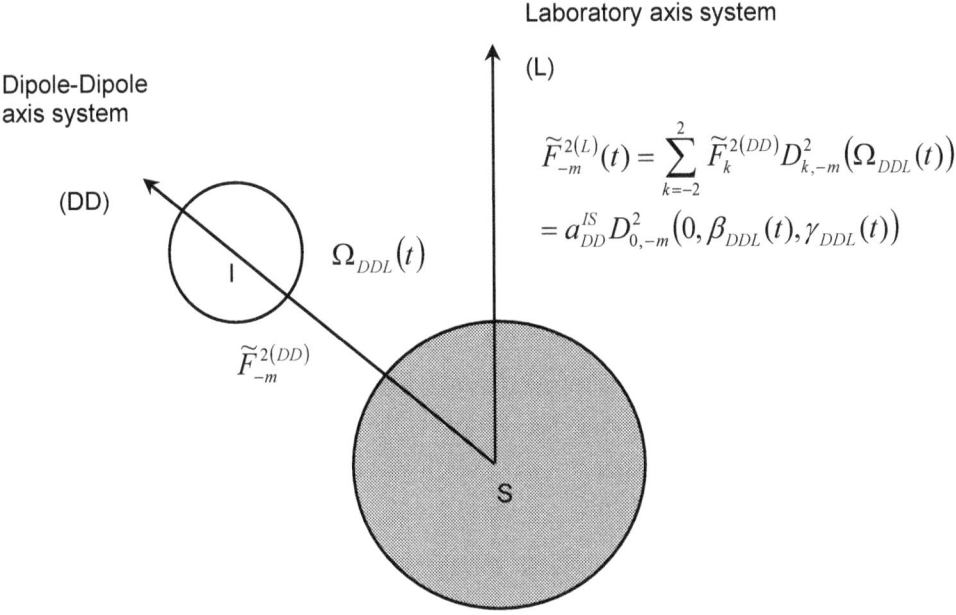

Fig.3.1b
The spatial components of the dipole-dipole Hamiltonian expressed in the dipole-dipole axis system of and in the laboratory frame.

Due to various dynamic processes, which can occur in molecular systems, the spatial tensor components associated with the discussed interactions can depend stochastically on time. It implies that, in general, we should consider the Hamiltonians as time dependent. This time dependence should be considered in a 'Galilean sense', *i.e.* if a given Hamiltonian is time independent in the reference frame denoted as (P_1), but this frame moves (fluctuates) with respect to a reference coordinate system denoted as (P_2), then the Hamiltonian considered in the (P_2) frame becomes time dependent. This issue is of fundamental importance for the problems presented in this book and therefore I shall discuss it in detail in Section III.3.

Looking carefully at Eq.3.1 and Eq.3.6 one concludes immediately that the representations of the quadrupolar and the ZFS interactions are equivalent. One can easily establish a line of analogy between the two interactions by replacing formally the quadrupolar parameters by their ZFS counterparts: $D \to \frac{3}{4}\frac{a_Q}{S(2S-1)}$ and $E \to \frac{1}{4}\eta\frac{a_Q}{S(2S-1)}$ [2, 29]. I shall use the analogous Hamiltonian representations of the two interactions very intensively in this book. Because of the equivalent representations of the quadrupolar and ZFS interactions we can apply the theories and models presented in the forthcoming chapters to molecular systems containing electron as well as quadrupolar spins; the required modifications are straightforward.

III.2. Examples of spin systems and motional models

An arbitrary spin interaction, described by Hamiltonian $H(t)$, can be expressed as a sum of two components: $H(t) = \langle H(t) \rangle + (H(t) - \langle H(t) \rangle)$. The first term represents an averaged part of the considered interaction: $\langle H(t) \rangle \equiv H_0$, while the second one describes stochastic fluctuations of the $H(t)$ Hamiltonian around this averaged value. In particular the averaged part of the Hamiltonian can vanish because of symmetry reasons, however, if an electron spin or a quadrupolar spin $S \geq 1$ is placed in a low-symmetry environment, one has to expect a large averaged ZFS or a large averaged quadrupolar coupling, respectively. At this moment it is essential to understand that the decomposition of the total Hamiltonian must be performed in a molecular frame, *i.e.* in any coordinate system fixed in the molecule. The averaged (permanent) part of the Hamiltonian is time independent (static) only in a molecule fixed frame. In particular, one can treat the principal axis system of the electric field gradient tensor or crystal field (in the case of quadrupolar or ZFS interactions, respectively) as a molecular frame. If for some reasons the molecular frame is associated with different interactions (for instance one can choose a dipole-dipole axis as a molecular axis) or chosen due to some symmetry properties of the considered molecule, the principal axis system of the averaged quadrupolar (ZFS) interaction does not coincide with the molecular frame, but remains fixed with respect to it. The averaged (permanent) interactions are often referred to in the literature as static interactions, and I shall use the same terminology (even though it might be more appropriate to call them 'permanent'). Therefore, to avoid confusion, which can lead to a situation that the further considerations are understandable on the level of classifying interactions as main and perturbing ones, I would like to stress once again the most important aspect of the current discussion. If one talks about static interactions it means that they are static only in a molecular frame, one may not think that they are just always time independent. The molecular frame can change its orientation with respect to the laboratory axis systems due to, for example, fast molecular tumbling. In this situation the permanent (static) interactions fluctuate stochastically in time with respect to the laboratory frame, and therefore under some circumstances (discussed in the next chapters) they can cause relaxation processes. The deviation between the momentary Hamiltonian and its averaged value $(H(t) - \langle H(t) \rangle)$, in the case of the ZFS interaction is referred to in the literature as the transient ZFS. One can use exactly the same terminology for the quadrupolar interaction. This terminology is general and can be applied to an arbitrary molecular system. Nevertheless, physical origins of the stochastic fluctuations of the total Hamiltonian around the averaged value (considered in a molecular frame) can be completely different, depending on the system under interest. From the point of view of a molecule-fixed frame, the static and transient parts represent the mean and the spread, respectively, of the total interactions. The transient interactions do depend on time in a molecule frame, in contradiction to their static counterparts.

In this book I shall deal with various types of molecular systems, liquids as well as solids, treating them as examples of applications of the presented theoretical models. At this moment I wish to comment more on one of them, to which I shall refer very often. Let us consider water solutions of transition metal ion complexes. Such molecules have well-known applications as contrast agents in magnetic resonance imaging. In Fig.3.2 a schematic view of such complexes surrounded by directly coordinated water molecules is shown.

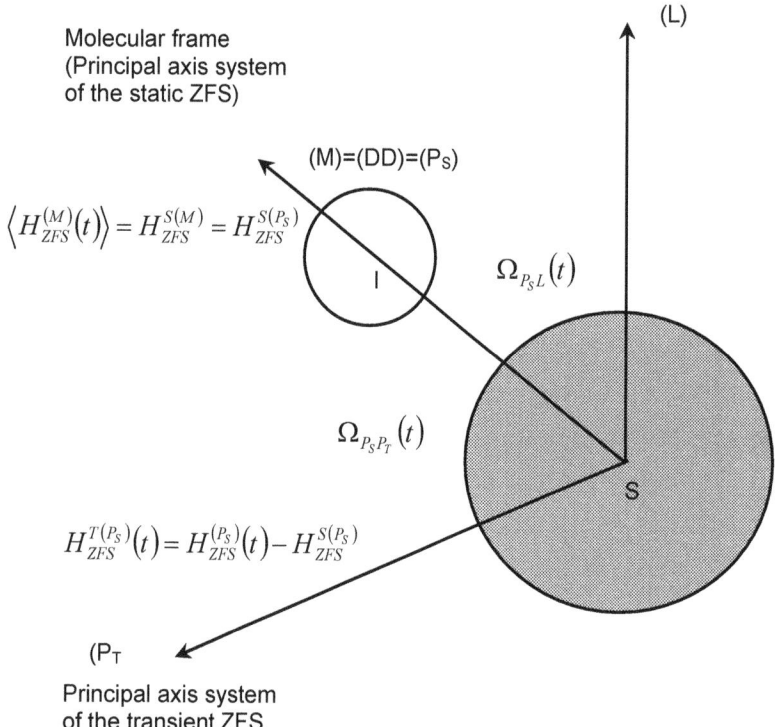

Fig.3.2

A schematic view of nuclear spin – electron spin systems, such as transition metal complexes in water solution. The static ZFS interaction is defined as the averaged part of the entire ZFS interaction considered in a molecule fixed frame (M). For simplicity we have assumed that the molecular frame is defined as the dipole-dipole frame (which coincides, in addition, with the principal axis system of the static ZFS). The transient ZFS represents the momentary deviation of the ZFS interaction from its averaged value.

The nuclear spin of water protons is here denoted as I, while the electron spin of the paramagnetic ion as S. Let us choose the $I-S$ dipole-dipole axis as a molecular axis. Because of low symmetry of the ligand shell there is a non-zero averaged (static) ZFS interaction $\langle H_{ZFS}^{(M)}(t) \rangle = H_{ZFS}^{S(M)}$. The index 'S' refers to the terminology 'static' (it should not be confused with the spin S); it is denoted here explicitly that this averaging (over time) has been performed in a molecular frame (M). To simplify the physical picture of this system as far as possible let us assume that the principal axis system of the static ZFS Hamiltonian, H_{ZFS}^{S}, coincides with the $I-S$ dipole-dipole axis. The static ZFS Hamiltonian expressed in its principal axis system (P_S) takes a form, which results directly from Eq.3.6:

$$H_{ZFS}^{S(P_S)}(S) = \sqrt{\frac{2}{3}} D_s T_0^2(S) + E_S \left(T_{-2}^2(S) + T_2^2(S) \right) \tag{3.9}$$

where the quantities D_S and E_S describe the axial and rhombic components of the static ZFS tensor, respectively. Since the entire molecule rotates, the static ZFS fluctuates in time with respect to the laboratory axis system. The same Hamiltonian expressed in the laboratory frame takes the form:

$$H_{ZFS}^{S(L)}(S)(t) = \sqrt{\frac{2}{3}} D_S \sum_{m=-2}^{2} (-1)^m \left(\sum_{k=-2}^{2} V_k^{2S(P_S)} D_{k,-m}^2 (\Omega_{P_S L}(t)) \right) T_m^2(S)$$
$$= \sum_{m=-2}^{2} (-1)^m \tilde{V}_{-m}^{2S(L)}(t) T_m^2(S) \tag{3.10}$$

where $V_0^{2S(P_S)} = 1$, $V_{\pm 1}^{2S(P_S)} = 0$, $V_{\pm 2}^{2S(P_S)} = \frac{4}{\sqrt{6}} \frac{E_S}{D_S}$, as already defined above, while

$\tilde{V}_{-m}^{2S(L)} = \sqrt{\frac{2}{3}} D_S D_{0,-m}^2 (\Omega_{P_S L}(t)) + E_S \left[D_{-2,-m}^2 (\Omega_{P_S L}(t)) + D_{2,-m}^2 (\Omega_{P_S L}(t)) \right]$. Setting up this expression we have applied the transformation rule of Eq.3.3. The relative orientation of the (P_S) and (L) frames is described by the angle $\Omega_{P_S L}(t)$, which varies in time due to the molecular tumbling.

It is highly appropriate to comment now more on the transient ZFS interaction. The spread of the ZFS interaction around its averaged value is caused in this case by stochastic fluctuations of the ligand framework. The transient ZFS Hamiltonian representing this spread in the molecular frame, $H_{ZFS}^{T(P_S)}(t)$, depends on time (the index 'T' refers to the attribute 'transient'). This statement should be obvious in the context of the explanations given so far. Nevertheless, the form of this Hamiltonian depends on the applied motional model describing the dynamic processes, which lead to the spread of the ZFS interaction. The simplest possible model, and therefore the most popular one, assumes that the transient ZFS has a constant magnitude and a principal direction (a principal axis system (P_T)), which is not fixed in the molecule [23, 24, 30-35]. The (P_T) frame changes its orientation relative to the (P_S) frame according to the rotational diffusion equation [1, 8, 9, 37-41]. Therefore this model is referred to in the literature as the 'pseudorotational model'. Thus, by analogy with the static ZFS interaction, the Hamiltonian $H_{ZFS}^{T(P_T)}$, written in its principal axis system, is time independent and takes the form:

$$H_{ZFS}^{T(P_T)}(S) = \sqrt{\frac{2}{3}} D_T T_0^2(S) + E_T \left(T_{-2}^2(S) + T_2^2(S) \right) \tag{3.11}$$

where D_T and E_T are the transient counterparts of the D_S and E_S parameters. I would like to remind once again at this stage that the (P_T) frame is not fixed in the molecule. Therefore the transient ZFS Hamiltonian expressed in the (P_S) frame fluctuates in time:

$$H_{ZFS}^{T(P_S)}(S)(t) = \sqrt{\frac{2}{3}} D_T \sum_{m=-2}^{2} (-1)^m \left(\sum_{k=-2}^{2} V_k^{2T(P_T)} D_{k,-m}^2 \left(\Omega_{P_T P_S}(t) \right) \right) T_m^2(S)$$
$$= \sum_{m=-2}^{2} (-1)^m \tilde{V}_{-m}^{2T(P_S)} T_m^2(S) \tag{3.12}$$

where $V_0^{2T(P_T)} = 1$, $V_{\pm 1}^{2T(P_T)} = 0$, $V_{\pm 2}^{2T(P_T)} = \frac{4}{\sqrt{6}} \frac{E_T}{D_T}$, while

$\tilde{V}_{-m}^{2T(P_S)} = \sqrt{\frac{2}{3}} D_T D_{0,-m}^2 \left(\Omega_{P_S P_T}(t) \right) + E_T \left[D_{-2,-m}^2 \left(\Omega_{P_S P_T}(t) \right) + D_{2,-m}^2 \left(\Omega_{P_S P_T}(t) \right) \right]$. The relative orientation of the (P_T) and (P_S) frames fluctuates in time because of distortions of the ligand framework. One can argue that the pseudorotational model of the ZFS is a rather crude simplification, because it does not relate the parameters of the ligand field to the internal geometry of the complex. Yes, in fact it is so. However, this model captures the essential physics of the transient ZFS interaction and involves only two parameters, namely its amplitude and a time constant describing fluctuations in the relative orientation of the (P_T) and (P_S) frames, encoded in the Euler angle $\Omega_{P_T P_S}(t)$.

If one wishes to consider the transient ZFS interaction in the laboratory frame, one has to perform a two step transformation, which leads to the formula:

$$H_{ZFS}^{T(L)}(S)(t) = \sqrt{\frac{2}{3}} D_T \sum_{m=-2}^{2} (-1)^m \left[\sum_{n=-2}^{2} \left(\sum_{k=-2}^{2} V_k^{2T(P_T)} D_{k,-n}^2 (\Omega_{P_T P_S}(t)) \right) D_{-n,-m}^2 (\Omega_{P_S L}(t)) \right] T_m^2(S) = \sum_{m=-2}^{2} (-1)^m \widetilde{V}_{-m}^{2T(L)} T_m^2(S) \quad (3.13)$$

where

$$\widetilde{V}_{-m}^{T(L)} = \sum_{n=-2}^{2} \left\{ \sqrt{\frac{2}{3}} D_T D_{0,-n}^2 (\Omega_{P_S P_T}(t)) + E_T \left[D_{-2,-n}^2 (\Omega_{P_S P_T})(t) + D_{2,-n}^2 (\Omega_{P_S P_T})(t) \right] \right\} D_{-n,-m}^2 (\Omega_{P_S L}(t))$$

The first transformation is between the (P_T) and (P_S) frames through the Euler angle $\Omega_{P_T P_S}(t)$, which is modulated in time by the distortional motion, while the second transformation occurs between the (P_S) and laboratory (L) frames via the Euler angle $\Omega_{P_S L}(t)$ modulated by the molecular tumbling. The three relevant coordinate systems and the corresponding representations of the static and transient interactions are presented in Fig.3.3.

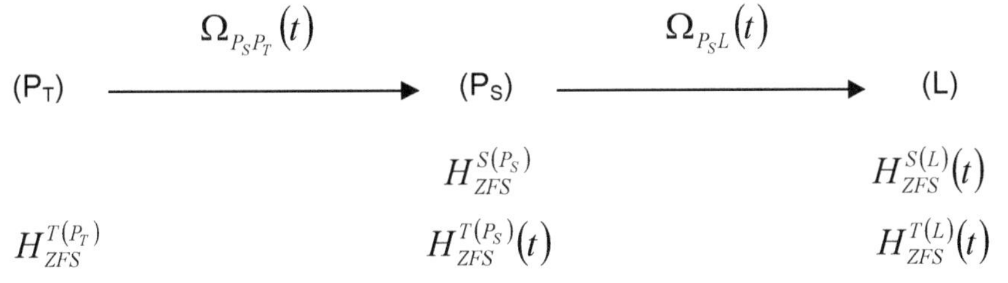

$$\begin{array}{ccccc}
 & \Omega_{P_S P_T}(t) & & \Omega_{P_S L}(t) & \\
(P_T) & \longrightarrow & (P_S) & \longrightarrow & (L) \\
 & & H_{ZFS}^{S(P_S)} & & H_{ZFS}^{S(L)}(t) \\
H_{ZFS}^{T(P_T)} & & H_{ZFS}^{T(P_S)}(t) & & H_{ZFS}^{T(L)}(t)
\end{array}$$

Fig.3.3
The representations of the static and transient interactions depending on the reference frames.

Since in this book the attention is rather focused on quantum-mechanical aspects of spin evolution and relaxation, in many cases the considerations are just based on the pseudorotational model of the ZFS interaction, which we shall adapt to the quadrupolar coupling. I just do not aim for a more specific discussion of possible mechanisms causing fluctuations of the electric field gradient or the ligand field, and in consequence for more sophisticated motional models. As has been stated above, the spin systems are treated as illustrations of various quantum-mechanical problems; a description of such motional mechanisms is a complicated issue by itself and is beyond the scope of this book. An exception is Chapter XIII, where we discuss a model of the ZFS interactions based on quantum vibrations of molecular complexes (lattices) leading to temporal changes of normal coordinates for the molecular systems of interest. Nevertheless, the purpose of this chapter is rather to demonstrate how to describe relaxation processes when the lattice must the treated quantum-mechanically, and the proposed motional model is just an example.

We have learned from Chapter I and Chapter II that correlation functions and their spectral densities are fundamental quantities for relaxation processes. Therefore, it is worthwhile and very helpful for further studying of specific relaxation theories presented in the forthcoming chapters to evaluate the correlation functions (spectral densities) associated with the ZFS (quadrupolar) interactions. We need for this purpose a solution of the isotropic rotational diffusion equation (Eq.1.8a). The solution is obviously well known and has a form of a superposition of Wigner rotation matrices [1, 9, 36]:

$$P(\Omega,\Omega_0,\tau) = \frac{2l+1}{4\pi} \sum_{l,m} D_{0,m}^{l*}(\Omega_0) D_{0,m}^{l}(\Omega) \exp\left(-\frac{\tau}{\tau_R}\right) \quad (3.14)$$

where the characteristic time constant τ_R is the rotational correlation time for second rank Wigner rotational matrices. Substituting this series into the general definition of Eq.1.6, taking into account that $P_{eq}(\Omega_0) = \frac{1}{4\pi}$ and making use of orthogonal properties of the Wigner rotation matrices [11-14]:

$$\int D_{k,m}^{l*}(\Omega) D_{k',m'}^{l'}(\Omega) d\Omega = \frac{8\pi^2}{2l+1} \delta_{ll'} \delta_{kk'} \delta_{mm'} \quad (3.15)$$

one can obtain closed form expressions for the relevant correlation functions. In the case of the static ZFS interaction considered in the laboratory frame one gets:

$$\left\langle \widetilde{V}_m^{2S(L)*}(\tau)\widetilde{V}_m^{2S(L)}(0)\right\rangle = \frac{\Delta_S^2}{5}\exp\left(-\frac{\tau}{\tau_R}\right) \quad (3.16)$$

where the amplitude Δ_S of the static ZFS is defined as $\Delta_S^2 = \frac{2}{3}D_S^2 + 2E_S^2$. We have explained in Chapter II that the correlation function describes the way in which the molecular system loses its memory of the initial configuration when time passes. Fig.3.4 illustrates this statement.

We can see progressively less influence of the initial orientation of the molecular frame on its position with respect to the laboratory axis after time τ.

In the same way we can evaluate the correlation function for the transient ZFS when considered in the (P_S) frame:

$$\left\langle \widetilde{V}_m^{2T(P_S)*}(\tau)\widetilde{V}_m^{2T(P_S)}(0)\right\rangle = \frac{\Delta_T^2}{5}\exp\left(-\frac{\tau}{\tau_D}\right) \quad (3.17)$$

The amplitude of the transient ZFS is defined by full analogy with its static counterpart, i.e. $\Delta_T^2 = \frac{2}{3}D_T^2 + 2E_T^2$, but the correlation time τ_D characterizes now the distortional (vibrational) motion leading to the spread of the ZFS interaction, assuming that this motion can be described by the pseudorotational model. It can be interesting to evaluate also the correlation function associated with the modulations of the transient ZFS with respect to the laboratory frame. The correlation function decays with the effective correlation time τ_{eff}:

$$\left\langle \widetilde{V}_m^{2T(L)*}(\tau)\widetilde{V}_m^{2T(L)}(0)\right\rangle = \frac{\Delta_T^2}{5}\exp\left(-\frac{\tau}{\tau_{eff}}\right) \quad (3.18)$$

resulting from the two-step transformation and therefore containing the rotational as well as the distortional contributions: $\tau_{eff}^{-1} = \tau_R^{-1} + \tau_D^{-1}$. This is so if the distortional and rotational motions are stochastically independent, but this problem will be considered in the next chapters.

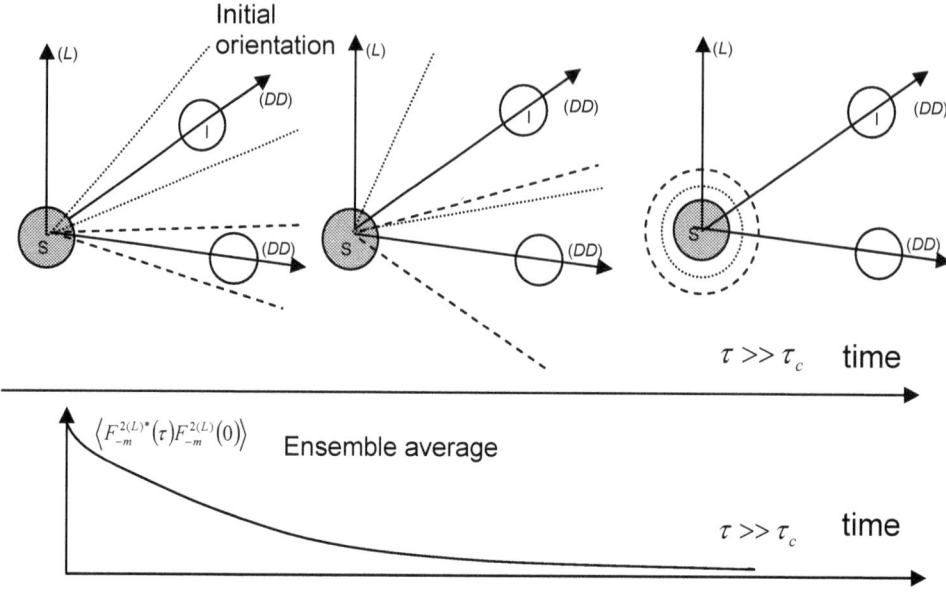

Fig.3.4
*The concept of the correlation function.
After time τ significantly shorter than the correlation time τ_c (the rotational correlation time in this case) the orientation of the dipole-dipole axis does not differ very much from its initial orientation. The available positions of the dipole-dipole axis are restricted to the areas between dotted or dashed lines, depending on the corresponding initial orientations. The ranges of available orientations become broader when the time interval, τ, increases. In consequence they overlap partially. This means that there are orientations of the dipole-dipole axis which can be reached after time τ starting from different initial orientations. This indicates that the system continuously loses its memory. After time τ much longer than the correlation time all orientations are available with the same probability. The memory is completely lost and therefore the correlation function $\langle F_{-m}^{2(L)*}(\tau) F_{-m}^{2(L)}(0) \rangle$ decays to zero.*

References:

69. A. Abragam, The principles of nuclear magnetism, *Oxford University Press, Oxford* (1961)
70. Ch. P. Poole, Jr, H.A. Farach, Theory of magnetic resonance, *A Wiley – Interscience publication, John Wiley & Sons* (1987)
71. C.P. Slichter, Principles of magnetic resonance, *Springer-Verlag, Berlin* (1990)
72. R.R. Ernst, G. Bodenhausen, A, Wokaun, Principles of nuclear magnetic resonance in one and two dimensions, *Clarendon Press, Oxford* (1994)
73. S.A. Smith, W.E. Palke, J.T. Gerig, The Hamiltonians of NMR. Part I, *Concepts Magn. Reson.* **4** (1992) 107-144
74. S.A. Smith, W.E. Palke, J.T. Gerig, The Hamiltonians of NMR. Part II, *Concepts Magn. Reson.* **4** (1992) 181-204
75. S.A. Smith, W.E. Palke, J.T. Gerig, The Hamiltonians of NMR. Part III, *Concepts Magn. Reson.* **4** (1993) 151-177
76. R. Kimmich, NMR - Tomography, Diffusometry, Relaxometry, *Springer – Verlag Berlin* (1997)
77. J. Kowalewski, L. Mäler, Nuclear spin relaxation in liquids: theory, experiments and applications, Series in Chemical Physics, *Taylor & Francis Group* (2006)
78. P.P. Man, Quadrupolar interactions, in Encyclopedia of Nuclear Magnetic Resonance, D.M. Grant, R.K. Harris (Eds.) *Chichester; Willey* (1996) 3838-3847
79. D.A. Varshalovich, A.N. Moskalev, V.K. Khersonkii, Quantum theory of angular momentum, *Word Scientific Publishing, Singapore* (1988)
80. D. M. Brink, G.R. Satchler, Angular momentum, *Clarendon Press, Oxford* (1979)
81. A. R. Edmunds, Angular momentum in quantum mechanics, *Princeton University Press, Princeton* (1974)
82. M. E. Rose, Elementary theory of angular momentum, *Wiley, New York* (1957)
83. W. Low, Paramagnetic resonances in solids, in Solid State Physics, Supplement 2, F. Seitz, D. Turnbull (Ed.), *Academic Press, New York* (1960)

84. J.S. Griffith, The Theory of Transition metal ions, *Cambridge University Press, Cambridge* (1961)
85. A. Abragam, B. Bleaney, Electron paramagnetic resonance of transition ions, *Clarendon Press, Oxford* (1970)
86. J. E. Wertz, J. R. Bolton, Electron Spin Resonance: Elementary theory and practical applications, *McGraw-Hill, New York* (1972)
87. S. R. Langhoff, C. W. Kern, Applications of electronic structure theory, H.F. Schaefer (Ed.), *Plenum Press, New York* (1977) 381-437
88. M. Bersohn, J.C. Baird, An introduction to electron paramagnetic resonance, *Benjamin, New York* (1996)
89. I. Bertini, C. Luchinat, G. Parigi, Solution NMR of paramagnetic molecules, *Elsevier: Amsterdam* (2001)
90. R. Sharp, L. Lohr, J. Miller, Paramagnetic NMR relaxation enhancement: recent advances in theory, *Prog. Nucl. Magn. Reson. Spectr.* **38** (2001) 115-158
91. P.-O. Westlund, Nuclear paramagnetic spin relaxation theory: Paramagnetic spin probes in homogenous and microheterogenous solutions, in Dynamics of solutions and fluid mixtures by NMR, J. J. Delpuech (Ed.), *Wiley, Chichester* (1995) 173-229
92. J. Kowalewski, Paramagnetic relaxation in solution, in Encyclopedia of Nuclear Magnetic Resonance, D.M. Grant, R.K. Harris (Ed.), *Wiley, Chichester* (1996) 3455-3462
93. N. Benetis, J. Kowalewski, L. Nordenskiöld, H. Wennerström, P.-O. Westlund, *Mol. Phys.* **48** (1983) 329-346
94. N. Benetis, J. Kowalewski, L. Nordenskiöld, H. Wennerström, P.-O. Westlund, *Mol. Phys.* **50** (1983) 515-530
95. N. Benetis, J. Kowalewski, L. Nordenskiöld, H. Wennerström, P.-O. Westlund, *J. Magn. Reson.* **58** (1984) 261-281
96. P.-O. Westlund, H. Wennerström, L. Nordenskiöld, J. Kowalewski, N. Benetis, *J. Magn. Reson.* **59** (1984) *91-109*
97. D. Kruk, Field dependent electron and quadrupole spin relaxation: A unified treatment, *Mol. Phys. Rep.* – in press

98. M. Rubinstein, A. Baram, Z. Luz, Electronic and nuclear relaxation in solutions of transition metal ions with spin S=3/2 and 5/2, *Mol. Phys.* **20** (1971) 67-80
99. N. Bloembergen, L.O. Morgan, Proton relaxation times in paramagnetic solutions: effects of electron spin relaxation, *J. Chem. Phys.* **34** (1961) 842-850
100. P.-O. Westlund, N. Benetis, H. Wennerström, Paramagnetic proton nuclear magnetic relaxation in the Ni^{2+} hexa-aquo complex: A theoretical study, *Mol. Phys.* **61** (1987) 177-194
101. P.-O. Westlund, P. T. Larsson, O. Teleman, Paramagnetic enhanced proton spin – lattice relaxation in the Ni^{2+} hexa-aquo complex: A theoretical and molecular dynamics simulation study of the Bloembergen – Morgan decomposition approach, *Mol. Phys.* **78** (1983) 1365 – 1384
102. J. Svoboda, T. Nilsson, J. Kowalewski, P.-O. Westlund, P. T. Larsson, Field – dependent proton NMR relaxation in aqueous solutions of Ni (II) ions. A new interpretation, *J. Magn. Reson. A* **121** (1996) 108-113
103. T. Larsson, P.-O. Westlund, J. Kowalewski, S. H. Koenig, Nuclear – spin relaxation in paramagnetic complexes in the slow-motion regime for the electron spin – the anisotropic pseudorotational model for S=1 and the interpretation of nuclear magnetic-relaxation dispersion results for a low-symmetry Ni(II) complex, *J. Chem. Phys.* **101** (1994) 1116 – 1128
104. M. Doi, S. F. Edwards, The theory of polymer dynamics, *Clarendon Press, Oxford* (1986)
105. H. W. Spiess, in NMR basic principles and progress, 15, *Springer – Verlag, Berlin Heidelberg New York* (1978) 56-214
106. D. E. Woessner, Brownian motion and correlation times, in Encyclopedia of Nuclear Magnetic Resonance, D.M. Grant, R.K. Harris, (Eds.) *Chichester: Wiley* (1996) 1068-1084
107. D. E. Woessner, Nuclear spin relaxation in ellipsoids undergoing rotational Brownian motion, *J. Chem. Phys.* **37** (1962) 647
108. D.M. Grant, R.A. Brown, Relaxation of coupled spin systems from rotational diffusion, in Encyclopedia of Nuclear Magnetic Resonance, D.M. Grant, R.K. Harris, (Eds.) *Chichester: Wiley* (1996) 4003-4018

CHAPTER IV

Examples of relaxation processes treated within the perturbation approach

In this chapter I shall use the relaxation theory, presented in Chapter II in a general and formal manner, to obtain closed form expressions for relaxation rates characterizing 'typical' spin systems like, for example two equivalent or non-equivalent spins 1/2. For some selected cases I shall describe the relaxation processes by applying the relaxation matrix formalism as well as the Liouville operator approach and demonstrate that they are fully equivalent. This chapter is meant to provide a practical background for further more advanced considerations.

IV.1 Two equivalent spins 1/2 relaxing via mutual dipole-dipole coupling

Consider two equivalent spins I_1 and I_2 of the spin quantum number 1/2 coupled by a dipole-dipole interaction. Let us assume that the inter-spin vector $\vec{r}_{12}$, determining the principal axis of the dipole-dipole interaction, changes its orientation with respect to the direction of the external magnetic field due to rotational motion. The Hamiltonian H for the entire system consists of the pure spin part H_I containing Zeeman couplings of the participating spins:

$H_I = H_z(I_1) + H_z(I_2)$, the perturbing part $H_{IL}(t) = H_{DD}(I_1, I_2)(t)$ provided by the dipole-dipole interactions fluctuating in time due to the molecular tumbling and the pure lattice part H_L describing the classical continuum of rotational states of the lattice. Fig.4.1 shows a schematic view of the system.

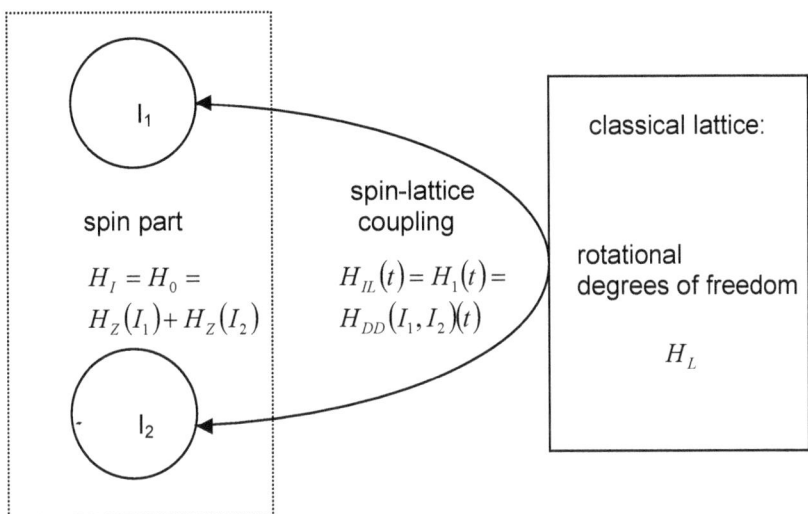

Fig.4.1
Schematic view of a spin system containing two spins (I_1 and I_2) coupled by a dipole-dipole interaction, which fluctuates in time due to rotational motion. The dipole-dipole interaction couples mutually the spins via the classical lattice including the rotational degrees of freedom.

We shall denote the main, unperturbed Hamiltonian of the spin system as $H_0 = H_I = \omega_0(I_{1z} + I_{2z})$, and the perturbing, spin-lattice coupling as $H_1(t) = H_{IL}(t)$. Such a notation is common for calculations based on the perturbation theory and we shall use it as well. The frequency $\omega_0 = \omega_I$ is the characteristic Larmor frequency of the spins under interest $\omega_0 = \gamma_I B_0$ ($\gamma_I = \gamma_{I_1} = \gamma_{I_2}$ is the gyromagnetic factor) while I_{1z} and I_{2z} are z-components of the spin operators. The Hilbert basis $\{|n\rangle\}$, appropriate for the case of two spins

$\frac{1}{2}$, can be formed by four functions, $|n\rangle = |m_1, m_2\rangle$, where m_1 and m_2 are the magnetic quantum numbers of the spins I_1 and I_2, respectively. The basis $\{|m_1, m_2\rangle\}$ is obviously the eigenbasis of the main Hamiltonian H_0. In the further description of the $I_1 - I_2$ spin system we shall use the following numbering of the eigenfunctions $|m_1, m_2\rangle$: $|1\rangle = \left|\frac{1}{2}, \frac{1}{2}\right\rangle$, $|2\rangle = \left|\frac{1}{2}, -\frac{1}{2}\right\rangle$, $|3\rangle = \left|-\frac{1}{2}, \frac{1}{2}\right\rangle$ and $|4\rangle = \left|-\frac{1}{2}, -\frac{1}{2}\right\rangle$. The corresponding energy levels are given as: $E(m_1, m_2) = \omega_0(m_1 + m_2)$, that $E_1 = \omega_0$, $E_2 = E_3 = 0$ (the two energy levels are degenerate) and $E_4 = -\omega_0$. The dipole-dipole Hamiltonian, expressed in the laboratory frame in terms of the second rank spin tensor operators $T_m^2(I_1, I_2)$ and the corresponding lattice functions $F_{-m}^2(t) \equiv F_{-m}^{2(L)}(t)$ (Eq. 3.7) can be represented conveniently by its matrix elements in this Zeeman basis $\{|m_1, m_2\rangle\}$. The matrix representation of the dipole-dipole Hamiltonian $\langle \alpha | H_{DD}^{(L)}(I_1, I_2) | \beta \rangle$ results from the properties of the raising and lowering operators [1-2] and has the form:

$$\left[\frac{H_{DD}^{(L)}(I_1, I_2)}{a_{DD}^{I_1 I_2}}\right] = \begin{array}{c} \\ \langle 1| \\ \langle 2| \\ \langle 3| \\ \langle 4| \end{array} \begin{bmatrix} |1\rangle & |2\rangle & |3\rangle & |4\rangle \\ \frac{1}{2\sqrt{6}} F_0^{2(L)} & -\frac{1}{4} F_{-1}^{2(L)} & -\frac{1}{4} F_{-1}^{2(L)} & \frac{1}{2} F_{-2}^{2(L)} \\ \frac{1}{4} F_1^{2(L)} & -\frac{1}{2\sqrt{6}} F_0^{2(L)} & -\frac{1}{2\sqrt{6}} F_0^{2(L)} & \frac{1}{4} F_{-1}^{2(L)} \\ \frac{1}{4} F_1^{2(L)} & -\frac{1}{2\sqrt{6}} F_0^{2(L)} & -\frac{1}{2\sqrt{6}} F_0^{2(L)} & \frac{1}{4} F_{-1}^{2(L)} \\ \frac{1}{2} F_2^{2(L)} & -\frac{1}{4} F_1^{2(L)} & -\frac{1}{4} F_1^{2(L)} & \frac{1}{2\sqrt{6}} F_0^{2(L)} \end{bmatrix} \quad (4.1)$$

Even though the calculations are very simple, we provide the explicit expression for convenience of the reader and clarity of the further calculations. The lattice functions $F_{-m}^{2(L)}(t)$ are Wigner rotation matrices of the angles Ω_{DDL} specifying the direction of the dipole-dipole axis with respect to the laboratory frame, $F_{-m}^{2(L)}(t) = D_{0,-m}^2[\Omega_{DDL}(t)]$ (see Section III.2). We shall calculate the relaxation matrix elements $R_{\alpha\alpha'\beta\beta'}$ for this spin system applying the WBR relaxation theory presented in Section II.3; Eq.2.14 provides the computational framework for this purpose. The relaxation matrix elements can be obtained straightforwardly utilizing the representation of the perturbing interaction in the eigenbasis of the main Hamiltonian (Eq.4.1) and keeping in mind the simple energy level structure for this system of two spins. The dimension of the relaxation matrix is 16 while the population, zero-quantum block is of the dimension 4. The individual relaxation matrix elements $R_{\alpha\alpha\beta\beta}$ connecting the time evolution of the populations $\rho_{\alpha\alpha}$ and $\rho_{\beta\beta}$ have the explicit form:

$$\frac{d}{dt}\begin{bmatrix}\rho_{11}\\ \rho_{22}\\ \rho_{33}\\ \rho_{44}\end{bmatrix}=$$

$$\begin{bmatrix} -\frac{1}{4}J_1(\omega_0)-\frac{1}{2}J_2(2\omega_0) & \frac{1}{8}J_1(\omega_0) & \frac{1}{8}J_1(\omega_0) & \frac{1}{2}J_2(2\omega_0) \\ \frac{1}{8}J_1(\omega_0) & -\frac{1}{12}J_0(0)-\frac{1}{4}J_1(\omega_0) & \frac{1}{12}J_0(0) & \frac{1}{8}J_1(\omega_0) \\ \frac{1}{8}J_1(\omega_0) & \frac{1}{12}J_0(0) & -\frac{1}{12}J_0(0)-\frac{1}{4}J_1(\omega_0) & \frac{1}{8}J_1(\omega_0) \\ \frac{1}{2}J_2(2\omega_0) & \frac{1}{8}J_1(\omega_0) & \frac{1}{8}J_1(\omega_0) & -\frac{1}{4}J_1(\omega_0)-\frac{1}{2}J_2(2\omega_0) \end{bmatrix} \times \begin{bmatrix}\rho_{11}\\ \rho_{22}\\ \rho_{33}\\ \rho_{44}\end{bmatrix}$$

(4.2)

The lattice spectral densities $J_{m,m}^2(\omega)$ are defined in a general manner by Eq.2.18. Since we assume isotropic molecular tumbling modulating the dipole-dipole interaction, the correlation function takes the form:

$$\langle F_m^{2(L)*}(t)F_m^{2(L)}(0)\rangle = \frac{1}{5}\exp\left(-\frac{t}{\tau_R}\right)$$ [3-5] (in analogy to Eqs.3.16-3.18, but the $F_m^{2(L)}$ quantities do not include the dipole-dipole constant $a_{DD}^{I_1 I_2}$); in consequence the spectral densities (see Eq.1.9) are given as:

$$J_m(\omega) = J_{m,m}^2(\omega) = \frac{1}{5}\frac{\tau_R}{1+\omega^2\tau_R^2} \quad (4.3)$$

where $\tau_R = \tau_R^{(2)}$ is the rotational correlation time for rank-two Wigner rotational matrices (see Eq. 3.16). Apparently the obtained expressions for the particular relaxation rates are valid for any form of spectral densities reflecting the type of motion occurring in the lattice. Noticing that $R_{1122} = R_{1133} = R_{4422} = R_{4433}$, one can easy conclude that the vector $[1,0,0,-1]$ corresponding to the combination of the density matrix elements $\rho_{11} - \rho_{44}$ (proportional to the expectation value $\langle I_z \rangle$ of the $I_z = I_{1z} + I_{2z}$ operator [6-8]; $\langle I_z \rangle = \frac{1}{2}(\rho_{11} - \rho_{44})$) is an eigenvector of the population (zero-quantum) block of the relaxation matrix. The corresponding eigenvalue multiplied by a square of the dipole-dipole constant gives the spin-lattice relaxation rate $R_1 = T_1^{-1}$ for the system of two equivalents spins 1/2:

$$R_1 = \left(a_{DD}^{I_1 I_2}\right)^2 \left(\frac{1}{4} J_1(\omega_0) + J_2(2\omega_0)\right) = 2\left(\frac{\mu_0}{4\pi} \frac{\gamma_I^2 \hbar}{r_{12}^3}\right)^2 I(I+1)[J_1(\omega_0) + 4J_2(2\omega_0)] \quad (4.4)$$

This expression as well as its generalization to an arbitrary value of the spin quantum number I is the most fundamental (and simplest) example of closed form descriptions of relaxation processes; one can find it everywhere in NMR literature, for example [3-5, 8-10]. It has been derived in 1948 by Bloembergen, Purcell and Pound [11] and is known as the classical BPP formula.

It is very important to realize at this moment that even though we have obtained the proper final expression our calculations were not absolutely correct. Setting up the relaxation matrix for systems of equivalent spins one has to notice that the population block (covering the density matrix elements $\rho_{\alpha\alpha}$) cannot be just decoupled from some of the single quantum coherences (spin transitions leading to the change of the magnetic spin quantum number of the spin system $\Delta m = \Delta m_{I_1} + \Delta m_{I_2}$ by ± 1, i.e. $|\Delta m| = 1$) represented by the density matrix elements $\rho_{\alpha\beta}$. Actually in the considered case of two equivalent spins $\frac{1}{2}$ the evolution of the populations is coupled to the coherences ρ_{23} and ρ_{32}, because of the degeneracy of the energy levels $E_2 = E_3$, that $\omega_{\alpha\alpha} = \omega_{23} = \omega_{32}$. Therefore, the proper equations for the evolution of the populations ρ_{11} and ρ_{44} have the form:

$$\frac{d\rho_{11}}{dt} = -R_{1111}\rho_{11} - R_{1122}\rho_{22} - R_{1133}\rho_{33} - R_{1144}\rho_{44} - R_{1123}\rho_{23} - R_{1132}\rho_{32} \quad (4.5a)$$

$$\frac{d\rho_{44}}{dt} = -R_{1144}\rho_{11} - R_{2244}\rho_{22} - R_{3344}\rho_{33} - R_{4444}\rho_{44} - R_{2344}\rho_{23} - R_{3244}\rho_{32} \quad (4.5b)$$

We have not included into the calculations the couplings provided by the relaxation matrix elements R_{1123}, R_{1132} and R_{4423}, R_{4432}. Nevertheless, since $R_{1123} = R_{2344}$ and $R_{3211} = R_{3244}$, terms containing the coherences ρ_{23} and ρ_{32} do not appear in the expression for the population difference: $\langle I_z \rangle \propto \frac{d\rho_{11}}{dt} - \frac{d\rho_{44}}{dt}$. In other words, the vector $[1,0,0,-1]$ is really the eigenvector of the whole relaxation matrix (of the dimension 16) covering all coherences available for the $I_1 - I_2$ system. Therefore, we have obtained the proper formulation for the spin-lattice relaxation time neglecting the couplings to the coherences ρ_{23} and ρ_{32}. Nevertheless, this example can be treated as an illustration of the secular approximation condition, ($\omega_I >> \omega_{IL}^2 \tau_c$) which must be treated always with high caution.

IV.2 Dipolar relaxation in systems of two non-equivalent spins 1/2

In the more general case of two different spins I and S mutually coupled by a dipole-dipole interaction the spin-lattice relaxation process has a bi-exponential character and must be described by a set of two differential equations. It has been demonstrated by Solomon [12]. This result, based this time on a simple kinetic model can be obviously obtained within the density matrix formalism. I shall illustrate this for the simplest case of $I = S = \frac{1}{2}$ (we have escape for a while from the convention of this book that S denotes spins of the spin quantum number greater than $1/2$), generalizing the considerations from the previous chapter. The main Hamiltonian H_0 for the $I - S$ spin system must be now adjusted for different Larmor frequencies ω_I and ω_S of the spins I and S, respectively, that $H_0 = \omega_I I_z + \omega_S S_z$. It acts on the basis functions $|m_I, m_S\rangle$ (m_I and m_S are the magnetic quantum numbers) and gives the corresponding energy levels $E(m_I, m_S) = \omega_I m_I + \omega_S m_S$. The fact that the spins exhibit different Larmor frequencies removes the degeneracy of the E_2 and E_3 energy levels. The $R_{\alpha\alpha\beta\beta}$ relaxation matrix elements have to be adjusted for the new energy level structure of the present spin system. The relaxation rates appropriate for this case are collected in the matrix below:

$$\frac{d}{dt}\begin{bmatrix}\rho_{11}\\\rho_{22}\\\rho_{33}\\\rho_{44}\end{bmatrix} =$$

$$\begin{bmatrix} \Lambda & \frac{1}{8}J_1(\omega_S) & \frac{1}{8}J_1(\omega_I) & \frac{1}{2}J_2(\omega_I+\omega_S) \\ \frac{1}{8}J_1(\omega_S) & \Lambda & \frac{1}{12}J_0(\omega_I-\omega_S) & \frac{1}{8}J_1(\omega_I) \\ \frac{1}{8}J_1(\omega_I) & \frac{1}{12}J_0(\omega_I-\omega_S) & \Lambda & \frac{1}{8}J_1(\omega_S) \\ \frac{1}{2}J_2(\omega_I+\omega_S) & \frac{1}{8}J_1(\omega_I) & \frac{1}{8}J_1(\omega_S) & \Lambda \end{bmatrix} \times \begin{bmatrix}\rho_{11}\\\rho_{22}\\\rho_{33}\\\rho_{44}\end{bmatrix} \quad (4.6)$$

where the diagonal elements are given as:

$R_{\alpha\alpha\alpha\alpha} = -\sum_{\gamma\neq\alpha} R_{\alpha\alpha\gamma\gamma} = \Lambda = -\frac{1}{8}J_1(\omega_I)-\frac{1}{8}J_1(\omega_S)-\frac{1}{2}J_2(\omega_I+\omega_S)$. Since there is no degeneracy of the energy levels, the zero-quantum (population) block of the relaxation matrix ($\Delta m = 0$) is decoupled from other coherences and the set of differential equations given by Eq.4.6 is complete and correct. In particular, it leads to the relations:

$$\frac{d}{dt}(\rho_{11}+\rho_{22}-\rho_{33}-\rho_{44}) = -\left[\frac{1}{12}J_0(\omega_I-\omega_S)+\frac{1}{4}J_1(\omega_I)+\frac{1}{2}J_2(\omega_I+\omega_S)\right](\rho_{11}+\rho_{22}-\rho_{33}-\rho_{44})$$
$$+\left[\frac{1}{2}J_2(\omega_I+\omega_S)-\frac{1}{12}J_0(\omega_I-\omega_S)\right](\rho_{11}+\rho_{33}-\rho_{22}-\rho_{44}) \quad (4.7a)$$

$$\frac{d}{dt}(\rho_{11}+\rho_{33}-\rho_{22}-\rho_{44}) = -\left[\frac{1}{12}J_0(\omega_I-\omega_S)+\frac{1}{4}J_1(\omega_S)+\frac{1}{2}J_2(\omega_I+\omega_S)\right](\rho_{11}+\rho_{33}-\rho_{22}-\rho_{44})$$
$$+\left[\frac{1}{2}J_2(\omega_I+\omega_S)-\frac{1}{12}J_0(\omega_I-\omega_S)\right](\rho_{11}+\rho_{22}-\rho_{33}-\rho_{44}) \quad (4.7b)$$

The linear combinations of the density matrix elements $\rho_{11}+\rho_{22}-\rho_{33}-\rho_{44}$ and $\rho_{11}+\rho_{33}-\rho_{22}-\rho_{44}$ represent expectation values of the operators I_z and S_z, respectively. Thus Eqs.4.7a,b demonstrate the Solomon statement that the spin-lattice relaxation of two non-equivalent spins does not occur independently. The set of coupled equations can be written in the form being in a common use in the NMR literature, (for example [3 - 5 ,8]):

$$\frac{d\langle I_z \rangle}{dt} = -R_{II}\left(\langle I_z \rangle - \langle I_z \rangle_{eq}\right) - R_{IS}\left(\langle S_z \rangle - \langle S_z \rangle_{eq}\right) \tag{4.8a}$$

$$\frac{d\langle S_z \rangle}{dt} = -R_{SI}\left(\langle I_z \rangle - \langle I_z \rangle_{eq}\right) - R_{SS}\left(\langle S_z \rangle - \langle S_z \rangle_{eq}\right) \tag{4.8b}$$

where the coefficients $R_{IS} = R_{SI}$ are called cross-relaxation rates; the quantities $\langle I_z \rangle_{eq}$ and $\langle S_z \rangle_{eq}$ refer to the equilibrium state. The bi-exponential evolution of the $\langle I_z \rangle$ value is described by the two time constants $R_{1I}^{\pm}$ obtained from the solution Eqs.4.8a,b:

$$R_{1I}^{\pm} = -\frac{1}{2}\left[(R_{II} + R_{SS}) \pm \sqrt{(R_{II} - R_{SS})^2 + 4R_{IS}R_{SI}}\right] \tag{4.9}$$

The relaxation of the spin I is single exponential only if the spin S has an independent and highly efficient relaxation pathway. In this case the second term in Eq.4.8a becomes zero (the magnetization of the S spin disappears much faster than the magnetization of the I spin) and the spin-lattice relaxation of the I spin is described by the well known formula (for example [3-5, 8-10]):

$$T_{1I}^{-1} = R_{II} = \left(a_{DD}^{IS}\right)^2\left[\frac{1}{12}J_0(\omega_I - \omega_S) + \frac{1}{4}J_1(\omega_I) + \frac{1}{2}J_2(\omega_I + \omega_S)\right]$$

$$= \left(\frac{\mu_0}{4\pi}\frac{\gamma_I \gamma_S \hbar}{r_{IS}}\right)^2 \frac{2}{3}S(S+1)\left[J_0(\omega_I - \omega_S) + 3J_1(\omega_I) + 6J_2(\omega_I + \omega_S)\right] \tag{4.10}$$

From the perspective of the spin I the fast relaxation processes of the spin S contribute to time fluctuations of the mutual $I-S$ dipole-dipole coupling in a manner similar to other stochastic processes like for example molecular tumbling. In other words, the S spin can provide through its own relaxation mechanism an additional source of relaxation for the I spin. Therefore the S spin subsystem can be treated as a part of the lattice of the I spin. This is in fact the idea of the composite lattice (including classical as well as quantum mechanical degrees of freedom) introduced in Section II.5. In Section IV.4 we shall discuss the I spin relaxation once again invoking this concept and applying the Liouville operator formalism. If all dynamic processes relevant for the effective fluctuations of the *I-S* coupling (rotational diffusion, the *S* spin relaxation, etc.) can be treated as uncorrelated, one can formulate the description of the *I* spin relaxation by introducing a set of correlation times (rates) $\tau_{c,i}^{-1}$, which include the individual motional processes as a superposition of their characteristic time constants: $\tau_{c,i}^{-1} = \tau_R^{-1} + R_{iS}$. This entails that by adjusting the spectral densities of Eq.4.10 (determined so far by the single correlation time τ_R, Eq.4.3) in an appropriate manner one can get an expression for the *I* spin relaxation originating from rapid relaxation of the *S* spin. This treatment is particularly straightforward if the *S* spin relaxation can be described by a single spin-lattice relaxation rate $R_{1S} = T_{1S}^{-1}$ and a single spin-spin relaxation rate $R_{2S} = T_{2S}^{-1}$. Looking at Eq.4.10, it is apparent that transitions of the spin *I* leading to its relaxation are associated through the dipole-dipole coupling to certain transitions of the spin *S*. The connection between the dipolar spectral densities and the *S* spin relaxation is relatively simple: the spectral densities J_0 and J_2 corresponding to the transitions with $|\Delta m_S| = 1$ and therefore including the frequency ω_S are associated with the T_{S2} relaxation time, while the spectral density J_1 contains the T_{1S} time constant. This treatment leads to the well known Solomon-Bloembergen-Morgan formulation [12-16] of the *I* spin relaxation (originally formulated for the relaxation of a nuclear spin $I = \frac{1}{2}$ induced by a dipole-dipole coupling to a fast relaxing electron spin $S = 1$):

$$R_{1I} = R_{II} =$$

$$\left(\frac{\mu_0}{4\pi}\frac{\gamma_I \gamma_S \hbar}{r_{IS}^3}\right)^2 \frac{2}{15} S(S+1) \left[\frac{\tau_{c2}}{1+(\omega_I - \omega_S)^2 \tau_{c2}^2} + \frac{3\tau_{c1}}{1+\omega_I^2 \tau_{c1}^2} + \frac{6\tau_{c2}}{1+(\omega_I + \omega_S)^2 \tau_{c2}^2}\right] \quad (4.11)$$

where $\tau_{ci}^{-1} = \tau_R^{-1} + R_{iS}$, ($i = 1,2$). In particular, when the molecular tumbling is slow the electron spin dynamics is the only source of the relaxation processes of the *I* spin.

At this instant it is important to notice the hierarchy of events: the essence of the already presented treatments is the assumption that the relaxation dynamics of the spin *S* is not affected by the presence of the spin *I*. Actually, electron spins

Chapter IV – Examples of relaxtion processes treated within the perturbation approach

of $S \geq 1$ have a rapid relaxation pathway of their own, provided by the fluctuating part of the zero field splitting tensor (see Chapter III). As an illustration of the ZFS relaxation mechanism, we shall derive explicit expressions for the electron spin relaxation rates R_{1S} and R_{2S} in the case of $S = 1$ and high molecular symmetry, that there is no static (long-time averaged) component of the ZFS tensor. Most importantly, this example provides an introduction to relaxation processes in coupled, multi-spin systems and we shall profit from this detailed description in the forthcoming sections dealing with much more complicated systems.

IV.3. Electron spin relaxation caused by fluctuations of zero field splitting (ZFS) tensor

The Zeeman energy splitting of the spin $S = 1$ consists of three energy levels, $E(m_S) = \omega_S m_S$, corresponding to the three eigenstates $|m_S\rangle$ labeled as follows: $|1\rangle = |1\rangle, |2\rangle = |0\rangle$ and $|3\rangle = |-1\rangle$. Relaxation matrix elements for the electron spin can be obtained in the same manner as in Sections IV.2 And IV.3, *i.e.* from Eq.2.14, but with the ZFS Hamiltonian, expressed in the laboratory frame (Eq.3.13) as the perturbing interaction. To simplify the physical picture of the spin system as much as possible we assume that the static (permanent) ZFS vanishes. For convenience of the reader we collect below the matrix elements of the transient ZFS Hamiltonian $H_{ZFS}^{T(L)}$ in the Zeeman basis $\{|m_S\rangle\}$:

$$\left[H_{ZFS}^{T(L)}\right] = \begin{array}{c} \\ \langle 1| \\ \langle 2| \\ \langle 3| \end{array} \begin{bmatrix} |1\rangle & |2\rangle & |3\rangle \\ \frac{1}{\sqrt{6}}\widetilde{V}_0^{2T(L)} & \frac{1}{\sqrt{2}}\widetilde{V}_{-1}^{2T(L)} & \widetilde{V}_{-2}^{2T(L)} \\ -\frac{1}{\sqrt{2}}\widetilde{V}_1^{2T(L)} & -\frac{2}{\sqrt{6}}\widetilde{V}_0^{2T(L)} & -\frac{1}{\sqrt{2}}\widetilde{V}_{-1}^{2T(L)} \\ \widetilde{V}_2^{2T(L)} & \frac{1}{\sqrt{2}}\widetilde{V}_1^{2T(L)} & \frac{1}{\sqrt{6}}\widetilde{V}_0^{2T(L)} \end{bmatrix} \quad (4.12)$$

Setting up this matrix we have included the interaction constant into the spatial tensor functions (we use the $\widetilde{V}_m^{2T(L)}$ quantities instead of $V_m^{2T(L)}$); it is just the matter of convenience. Utilizing the Redfield formulation (Eq.2.14) one can derive the set of coupled equations for the time evolution of the populations $\rho_{\alpha\alpha}^S$ for the spin S:

$$\frac{d}{dt}\begin{bmatrix}\rho_{11}^S \\ \rho_{22}^S \\ \rho_{33}^S\end{bmatrix} = \begin{bmatrix} -J_1^S(\omega_S)-2J_2^S(2\omega_S) & J_1^S(\omega_S) & 2J_2^S(2\omega_S) \\ J_1^S(\omega_S) & -2J_1^S(\omega_S) & J_1^S(\omega_S) \\ 2J_2^S(2\omega_S) & J_1^S(\omega_S) & -J_1^S(\omega_S)-2J_2^S(2\omega_S) \end{bmatrix} \times \begin{bmatrix}\rho_{11}^S \\ \rho_{22}^S \\ \rho_{33}^S\end{bmatrix} \quad (4.13)$$

where $J_m^S(m\omega_S)$ are the S spin spectral densities resulting from the fluctuations of the zero field splitting tensor (and including the amplitude of the ZFS tensor). The expectation value $\langle S_z \rangle$ is represented by the difference between the populations of the states of $m_S = 1$ and $m_S = -1$, $\langle S_z \rangle = \rho_{11} - \rho_{33}$ and evolves exponentially in time according to the equation:

$$\frac{d}{dt}(\rho_{11}-\rho_{33}) = -[J_1^S(\omega_S)+4J_2^S(2\omega_S)](\rho_{11}-\rho_{33}) \quad (4.14)$$

In the similar manner one can set up the multi-quantum block of the relaxation matrix, yielding:

$$\frac{d}{dt}\begin{bmatrix}\rho_{12}^S \\ \rho_{23}^S \\ \rho_{13}^S\end{bmatrix} = \begin{bmatrix} -i\omega_S-\frac{3}{2}J_0^S(0)-\frac{3}{2}J_1^S(\omega_S)-J_2^S(2\omega_S) & -J_1^S(\omega_S) & 0 \\ -J_S^1(\omega_S) & -i\omega_S-\frac{3}{2}J_0^S(0)-\frac{3}{2}J_1^S(\omega_S)-J_2^S(2\omega_S) & 0 \\ 0 & 0 & -i\omega_S-J_1^S(\omega_s)-2J_2^S(2\omega_S) \end{bmatrix}$$

$$\times \begin{bmatrix}\rho_{12}^S \\ \rho_{23}^S \\ \rho_{13}^S\end{bmatrix} \quad (4.15)$$

The single-quantum coherences ($|\Delta m|=1$) ρ_{12}^S and ρ_{23}^S are coupled, because they are characterized by the same frequency ω_S, while the double-quantum coherence

($|\Delta m| = 2$) ρ_{13}^S evolves in time independently. The expectation value of the S_- operator, $\langle S_- \rangle \propto \rho_{12}^S + \rho_{23}^S$, follows the equation:

$$\frac{d}{dt}\left(\rho_{12}^S + \rho_{23}^S\right) = -\left[i\omega_S + \left(\frac{3}{2}J_0^S(0) + \frac{5}{2}J_1^S(\omega_S) + J_2^S(2\omega_S)\right)\right]\left(\rho_{12}^S + \rho_{23}^S\right) \quad (4.16)$$

It is apparent from Eq.4.14 and Eq.4.16 that for the $S = 1$ electron spin system the spin-lattice and spin-spin relaxation rates (R_{1S} and R_{2S}) caused by fluctuations of the ZFS tensor are given as:

$$R_{1S} = \frac{1}{5}\Delta_T^2\left[\frac{\tau_{eff}}{1+\omega_S^2\tau_{eff}^2} + \frac{4\tau_{eff}}{1+(2\omega_S\tau_{eff})^2}\right] \quad (4.17a)$$

$$R_{2S} = \frac{1}{10}\Delta_T^2\left[3\tau_{eff} + \frac{5\tau_{eff}}{1+\omega_S^2\tau_{eff}^2} + \frac{2\tau_{eff}}{1+(2\omega_S\tau_{eff})^2}\right] \quad (4.17b)$$

We have used here the explicit form of the spectral densities $J_m^S(\omega) = J^S(\omega) = \frac{1}{5}\Delta_T^2\frac{\tau_{eff}}{1+\omega^2\tau_{eff}^2}$ resulting from the correlation function of Eq.3.18. This has been recognized by Bloembergen and Morgan [14, 15] in early 1960s. In their original formulation instead of the product $\Delta_T^2\tau_D$ an electron spin relaxation time at zero magnetic field, τ_{S0} is used, that $\Delta_T^2\tau_D = \tau_{S0}^{-1}$. The Bloembergen-Morgan theory of the electron spin relaxation combined with the Solomon–Bloembergen treatment applied to a nuclear spin of $I = \frac{1}{2}$ coupled to rapidly relaxing electron spin of $S = 1$ yields a first, landmark theory for field dependent relaxation processes in multispin systems, known as the SBM model.

In the next section we shall examine in detail the multipole representation approach employing the concept of a composite lattice containing quantum-mechanical degrees of freedom. The same $I - S$ spin system will be used as a guide for understanding some aspects of practical evaluations of the quantum-mechanical correlation functions. Obviously, the final result must be the same. Since, we discuss in next chapters of this book relaxation processes in systems where such a treatment is highly recommendable, even mandatory, it is of primary importance to provide simple examples illustrating it in great detail.

IV.4. Quantum-mechanical correlation function and the Solomon-Bloembergen-Morgan (SBM) theory

The idea of this approach applied to the $I-S$ spin system is to let the electron spin S to be a part of a composite lattice of the nuclear spin I. The composite lattice contains quantum mechanical degrees of freedom related to the S spin subsystem as well as classical degrees of freedom in terms of the rotational motion modulating the orientation of the $I-S$ dipole-dipole axis and the distortions of the ligand field leading to time fluctuations of the ZFS tensor [17-20]. The number of degrees of freedom of the lattice is so great as to be effectively infinite, with essentially a continuum of eigenstates. Since we aim to extend (compared to the examples considered so far) the definition of the lattice we must also reconsider the form of the spin-lattice coupling provided by the dipole-dipole interaction. The present treatment requires separating the I spin from the lattice enclosing now explicitly the S spin subsystem. Thus, to follow the general idea of this approach, we need to separate in the representation of the $I-S$ dipole-dipole Hamiltonian (Eq.3.7) the part, which is dependent only on the I spin operators from other quantities characterizing the composite lattice. This problem has been already anticipated in Section III.1. One can achieve this expressing the second order $I-S$ tensor operators $T_m^2(I,S)$ in terms of the first order tensors: I_n^1 and S_n^1 for the I and S spins, respectively [21-24]. Then the dipole-dipole Hamiltonian yields the form [25-28]:

$$H_{DD}^{(L)}(I,S) = a_{DD}^{IS} \sum_{n=-1}^{1}(-1)^n I_n^1 T_{-n}^{1(DD)(L)} \\ = a_{DD}^{IS}\sum_{n=-1}^{1}(-1)^n I_{-n}^1\left\{\sqrt{5}\sum_{q=-1}^{1}\begin{pmatrix}2 & 1 & 1 \\ n-q & q & -n\end{pmatrix}S_q^1 D_{0,n-q}^2(\Omega_{DDL})\right\} \quad (4.18)$$

where the last equality gives explicitly the tensor operators $T_{-n}^{1(DD)(L)}$ of the composite lattice. Since now the lattice contains quantum mechanical degrees of freedom (represented explicitly by the S spin tensor components S_q^1, the lattice functions (operators) are denoted as $T_{-n}^{1(DD)(L)}$, not $F_{-n}^{1(DD)(L)}$. This is, in fact, important only for keeping a consistent labeling in this book. The tensor operators

I_n^1 (S_q^1) are related to the angular momentum operators: $P_0^1 = P_z$, $P_{\pm 1}^1 = \frac{1}{\sqrt{2}} P_\pm$, $P = I, S$ (see Eq.2.29a), while $\begin{pmatrix} 2 & 1 & 1 \\ n-q & q & -n \end{pmatrix}$ are the appropriate 3j symbols [21-24, 29]. The lattice tensors are of the rank one (like it has been anticipated in Chapter II) and they are written as a scalar contraction of the spherical tensor operators S_q^1 and the Wigner rotation matrices $D_{0,n-q}^2 [\Omega_{DDL}(t)]$. The two elements of the lattice operators $T_{-n}^{1(DD)(L)}$ reflect the main components of the lattice: the rotational degrees of freedom are encoded directly into the Wigner rotation matrices, while the S_q^1 operators represent the contribution of the S spin system. The dynamics of the S spin results from its energy level structure and relaxation processes within these eigenstates, caused by the zero field splitting affected by the distortional motion. A pictorial view the entire system is presented in Fig.4.2, which shows the partition of the lattice into the classical, rotational component and the quantum-mechanical part of the spin S connected, in turn, with the classical, distortional component.

Now a computational framework has to be built around the evaluation of the quantum-mechanical spectral density involving the three relevant degrees of freedom: rotation, electron spin relaxation and distortion. However, before we shall begin the calculations, it is very useful for their clarity to present the geometry of the entire system. Fig.3.2 illustrates the concept of the ZFS tensor decomposed into the static and fluctuating parts. In fact, so far we have not considered any effects of the static ZFS, assuming that it vanishes.

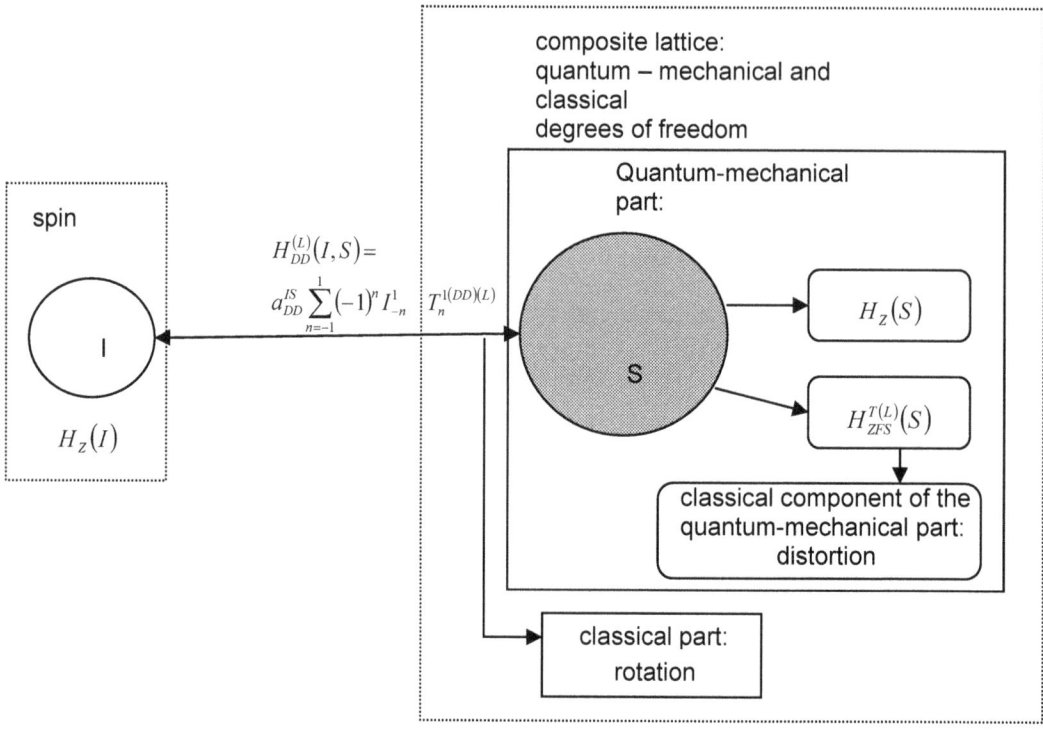

Fig.4.2

The concept of the composite lattice for a nuclear spin $I = \frac{1}{2}$ coupled by a dipole – dipole interaction to an electron spin S. The lattice contains classical as well as quantum-mechanical degrees of freedom. It includes the classical rotational motion modulating the dipole-dipole coupling as well as the quantum-mechanical dynamics of the spin S. From the perspective of the spin I the spin S subsystem is just a part of the lattice. The quantum-mechanical part of the composite lattice can include also classical degrees of freedom; in this case the distortional motion modulating the transient ZFS interaction.

In Fig.4.3 we show the laboratory, dipole-dipole and transient ZFS frames, pointing out the effects on rotational and distortional motions on their relative orientations. According to Eq.2.33 the spin-lattice relaxation rate of the spin I is determined by the spectral density functions $K_{1,1}(-\omega_I)$ taken at the 'minus' Larmor frequency ω_I. We begin with the general formulation of the dipole-dipole

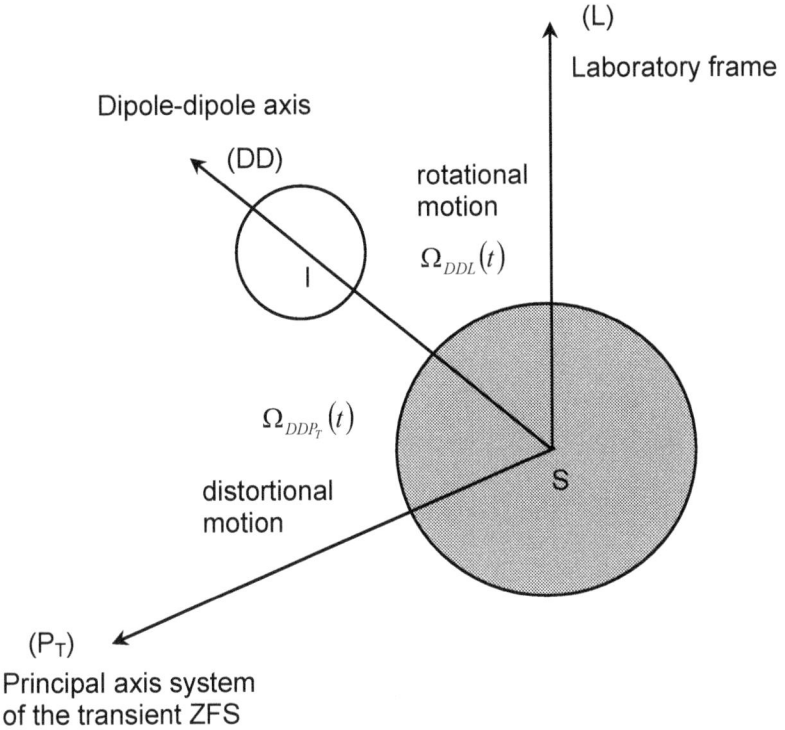

Fig.4.3
The relative orientation of the dipole-dipole axis with respect to the laboratory axis system (described by the angle $\Omega_{DDL}(t)$) changes in time due to the rotational motion, while the principal axis system of the transient ZFS fluctuates with respect to the dipole-dipole axis (the angle $\Omega_{DDP_T}(t)$) due to the distortional motion.

spectral density $K_{1,1}^{DD}(-\omega_I)$, which can be obtained as the Fourier transform of the correlation function for the composite lattice $G_{1,1}^{DD}(-\tau)$ (Eq.2.25b, Eq.2.26), which is represented by the lattice tensors $T_1^{1(DD)(L)}$:

$$K_{1,1}^{DD}(-\omega_I) = \int_0^\infty G_{1,1}^{DD}(-\tau)\exp(-i\omega_I\tau)d\tau$$

$$= \int_0^\infty Tr_L\left\{T_1^{1(DD)(L)+}\left[\exp(-i\hat{\hat{L}}_L\tau)T_1^{1(DD)(L)}\rho_L^{eq}\right]\right\}\exp(-i\omega_I\tau)d\tau \quad (4.19)$$

One should not be confuse here by the notation; the subscript L denotes the lattice, while (L) refers to the laboratory frame. The lattice Liouville operator $\hat{\hat{L}}_L$ contains terms describing the already discussed components:

$$\hat{\hat{L}}_L = \hat{\hat{L}}_Z(S) + \hat{\hat{L}}_{ZFS}^{T(L)}(S) + \hat{\hat{L}}_D + \hat{\hat{L}}_R \quad (4.20)$$

The first term $\hat{\hat{L}}_Z(S)$ is the Liouvilian generated by the Zeeman Hamiltonian $H_Z(S)$: $\hat{\hat{L}}_Z(S) = [H_Z(S),...]$, while the second Liouville operator $\hat{\hat{L}}_{ZFS}^{T(L)}(S)$ is associated with the fluctuating part of the ZFS tensor, $H_{ZFS}^{T(L)}(S)$. The two remaining terms, $\hat{\hat{L}}_D$ and $\hat{\hat{L}}_R$, describe the rotation of the $I-S$ dipole-dipole axis and the distortion of the ligand framework, respectively. Although, for the purpose of demonstrating the computational procedure based on the concept of the multipart lattice it is not necessary to use explicit forms of these superoperators, we wish to provide them for completeness of the lattice description. They are related to the free diffusion operator $-\frac{1}{6\tau_c}\nabla_\Omega^2$ acting on the set of Euler angles Ω, which is relevant for the particular motions. The correlation time τ_c ($c = R, D$) corresponds to the rank-two spherical harmonics, thus one has for the rotation: $\hat{\hat{L}}_R = -\frac{i}{6\tau_R}\nabla^2(\Omega_{DD,L})$, and, in full analogy, for the distortional motion: $\hat{\hat{L}}_D = -\frac{i}{6\tau_D}\nabla^2(\Omega_{P_T,L})$. With the expression of the components of the lattice tensor operators $T_n^{1(DD)(L)}$, resulting from the dipole-dipole spin-lattice coupling given by Eq.4.18, the specific form of the spectral density $K_{1,1}^{DD}(-\omega_I)$ may be written as:

$$K_{1,1}^{DD}(-\omega_I) = 30\left(\frac{\mu_0 \hbar \gamma_I \gamma_S}{4\pi r_{IS}^3}\right)^2 \sum_{p,q=-1}^{1} \begin{pmatrix} 2 & 1 & 1 \\ 1-q & q & -1 \end{pmatrix}\begin{pmatrix} 2 & 1 & 1 \\ 1-p & p & -1 \end{pmatrix} \times \quad (4.21)$$

$$\int_0^\infty Tr_L\left\{S_q^{1(L)+} D_{0,1-q}^{2*}(\Omega_{DDL}(\tau))\left[\exp(-i\hat{L}_L \tau)\right] S_p^{1(L)} D_{0,1-p}^{2}(\Omega_{DDL}(0))\rho_L^{eq}\right\} \exp(-i\omega_I \tau)d\tau$$

The index (L) associated to the S spin tensor operators $S_q^{1(L)}$ reminds that we consider the laboratory representation the $I-S$ dipole-dipole interaction, $H_{DD}^{(L)}(I,S)$. Setting up Eq.4.21 the last equation we have used the explicit expression for the dipolar coupling constant a_{DD}^{IS} and brought together the factors $\sqrt{6}$ from a_{DD}^{IS} and $\sqrt{5}$ from Eq.4.18. The spectral density $K_{1,1}^{DD}(-\omega_I)$ of Eq.4.21 is not in a form that can be handled easily. We need to apply several additional approximations that will eventually lead to a manageable formulation. The first step is to separate the rotational motion and the dynamics of the S spin. Finishing Section IV.2 we have pointed out that Eq.4.11 has been derived under the assumption that the dynamics of the electron spin S is independent of the presence of the I spin. This assumption is, in fact, not sufficient to decompose in Eq.4.21 the molecular tumbling and the electron spin dynamics. A detailed discussion of this issue shall be given in Chapter IX. Nevertheless to perform this decomposition safely we can assume that the rotational fluctuations of the $I-S$ dipole-dipole axis do not affect the evolution of the spin S at all, since it does not contribute to the modulations of the transient ZFS tensor with respect to the laboratory frame (the static ZFS does not bother us; it vanishes because of the symmetry reasons). Actually, this assumption is easy to fulfill. The distortional motion gives usually the main contribution to the modulations of the transient ZFS interaction, since typically $\tau_R \gg \tau_D$ and therefore $\tau_{eff} \cong \tau_D$. This is a nice example that one should consider very carefully every step of such calculations and therefore it fits very well to the scope of this book. It is also worthwhile to notice that, in fact, there are two relaxation pathways for the electron spin: the transient ZFS modulated by the distortional motion and the mutual $I-S$ dipole-dipole coupling modulated by the rotation. Since the predominant relaxation channel is provided by the zero field splitting, the dipole-dipole contribution, involving the I spin, is negligible and therefore the electron spin dynamics is not affected by the presence of the nuclear spin.

The independence of the S spin subsystem of the rotational degrees of freedom leads to the separation of the correlation function $G_{1,1}^{DD}(\tau)$ into one function involving only the electron spin part and another including the reorientational part:

$$K_{1,1}^{DD}(-\omega_I) = 30\left(\frac{\mu_0 \hbar \gamma_I \gamma_S}{4\pi r_{IS}^3}\right)^2 \sum_{p,q=-1}^{1} \begin{pmatrix} 2 & 1 & 1 \\ 1-q & q & -1 \end{pmatrix}\begin{pmatrix} 2 & 1 & 1 \\ 1-p & p & -1 \end{pmatrix} \times$$

$$\int_0^\infty Tr_R \left\{ D_{0,1-q}^{2*}(\Omega_{DDL}(\tau)) \left[\exp(-i\hat{L}_R \tau) D_{0,1-p}^2(\Omega_{DDL}(0))\right] \rho_R^{eq} \right\} \times \qquad (4.22)$$

$$Tr_S \left\{ S_q^{1(L)+} \left[\exp(-i\hat{L}_S \tau) S_p^{1(L)} \rho_S^{eq}\right] \right\} \exp(-i\omega_I \tau) d\tau$$

In this equation we have separated also the equilibrium density operator ρ_L^{eq} into the rotational and spin parts. Since the two contributions to the lattice are independent, one can consider the equilibrium lattice density operator as an outer product of the particular density operators: $\rho_L^{eq} = \rho_R^{eq} \otimes \rho_S^{eq}$. The electron spin Liouvilian, $\hat{L}_S$, covers all terms of Eq.4.20 except the rotational superoperator, i.e. $\hat{L}_S = \hat{L}_L - \hat{L}_R = \hat{L}_Z(S) + \hat{L}_{ZFS}^T(S) + \hat{L}_D$. Since the reorientational correlation function is modeled as isotropic rotational diffusion it takes the simple form of Eq.4.3:

$$Tr_R \left\{ D_{0,1-q}^{2*}(\Omega_{DDL}(\tau))\exp(-i\hat{L}_R \tau) D_{0,1-p}^2(\Omega_{DDL}(0)) \rho_R^{eq} \right\}$$

$$\equiv \left\langle D_{0,1-q}^{2*}(\Omega_{DDL}(\tau)) D_{0,1-p}^2(\Omega_{DDL}(0)) \right\rangle = \delta_{pq} \frac{1}{5} \exp\left(-\frac{\tau}{\tau_R}\right) \qquad (4.23)$$

The equilibrium density operator for the rotation is $\rho_R^{eq} = 1$. It is so for all classical degrees of freedom. One can conclude this from Eq.2.21, taking into account that classical systems are characterized by a continuum of eigenstates ($\frac{\omega_L}{k_B T} \ll 1$, where ω_L is the amplitude of the H_L Hamiltonian). It is worthwhile to notice the advantages of introducing at this stage an explicit form of the correlation function. If we use the single exponential rotational correlation function of Eq.4.23 the spectral density $K_{1,1}^{DD}(-\omega_I)$ reaches the much simpler form:

$$K_{1,1}^{DD}(-\omega_I) = 6\left(\frac{\mu_0 \hbar \gamma_I \gamma_S}{4\pi r_{IS}^3}\right)^2 \sum_{q=-1}^{1}\begin{pmatrix} 2 & 1 & 1 \\ 1-q & q & -1 \end{pmatrix}^2 \times$$
$$\int_0^\infty Tr_S\left\{S_q^{1(L)+}\left[\exp\left(-i\hat{L}_S\tau\right)S_q^{1(L)}\rho_S^{eq}\right]\right\} \exp\left[-\left(i\omega_I + \frac{1}{\tau_R}\right)\tau\right]d\tau \qquad (4.24)$$

Now, we can focus attention on the second component of the composite lattice, namely the S spin subsystem. We have to evaluate explicitly the Fourier transforms of the quantum-mechanical electron spin correlation function $G_{q,q}(\tau) = Tr_S\left\{S_q^{1(L)+}\left[\exp\left(-i\hat{L}_S\tau\right)S_q^{1(L)}\rho_S^{eq}\right]\right\}$ multiplied by the rotational exponential decay. It is determined by the electron spin interaction, *i.e.* the Zeeman coupling and the transient ZFS, and the distortional motion:

$$G_{q,q}(\tau) = \frac{1}{2S+1}Tr_S\left\{S_q^{1(L)+}\left[\exp\left(-i\left(\hat{L}_Z(S)+\hat{L}_{ZFS}^{T(L)}(S)+\hat{L}_D\right)\tau\right)S_q^{1(L)}\right]\right\} \qquad (4.25)$$

The factor $(1/(2S+1))$ corresponds to the equilibrium electron spin density operator, ρ_S^{eq}, under the high temperature approximation [9] ($\frac{\omega_S}{k_B T} \ll 1$), which is easily fulfilled at the room temperature. This results directly from the expansion of the density operator given by Eq.2.30. The ZFS interaction ($\hat{L}_{ZFS}^{T(L)}$) in combination with the distortional motion ($\hat{L}_D$) leads to the electron spin relaxation. Therefore one can write:

$$i\left(\hat{L}_{ZFS}^{T(L)}(S)+\hat{L}_D\right) \equiv \hat{R}_S \qquad (4.26)$$

where $\hat{R}_S$ is the relaxation superoperator for the electron spin. Consequently, the electron spin correlation function can be rewritten as:

$$G_{q,q}(\tau) = \frac{1}{2S+1}Tr_S\left\{S_q^{1(L)}\exp\left[\left(-\left(i\hat{L}_Z(S)+\hat{R}_S\right)\tau\right)S_q^{1(L)}\right]\right\} \qquad (4.27)$$

Now, the link between the present treatment and the description of the electron spin relaxation presented in Section IV.2 becomes apparent. The expression $i\hat{L}_z(S) + \hat{R}_S$ is just the operator representation of the Redfield matrix. The Liouville space for the electron spin is constructed from the Zeeman eigenstates $|m_S\rangle$ and for the spin $S = 1$ contains 9 elements $|\alpha\rangle\langle\beta|$ ($\alpha, \beta = 1, 2, 3$; we remind to the reader of the labeling: $|1\rangle = |1\rangle, |2\rangle = |0\rangle, |3\rangle = |-1\rangle$), which correspond to the density matrix elements $\rho_{\alpha\beta}$. According to the convention introduced in Section II.1 we denote the basis vectors of the Liouville space as $|\alpha\rangle\langle\beta| \equiv |m_S^\alpha, m_S^\beta\rangle \equiv |\alpha, \beta\rangle$. The matrix elements of the operator $i\hat{L}_z(S) + \hat{R}_S$ in the basis $|\alpha, \beta\rangle$ are given by:

$$(\alpha', \beta'|(i\hat{L}_z(S) + \hat{R}_S)|\alpha, \beta) = i\omega_{\alpha\beta}\delta_{\alpha'\alpha}\delta_{\beta'\beta} + R_{\alpha'\beta'\alpha\beta}\big|_{\omega_{\alpha\alpha'} = \omega_{\beta\beta'}} \quad (4.28)$$

where the mark $\big|_{\omega_{\alpha\alpha'} = \omega_{\beta\beta'}}$ refers to the secular approximation (Section II.3) and indicates that only the relaxation matrix elements connecting the coherences, which fulfill this condition are relevant. Obviously one can expand also the electron spin tensor operators S_q^1 in the Liouville basis set $|\alpha\rangle\langle\beta|$. For the spin quantum number $S = 1$, the relations are very simple [25-27]:

$$S_0^1 = S_z = |1\rangle\langle1| - |-1\rangle\langle-1| \quad (4.29a)$$

$$S_1^1 = \frac{1}{\sqrt{2}} S_+ = |0\rangle\langle1| + |-1\rangle\langle0| \quad (4.29b)$$

$$S_{-1}^1 = -\frac{1}{\sqrt{2}} S_- = -|1\rangle\langle0| - |0\rangle\langle-1| \quad (4.29c)$$

Nevertheless, the explicit calculations of the electron spin correlation functions $G_{q,q}(\tau)$ are still a difficult task. Even though we have set up the matrix representations of the operators $i\hat{L}_z(S) + \hat{R}_S$ and S_q^1, the fact that the former operator appears in Eq.4.27 as an exponential factor makes the evaluations complicated. Actually, one can write the correlation function $G_{q,q}(\tau)$ in the equivalent form: $G_{q,q}(\tau) = \left\langle S_q^{1(L)+}(\tau) S_q^{1(L)}(0) \right\rangle$ and evaluate it straightforwardly from the expressions:

$$G_{0,0}(\tau) = \left\langle S_0^{1(L)+}(\tau) S_0^{1(L)}(0) \right\rangle = \left[\sum_{m=-1,0,1} \left\langle m \left| S_0^1 \right| m \right\rangle \right]^2 \exp(-R_{1S}\tau) \quad (4.30a)$$

$$G_{\pm 1,\pm 1}(\tau) = \left\langle S_\pm^{1(L)+}(\tau) S_\pm^{1(L)}(0) \right\rangle = \left| \sum_{m,n=-1,0,1} \left\langle m \left| S_\mp^1 \right| n \right\rangle \left\langle n \left| S_\pm^1 \right| m \right\rangle \right| \exp(-R_{2S}\tau) \quad (4.30b)$$

where the relaxation rates R_{ie} are given by Eq.4.17a,b. Therefore it is rather difficult to appreciate at this stage advantages of the concept of the composite lattice and the Liouville superoperator formalism. However, to get a closed form of the dipolar spectral density $K_{1,1}^{DD}(-\omega_I)$ fortunately we do not need to evaluate explicitly the correlation functions $G_{q,q}(t)$; the relevant quantities are the spectral densities:

$$s_{q,q}(-\omega_I) = \int_0^\infty G_{q,q}(\tau) \exp\left[-i\left(\omega_I + \tau_R^{-1}\right)\tau\right] d\tau \quad (4.31)$$

One can write them in a form much more suitable for the computation by introducing the superoperator $\hat{\hat{M}}$ covering the electron spin operator, $i\hat{\hat{L}}_Z(S) + \hat{\hat{R}}_S$, and the term $i\omega_I + \tau_R^{-1}$:

$$\hat{\hat{M}} = i\hat{\hat{L}}_Z(S) + \hat{\hat{R}}_S + \left(i\omega_I + \tau_R^{-1}\right)\hat{\hat{1}} \quad (4.32)$$

where $\hat{\hat{1}}$ is the unit superoperator. Consequently the spectral densities may be evaluated as [23-26]:

$$s_{q,q}(-\omega_I) = \frac{1}{2S+1} \int_0^\infty Tr_S\left\{ S_q^{1(L)+} \left[\exp\left(-\hat{\hat{M}}\tau\right) S_q^{1(L)} \right] \right\} d\tau = \frac{1}{2S+1} \left[S_q^{1(L)} \right]^+ \left[\hat{\hat{M}} \right]^{-1} \left[S_q^{1(L)} \right] \quad (4.33)$$

The last equality implies that the actual computation of the spectral densities $s_{q,q}(-\omega_I)$ can be performed by setting up and inverting the matrix representation $\left[\hat{\hat{M}}\right]$ of the operator $\hat{\hat{M}}$ in the Liouville basis constructed from the Zeeman eigenstates of the electron spin. The vectors $\left[S_q^{1(L)}\right]$ represent the tensor operators S_q^1 in the same basis and are determined by Eqs.4.29a,b,c, while the matrix $\left[\hat{M}\right]$ can be obtained in a straightforward manner by adding to the diagonal elements of Redfield matrix the term $i\omega_I + \tau_R^{-1}$. The Zeeman representation of the operators S_q^1 shows clearly that the spectral density $s_{0,0}$ is determined by the zero-quantum (population) block of the matrix $\left[\hat{M}\right]$, while the $s_{\pm 1,\pm 1}$ spectral densities correspond to the single-quantum transitions.

Since we aim in this chapter for a detailed presentation of the computational procedure based on the concept of a composite lattice and a careful discussion of the equivalence of this approach and the traditional Wangsness – Bloch – Redfield [5, 8-10] treatment, we shall present explicitly all relevant steps of the calculations. To obtain a closed analytical form of the spectral density $s_{0,0}(-\omega_I)$ one has to perform the following matrix operations:

$$s_{0,0}(-\omega_I) = \frac{1}{3}\left[S_0^{1(L)}\right]^+ \left[\hat{M}\right]^{-1} \left[S_0^{1(L)}\right] =$$

$$\frac{1}{3}[1,0,-1]\begin{bmatrix} i\omega_I + R_{1111}^S + \tau_R^{-1} & R_{1122}^S & R_{1133}^S \\ R_{2211}^S & i\omega_I + R_{2222}^S + \tau_R^{-1} & R_{2223}^S \\ R_{3311}^S & R_{3322}^S & i\omega_I + R_{3333}^S + \tau_R^{-1} \end{bmatrix}^{-1} \begin{bmatrix} 1 \\ 0 \\ -1 \end{bmatrix} \quad (4.34)$$

Before we invert the population block of the matrix $\left[\hat{M}\right]$ it is worthwhile to realize that actually we need only two elements of the inverted matrix projected out by the S_0 operator:

$$s_{0,0}(-\omega_I) = \frac{1}{3}\left\{\left[\hat{M}\right]^{-1}_{1,1} + \left[\hat{M}\right]^{-1}_{3,3}\right\} \quad (4.35)$$

The factor 1/3 in Eq.4.34 and Eq.4.35 comes from the equilibrium density operator $1/(2S+1)$. Employing the explicit expressions for the relaxation matrix elements in terms of the quantities $J_m^S(\omega)$ (Eq.4.13) one obtains the final form of the spectral density $s_{0,0}(-\omega_I)$:

$$s_{0,0}(-\omega_I) = \frac{2}{3} \frac{\left(\tau_R^{-1} + J_1^S(\omega_S) + 4J_2^S(2\omega_S)\right)^{-1} - i\omega_I \left(\tau_R^{-1} + J_1^S(\omega_S) + 4J_2^S(2\omega_S)\right)^{-1}}{1 + \omega_I^2 \left[\left(\tau_R^{-1} + J_1^S(\omega_S) + 4J_2^S(2\omega_S)\right)^{-1}\right]^2} \quad (4.36)$$

Looking at this expression it is apparent that it contains the correlation time $\tau_{c,1}$, defined as $\tau_{c,1}^{-1} = J_1^S(\omega_S) + 4J_2^S(2\omega_S) + \tau_R^{-1} = R_{1S} + \tau_R^{-1}$, introduced in Section IV.2 as the effective time constant describing the modulation of the $I-S$ dipole-dipole axis due to the two independent dynamic processes: the rotational motion and the electron spin-lattice relaxation. Following the considerations it is important to remark that the fundamental statement of the SBM approach that the two processes are treated as independent sources of the fluctuations of the dipole-dipole interactions has been invoked when we set up Eq.4.22 by separating the rotational and electron spin correlation functions. In a similar manner one can evaluate the $s_{\pm 1, \pm 1}(-\omega_I)$ spectral densities; in particular:

$$s_{1,1}(-\omega_I) = \frac{1}{3}\left[S_1^{1(L)}\right]^+ \left[\hat{M}\right]^{-1} \left[S_1^{1(L)}\right] =$$

$$\frac{1}{3}[1,1]\begin{bmatrix} i(\omega_I - \omega_S) + R_{1212}^S + \tau_R^{-1} & R_{1223}^S \\ R_{2312}^S & i(\omega_I - \omega_S) + R_{2323}^S + \tau_R^{-1} \end{bmatrix}^{-1} \begin{bmatrix} 1 \\ 1 \end{bmatrix} \quad (4.37)$$

The closed form expression yields:

$$s_{\pm 1, \pm 1}(-\omega_I) = \frac{2}{3} \frac{\tau_{c,2} - i(\omega_I \mp \omega_S)\tau_{c,2}}{1 + (\omega_I \mp \omega_S)^2 \tau_{c,2}^2} \quad (4.38)$$

where the correlation time $\tau_{c,2}$ includes the electron spin-spin relaxation rate: $\tau_{c,2}^{-1} = \frac{3}{2}J_0^S(0) + \frac{5}{2}J_1^S(\omega_S) + J_2^S(2\omega_S) + \tau_R^{-1} = R_{2,S} + \tau_R^{-1}$. The formulas for $s_{1,1}$ and $s_{-1,-1}$ differ only by the sign of the frequency ω_S. If one introduces the values of the 3j symbols $\begin{pmatrix} 2 & 1 & 1 \\ 0 & 1 & -1 \end{pmatrix}^2 = \frac{1}{30}$, $\begin{pmatrix} 2 & 1 & 1 \\ 1 & 0 & -1 \end{pmatrix}^2 = \frac{1}{10}$ and $\begin{pmatrix} 2 & 1 & 1 \\ 2 & -1 & -1 \end{pmatrix}^2 = \frac{1}{5}$ the nuclear spin–lattice relaxation rate $R_{1,I}$ expressed in terms of the spectral densities s_{qq} takes the form:

$$R_{1I} = 2\operatorname{Re}\{K_{1,1}^{DD}(-\omega_I)\} = \frac{1}{5}\left(\frac{\mu_0}{4\pi}\frac{\gamma_I \gamma_S \hbar}{r_{IS}}\right)^2 (s_{1,1} + 3s_{0,0} + 6s_{-1,-1}) \quad (4.39)$$

With the explicitly evaluated quantities $s_{q,q}$ (Eq.4.36 and Eq.4.38) this equation is identical to Eq.4.11 for the spin quantum number $S = 1$.

We are aware that even though we have proved that the two approaches lead to the same result, the multipole representation treatment involving the composite lattice seems to be more complicated than the 'traditional' one. This is due to the fact that in the considered case the energy level structures of both spins have been fully determined by the Zeeman couplings.

If the relaxation processes of the spin S are slow compared to other sources of the modulations of the dipole-dipole interaction, like for example the rotational motion in this case, one can neglect the relaxation term $\hat{R}_S$ altogether evaluating the spectral densities s_{qq}. It simplifies the calculations significantly. The correlation times $\tau_{c,1}$ and $\tau_{c,2}$ become just equal to the rotational correlation time $\tau_{c,1} = \tau_{c,2} = \tau_R$ and Eq.4.39 taken together with the simplified spectral densities converges to the Solomon formulation.

Finishing this section we would like to comment on a very important aspect of the composite lattice approach. The S spin does not need to relax rapidly to be treated as a part of the lattice for the spin I. The essence of the concept of the composite lattice is to separate the S spin dynamics from any degrees of freedom relevant for the spin I. In other words, the S spin can be treated as a part of the lattice if its dynamics is completely independent of the presence of the I spin. If this condition is fulfilled, we can apply the approach to the $I-S$ spin system and we will obtain correct results. However, it is definitely not obvious that we are allowed to separate the S spin from the I spin, if the own relaxation mechanism of the spin S is not very efficient. Starting the considerations we have assumed that the predominant relaxation mechanism for the spin S is provided by the transient ZFS. Let us assume initially that this relaxation channel is not efficient. As it has been mentioned in Section IV.2 the spin S possesses also the second relaxation pathway via the dipole-dipole coupling to the spin I, fluctuating in time due to the molecular tumbling. The dipolar relaxation mechanism is now non-negligible compared to the one provided by the ZFS. However, since the dipolar relaxation of the S spin originates from the rotational motion, the Redfield condition $|\omega_{DD}\tau_R| \ll 1$ (ω_{DD} is the amplitude of the H_{DD} Hamiltonian in the angular frequency units) ensures that the dipolar relaxation is much slower than the rotation and therefore it can be neglected as a source of the modulations of the dipole-dipole interaction. This argument explains only why one can obtain the proper expression for the R_{II} term (Eq.4.8a) utilizing the concept of the composite lattice, even though the S spin does not relax fast. It does not contradict the obvious fact, that considering the dipolar relaxation of two mutually coupled spins, I and S, in the absence of other relaxation mechanisms, one cannot establish any hierarchy of them. To describe relaxation processes in such a system one must consider the entire relaxation operator in the basis formed by the two spin functions: $\{|m_S, m_I\rangle\}$, as has been done in Section IV.2. Therefore, one must remember that if the S spin does not relax efficiently due to its own (independent of the I spin) mechanism, the relaxation of the spin I as well of the spin S is bi-exponential (Eq.4.8a,b), while the quantum-mechanical spectral density $K_{1,1}^{DD}(-\omega_I)$ gives only the R_{II} term. We shall come back to this subject in Chapter XI.

In the next chapter we shall consider a similar $I - S$ spin system ($I = \frac{1}{2}, S = 1$), however characterized by a non-zero static component of the zero field splitting. We shall discuss the electron as well as the nuclear spin relaxation at a low magnetic field, when the Zeeman coupling for the electron spin is much weaker than the static ZFS. We shall perform once again the complete relaxation calculations employing the multipole representation and the WBR formalisms. Anticipating conclusions coming from the comparison of the two approaches we claim that the superoperator treatment is in this case more straightforward.

IV.5. Relaxation processes in nuclear spin – electron spin system at low magnetic field. Effects of the zero field splitting

Let us remove, in contrast with previous sections, the assumption of a permanent (static) ZFS that vanishes by symmetry. We begin the considerations from the description of the electron spin relaxation within the matrix (WBR) formalism. In the first step we assume that the static ZFS tensor is axially symmetrical, so that $E_S = 0$ (see Eq.3.9). In this case the basis $\{|m_S\rangle\}$ still remains the eigenbasis of the electron spin. Since the magnetic field is low we neglect the electron spin Zeeman coupling altogether. Therefore the energy level structure is fully determined by the static part of the zero field splitting. It leads to two degenerate energy levels $E_1 = E_3 = \frac{D_S}{3}$, which correspond to the eigenstates $|1\rangle = |m_S = 1\rangle$ and $|3\rangle = |m_S = -1\rangle$, respectively, while the energy level for the eigenstate $|2\rangle = |m_S = 0\rangle$ has the magnitude $E_2 = -\frac{2D_S}{3}$. Once again, the fluctuating part of the ZFS provides the mechanism of the electron spin relaxation. Employing the representation of the ZFS Hamiltonian in the basis $\{|m_S\rangle\}$ (Eq.4.12) and Eq.2.14 one can calculate the relaxation matrix elements. The populations evolve in time according to the set of equations for the electron spin density operator $\rho^S(t)$:

$$\frac{d}{dt}\begin{bmatrix} \rho_{11}^S \\ \rho_{22}^S \\ \rho_{33}^S \end{bmatrix} = \begin{bmatrix} -J_1^S(\omega_D) - 2J_2^S(0) & J_1^S(\omega_D) & 2J_2^S(0) \\ J_1^S(\omega_D) & -2J_1^S(\omega_D) & J_1^S(\omega_D) \\ 2J_2^0(0) & J_1^S(\omega_D) & -J_1^S(\omega_D) - 2J_2^S(0) \end{bmatrix} \times \begin{bmatrix} \rho_{11}^S \\ \rho_{22}^S \\ \rho_{33}^S \end{bmatrix} \quad (4.40)$$

where ω_D denotes the axial parameter of the static (permanent) ZFS, D_S (the subscript 's' refers to the permanent (static) part of the ZFS tensor, rather than the S spin), expressed in the angular frequency units. In fact, the relaxation matrix elements corresponding to the energy levels determined by the static ZFS can be obtained easily from the previous ones derived for the Zeeman energy level structure of the electron spin (Eq.4.13) by adjusting the energy differences at which the individual spectral densities are taken, namely $E_1 - E_2 = E_2 - E_3 = \omega_D$ instead of ω_S and $E_1 - E_3 = 0$ instead of $2\omega_S$. The resulting electron spin-lattice relaxation rate R_{1S} is given as [25, 26]:

$$\frac{d}{dt}\langle S_Z \rangle = \frac{d}{dt}(\rho_{11}^S - \rho_{33}^S) = -[J_1^S(\omega_D) + 4J_2^S(0)](\rho_{11}^S - \rho_{33}^S) = -R_{1S}\langle S_Z \rangle \quad (4.41a)$$

In a similar manner we obtain the spin-spin relaxation rate R_{2S}:

$$\frac{d}{dt}\left(\rho_{12}^S + \rho_{23}^S\right) = -\left[i\omega_D + \left(\frac{3}{2}J_0^S(0) + \frac{5}{2}J_1^S(\omega_D) + J_2^S(0)\right)\right]\left(\rho_{12}^S + \rho_{23}^S\right) = -R_{2S}\langle S_+\rangle \qquad (4.41b)$$

Obviously, the expressions for the electron spin relaxation rates at the low and high magnetic fields are different; they reflect the different energy level structures in the two cases. Eqs.4.41a,b and Eqs.4.17a,b give the same values for the relaxation rates if the condition $\omega\tau_D \ll 1$ is fulfilled.

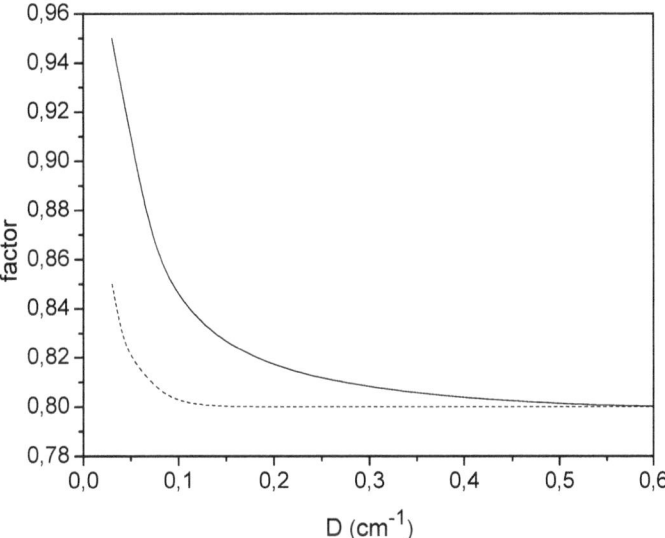

Fig.4.4

Discrepancies between the low field ($\omega_S \ll \omega_D$) electron spin-lattice relaxation rate obtained from Eq.4.41 and the predictions of the high field theory (Eq.4.17a) applied at low field. The ratio between the predicted values is equal to factor $= \frac{1}{5}\left(\frac{1}{1+\omega_D^2\tau_D^2} + 4\right)$, and this quantity is presented. For $\omega_D\tau_D \ll 1$ the two descriptions converge, but if $\omega_D\tau_D \gg 1$ the predictions of the Bloembergen - Morgan formulation are incorrect; they differ from the proper values by factor 0.8. Solid line corresponds to $\tau_D = 10\,ps$, while dashed line has been obtained for $\tau_D = 30\,ps$.

However, by comparing the two expressions one concludes also that if the condition $\omega_D \tau_D \ll 1$ does not hold, one may not use the high field formulations for the electron spin relaxation at the low field. Fig.4.4 shows the spin – lattice relaxation rates obtained from Eq.4.41a scaled by the factor $\Delta_T^2 \tau_D$ (assuming that $\tau_D \cong \tau_{eff}$) to which converges Eq.4.17a at the low magnetic field (because in the low field limit $\omega_S \tau_D \ll 1$). Dealing with the problem of a proper description of the electron spin relaxation in the presence of the static ZFS one should be aware that for higher spin quantum numbers the relaxation processes at high and low fields can be essentially different. We shall discuss this subject in the next chapter.

Now, we turn our attention to the nuclear spin $I = \frac{1}{2}$ coupled to the electron spin. The relaxation of the I spin is affected not only by the electron spin relaxation. Even if the rotational motion prevails and therefore the details of the S spin relaxation are not important (it can be negligible altogether compared to the rotation) the non-Zeeman energy level structure of the electron spin influences the I spin relaxation. The eigenbasis for the coupled $I-S$ spin system consists of the six vectors $|m_I, m_S\rangle$ labeled as follows: $|1\rangle = \left|\frac{1}{2},1\right\rangle$, $|2\rangle = \left|\frac{1}{2},0\right\rangle$, $|3\rangle = \left|\frac{1}{2},-1\right\rangle$, $|4\rangle = \left|-\frac{1}{2},1\right\rangle$, $|5\rangle = \left|-\frac{1}{2},0\right\rangle$ and $|6\rangle = \left|-\frac{1}{2},-1\right\rangle$. The fact that the presence of the static ZFS does not change the eigenbasis of the spin system does not imply that we can straightforwardly adapt the already formulated description of the electron spin relaxation under the condition that the Zeeman couplings provide the main Hamiltonian to the current situation. The relaxation theory requires one to represent the perturbing Hamiltonian in the principal axis system of the main interaction. The Redfield formula of Eq.2.14 has been derived in this way and can be applied only if one uses this representation. However, the principal axis systems of the main interactions of the nuclear and electron spins are now different. The first one is locked in the laboratory frame, while the second one processes around the principal axis of the static ZFS tensor. For that reason, we have to express the dipole interaction in terms of the electron and nuclear spin tensor operators, as it has been done in Section IV.3 and treat the two operators independently. It is convenient for the computations to start with the dipole-dipole Hamiltonian written in its principal axis system (DD) (the molecular frame), which coincides with the principal axis system of the static ZFS (P_S). The statement that the two frames coincide is just an assumption, which simplifies considerably the further calculations. The molecular frame representation of the dipole-dipole Hamiltonian contains only the $T_0^2(I,S)$ tensor, which can be easily decomposed into the first rank tensors for the electron and nuclear spins:

$$H_{DD}^{(DD)}(I,S) = a_{DD}^{IS} T_0^2(I,S) = a_{DD}^{IS} \frac{1}{\sqrt{6}} \left[2I_0^1 S_0^1 + I_1^1 S_{-1}^1 + I_{-1}^1 S_1^1 \right] \qquad (4.42)$$

The nuclear spin operators I_n^1 have to be transformed to the laboratory frame, according to the transformation rule of Eq.3.3: $I_n^{1(L)} = \sum_{k=0,\pm 1} I_k^{1(DD)} D_{k,n}^1(\Omega_{DD,L})$. The dipole-dipole Hamiltonian written in the 'mixed' representation takes then the form: $H_{DD}^{(L-P_S)}(I,S) = \sum_{n=-1}^{1} I_n^1 T_{-n}^1$, with the components T_{-n}^1 defined as:

$$T_{-n}^1 = a_{DD}^{IS} \frac{1}{\sqrt{6}} \left[2S_0^1 D_{-n,0}^1(\Omega_{DDL}) + S_1^1 D_{-n,-1}^1(\Omega_{DDL}) + S_{-1}^1 D_{-n,1}^1(\Omega_{DDL}) \right] \qquad (4.43)$$

It is worthwhile to realize at this moment that this treatment brings us, in fact, directly to the concept of the composite lattice. We have actually decomposed the electron spin variables from the entire Hamiltonian and formed the operators T_{-n}^1 including the electron spin and the rotational degrees of freedom. Next we set up the matrix representation of the dipole-dipole Hamiltonian $H_{DD}^{(L-P_S)}$ in the basis $\{|m_I, m_S\rangle\}$:

$$[H_{DD}^{(L-P_S)}] = $$

$$a_{DD}^{IS} \begin{bmatrix} & |1\rangle & |2\rangle & |3\rangle & |4\rangle & |5\rangle & |6\rangle \\ \frac{1}{\sqrt{6}}D_{0,0}^1 & \frac{1}{2\sqrt{6}}D_{0,-1}^1 & 0 & \frac{1}{\sqrt{3}}D_{1,0}^1 & \frac{1}{2\sqrt{3}}D_{1,-1}^1 & 0 \\ -\frac{1}{2\sqrt{6}}D_{0,1}^1 & 0 & \frac{1}{2\sqrt{6}}D_{0,-1}^1 & -\frac{1}{2\sqrt{3}}D_{1,1}^1 & 0 & \frac{1}{2\sqrt{3}}D_{1,-1}^1 \\ 0 & -\frac{1}{2\sqrt{6}}D_{0,1}^1 & -\frac{1}{\sqrt{6}}D_{0,0}^1 & 0 & -\frac{1}{2\sqrt{3}}D_{1,1}^1 & -\frac{1}{\sqrt{3}}D_{1,0}^1 \\ -\frac{1}{\sqrt{3}}D_{-1,0}^1 & -\frac{1}{2\sqrt{3}}D_{-1,-1}^1 & 0 & -\frac{1}{\sqrt{6}}D_{0,0}^1 & -\frac{1}{2\sqrt{6}}D_{0,-1}^1 & 0 \\ \frac{1}{2\sqrt{3}}D_{-1,1}^1 & 0 & -\frac{1}{2\sqrt{3}}D_{-1,-1}^1 & \frac{1}{2\sqrt{6}}D_{0,1}^1 & 0 & -\frac{1}{2\sqrt{6}}D_{0,-1}^1 \\ 0 & \frac{1}{2\sqrt{3}}D_{-1,1}^1 & \frac{1}{\sqrt{3}}D_{-1,0}^1 & 0 & \frac{1}{2\sqrt{6}}D_{0,1}^1 & \frac{1}{\sqrt{6}}D_{0,0}^1 \end{bmatrix} \quad (4.44)$$

Now we can use it for the evaluations of the individual relaxation matrix elements according to the Redfield formula of Eq.2.14. The orthogonality properties of the Wigner matrices simplify considerably the final formulas. The dipole-dipole coupling $H_{DD}^{(L-P_S)}$ leads to the following relaxation matrix for the populations:

$$\frac{d}{dt}\begin{bmatrix}\rho_{11}\\\rho_{22}\\\rho_{33}\\\rho_{44}\\\rho_{55}\\\rho_{66}\end{bmatrix} = $$

$$\begin{bmatrix} -\sum_{\gamma\neq 1}R_{11\gamma\gamma} & \frac{1}{12}J^{(1)}(\omega_D) & 0 & \frac{2}{3}J^{(1)}(\omega_I) & \frac{1}{6}J^{(1)}(\omega_I+\omega_D) & 0 \\ & -\sum_{\gamma\neq 2}R_{22\gamma\gamma} & \frac{1}{12}J^{(1)}(\omega_D) & \frac{1}{6}J^{(1)}(\omega_I-\omega_D) & 0 & \frac{1}{6}J^{(1)}(\omega_I-\omega_D) \\ & & -\sum_{\gamma\neq 3}R_{33\gamma\gamma} & 0 & \frac{1}{6}J^{(1)}(\omega_I+\omega_D) & \frac{2}{3}J^{(1)}(\omega_I) \\ & & & -\sum_{\gamma\neq 4}R_{44\gamma\gamma} & \frac{1}{12}J^{(1)}(\omega_D) & 0 \\ & & & & -\sum_{\gamma\neq 5}R_{55\gamma\gamma} & \frac{1}{12}J^{(1)}(\omega_D) \\ & & & & & -\sum_{\gamma\neq 6}R_{66\gamma\gamma} \end{bmatrix}\begin{bmatrix}\rho_{11}\\\rho_{22}\\\rho_{33}\\\rho_{44}\\\rho_{55}\\\rho_{66}\end{bmatrix}$$

$$(4.45)$$

Using this relaxation matrix one can set up, as usual, the equation for the expectation value of the I_z operator, $\langle I_z \rangle \propto \rho_{11} + \rho_{22} + \rho_{33} - \rho_{44} - \rho_{55} - \rho_{66}$ and extract the spin-lattice relaxation rate for the I spin:

$$\frac{d\langle I_z(t)\rangle}{dt} = -R_{1I}\langle I_z(t)\rangle = -\frac{8}{3}\left(\frac{\mu_0 \gamma_I \gamma_S \hbar}{4\pi r_{IS}^3}\right)^2 [2J(\omega_I) + J(\omega_I + \omega_D)]\langle I_z(t)\rangle \qquad (4.46)$$

The spectral densities $J(\omega)$ are given as Fourier transforms of the correlation functions corresponding to the rank one Wigner rotation matrices:

$$\langle D_{m,k}^{1*}(\Omega_{DDL}(\tau))D_{m',k'}^{1}(\Omega_{DDL}(0))\rangle = \delta_{m,m'}\delta_{k,k'}\frac{1}{3}\exp\left(-\frac{\tau}{3\tau_R}\right) \qquad (4.47)$$

and contain the corresponding correlation time $\tau_R^{(1)} = 3\tau_R^{(2)} = 3\tau_R$. Associating the spectral density $J(\omega_I)$ with the correlation time $\tau_{c,1}^{-1} = (3\tau_R)^{-1} + R_{1S}$ and the spectral density $J(\omega_I + \omega_D)$ with the correlation time $\tau_{c,2}^{-1} = (3\tau_R)^{-1} + R_{2S}$, including the spin-lattice and spin-spin relaxation rates of the electron spin, respectively, one gets [25,26]:

$$R_{1I} = \frac{8}{9}\left(\frac{\mu_0 \gamma_I \gamma_S \hbar}{4\pi}\right)^2 \left[\frac{2\tau_{c,1}}{1+\omega_I^2 \tau_{c,1}^2} + \frac{\tau_{c,2}}{1+(\omega_I + \omega_D)^2 \tau_{c,2}^2}\right] \qquad (4.48)$$

The obtained expression differs essentially from its high field counterpart. The quantization of the electron spin in the principal axis system of the ZFS tensor leads not only to the different transition frequencies for the spectral densities. It influences the effects of the rotational motion on the dipole-dipole coupling described now by the three times longer correlation time $\tau_R^{(1)}$.

We have combined the particular spectral densities with the correlation times $\tau_{c,1}$ and $\tau_{c,2}$ following the scheme: if the spectral density contains the transition frequency $\omega_D = E_1 - E_2 = E_2 - E_3$ it is related to the single-quantum transitions of the electron spin and therefore it should be expressed in terms of the spin-spin relaxation time, while the spectral density dependent only on the frequency ω_I refers to the spin-lattice relaxation of the electron spin. In a more general case when the S spin relaxation is multiexponential and the energy levels are degenerate this procedure becomes much more complicated and it requires very careful considerations of the coupled $I-S$ spin transitions. Fig.4.5 presents the ratio between the nuclear spin relaxation rate R_{1I} calculated from Eq.4.48 and the relaxation rate R_{1I} obtained from the SBM theory (Eq.4.11) for the same values of Δ_T and τ_D.

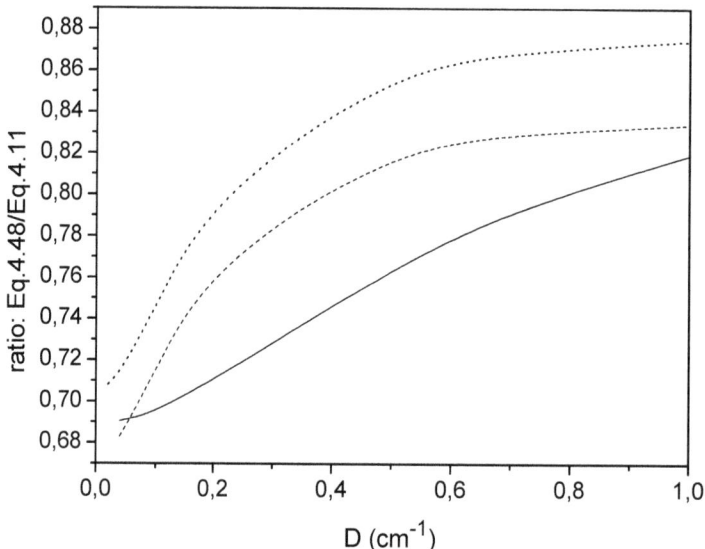

Fig.4.5

The ratio between the proton spin relaxation rate R_{1I} at low field ($\omega_S \ll \omega_D$) calculated from Eq.4.48 and the relaxation rate R_{1I} obtained from the SBM theory which does not include the ZFS. Solid line: $D_T = \sqrt{\frac{3}{2}}\Delta_T = 0.02 cm^{-1}, \tau_D = 10 ps, \tau_R = 1\mu s$,

dashed line: $D_T = 0.02 cm^{-1}, \tau_D = 30 ps, \tau_R = 1\mu s$,

dotted line: $D_T = 0.01 cm^{-1}, \tau_D = 30 ps, \tau_R = 1\mu s$.

Looking at the expressions for the nuclear spin relaxation rates one could expect that the discrepancies increase for higher ZFS. However, one should note that the electron spin relaxation rates determining the correlation times $\tau_{c,i}$ (since the rotation is slow) are different as well (see Eq.s4.41a,b and Eq.4.17a,b).

Chapter IV – Examples of relaxtion processes treated within the perturbation approach

To illustrate better the effects of the electron spin dynamics on the nuclear spin relaxation we have assumed that the rotational motion is very slow and its contribution to the effective correlation times $\tau_{c,1}$ and $\tau_{c,2}$ can be neglected. The energy level structure of the electron spin affects the nuclear spin relaxation by the electron spin relaxation rates, R_{1S} and R_{2S}, and by the frequencies of the joined $I-S$ spin transitions present in the dipole-dipole spectral densities. The final effects of the static ZFS on the electron as well as nuclear spin relaxation at low field are very significant [20, 25-28, 30-35].

In the next step we shall employ the multipole representation approach and evaluate the nuclear spin-lattice relaxation rate R_{1I} at low magnetic field treating the electron spin, characterized by a non-zero static ZFS, as a part of the lattice. We shall begin the considerations with the general form of the spectral density $K_{1,1}^{DD}(-\omega_I)$, Eq.4.21. First of all, it is very important to realize that in the formulation the electron spin tensor operators $S_q^{1(L)}$ refer to the laboratory frame. To evaluate properly the spectral density $K_{1,1}^{DD}(-\omega_I)$ at low magnetic field we have to represent the operators $S_q^{1(L)}$ in the dipole-dipole (molecular) (DD) frame, as it has been done above to illustrate the WBR formalism. Inserting the relationship $S_q^{1(L)} = \sum_{m=-1}^{1} S_m^{1(DD)} D_{m,q}^1(\Omega_{DDL})$ into Eq.4.21, one obtains:

$$K_{1,1}^{DD}(-\omega_I) = 30\left(\frac{\mu_0 \hbar \gamma_I \gamma_S}{4\pi r_{IS}^3}\right)^2 \sum_{p,q=-1}^{1} \begin{pmatrix} 2 & 1 & 1 \\ 1-q & q & -1 \end{pmatrix} \begin{pmatrix} 2 & 1 & 1 \\ 1-p & p & -1 \end{pmatrix} \times$$

$$\sum_{m,k=-1}^{1} \int_0^\infty Tr_L \left\{ \begin{bmatrix} S_q^{1(DD)+} D_{0,1-q}^{2*}(\Omega_{DDL}(\tau)) D_{m,q}^{1*}(\Omega_{DDL}(\tau)) \\ \exp(-i\hat{L}_L \tau) S_p^{1(DD)} D_{0,1-p}^2(\Omega_{DDL}(0)) D_{k,p}^1(\Omega_{DDL}(0)) \end{bmatrix} \rho_L^{eq} \right\} \exp(-i\omega_I \tau) d\tau \quad (4.49)$$

The expression can be appreciably simplified by contracting the Wigner rotation matrix elements according to the elementary angular momentum theory [21-23]:

$$D_{0,1-q}^2(\Omega_{DDL}) D_{mq}^1(\Omega_{DDL}) = (-1)^{m+1} \sum_{\lambda=1}^{3} (2\lambda+1) \begin{pmatrix} 2 & 1 & \lambda \\ 0 & m & -m \end{pmatrix} \begin{pmatrix} 2 & 1 & \lambda \\ 1-q & q & -1 \end{pmatrix} D_{m,1}^\lambda \quad (4.50)$$

This explains the statement that it is more convenient to start from the molecular representation of the dipole-dipole interaction; however the expression for the quantum-mechanical correlation function has been set up for the laboratory form of the dipole-dipole Hamiltonian. Properties of the 3j symbols restrict the summation to terms of $\lambda = 1$ and the dipolar spectral density $K_{1,1}^{DD}(-\omega_I)$ takes the form:

$$K_{1,1}^{DD}(-\omega_I) = \frac{10}{3}\left(\frac{\mu_0}{4\pi}\frac{\gamma_I\gamma_S\hbar}{r_{IS}^3}\right)^2 \sum_{p,q=-1}^{1}\begin{pmatrix}2 & 1 & 1\\ 1-q & q & -1\end{pmatrix}\begin{pmatrix}2 & 1 & 1\\ 1-p & p & -1\end{pmatrix} \times$$

$$\sum_{m,k=-1}^{1}\begin{pmatrix}2 & 1 & 1\\ 0 & m & -m\end{pmatrix}\begin{pmatrix}2 & 1 & 1\\ 0 & k & -k\end{pmatrix} \times \qquad (4.51)$$

$$\int_0^\infty Tr_L\left\{S_m^{1(DD)^+}D_{m,1}^{1*}(\Omega_{DDL}(\tau))\left[\exp(-i\hat{\hat{L}}_L\tau)S_k^{1(M)}D_{k,1}^1(\Omega_{DDL}(0))\right]\rho_L^{eq}\right\}\exp(-i\omega_I\tau)d\tau$$

The lattice Liouville operator $\hat{\hat{L}}_L$ contains now, instead of the electron spin Zeeman interaction, the static ZFS superoperator $\hat{\hat{L}}_{ZFS}^{S(P_S)}(S)$, generated by the static ZFS Hamiltonian written in its principal axis system (Eq.3.9): $\hat{\hat{L}}_L = \hat{\hat{L}}_{ZFS}^{S(P_S)}(S) + \hat{\hat{L}}_{ZFS}^{T(P_S)}(S) + \hat{\hat{L}}_D + \hat{\hat{L}}_R$. The operator $\hat{\hat{L}}_{ZFS}^{T(P_S)}(S)$, describing the fluctuating part of the ZFS interaction (generated by the transient ZFS Hamiltonian of Eq.3.12), forms together with the distortional Liouvilian, $\hat{\hat{L}}_D$, the electron spin relaxation superoperator: $i\left(\hat{\hat{L}}_{ZFS}^{T(P_S)}(S) + \hat{\hat{L}}_D\right) = \hat{\hat{R}}_S$. However, before we shall incorporate explicitly the electron spin dynamics into the spectral density $K_{1,1}^{DD}(-\omega_I)$, we separate once again the rotational correlation function: $\langle D_{m,1}^{1*}(\Omega_{DD,L}(\tau))D_{k,1}^1(\Omega_{DD,L}(0))\rangle = \frac{1}{3}\delta_{km}\exp\left(-\frac{\tau}{3\tau_R}\right)$ from the electron spin part. As a result, the dipolar spectral density takes the form:

$$K_{1,1}^{DD}(-\omega_I) = \frac{1}{3}\left(\frac{\mu_0}{4\pi}\frac{\gamma_I\gamma_S\hbar}{r_{IS}^3}\right)^2\left\{s_{1,1}^{LF}(-\omega_I) + 4s_{0,0}^{LF}(-\omega_I) + s_{-1,-1}^{LF}(-\omega_I)\right\} \qquad (4.52)$$

where the low filed (LF) spectral densities $s_{q,q}^{LF}(-\omega_I)$ can be evaluated employing the Redfield relaxation matrix $\hat{\hat{R}}_S$ for the electron spin:

$$s_{q,q}^{LF}(-\omega_I) =$$

$$\int_0^\infty Tr_S\left\{S_q^{1(DD)*}\exp\left[-\left(i\hat{L}_{ZFS}^{S(P_S)}(S)+\hat{R}_S\right)\tau\right]S_q^{1(DD)}\rho_S^{eq}\right\}\exp\left(-i\omega_I\tau+\frac{\tau}{3\tau_R}\right)d\tau = \qquad (4.53)$$

$$\frac{1}{2S+1}\left[S_q^{1(DD)}\right]^+\left[\hat{M}_{LF}\right]^{-1}\left[S_q^{1(DD)}\right]$$

The $\hat{M}_{LF}$ operator is defined as $\hat{M}_{LF} = i\hat{L}_{ZFS}^0(S)+\hat{R}_S+\left(\frac{1}{3\tau_R}+i\omega_I\right)\hat{1}$ and includes, in the same manner as the operator $\hat{M}$ of Eq.4.32, the electron spin dynamics, the rotational exponential decay and the nuclear spin transition frequency ω_I. Since the axially symmetric ZFS tensor is diagonal in the basis $\{|m_S\rangle\}$, the representations of the operators S_q^1 in the laboratory and in the molecular (dipole-dipole) frames are the same. Therefore the spectral densities $s_{q,q}^{LF}$ can be obtained in full analogy to the $s_{q,q}$ quantities (Eq.4.34 and Eq.4.37); in particular:

$$s_{0,0}^{LF}(-\omega_I) = \frac{1}{3}$$

$$[1,0,1]\begin{bmatrix} -J_1^S(\omega_D)-2J_2(0)-\frac{1}{3\tau_R}-i\omega_I & J_1^S(\omega_D) & 2J_2^S(0) \\ J_1^S(\omega_D) & -2J_1^S(\omega_D)-\frac{1}{3\tau_R}-i\omega_I & J_1^S(\omega_D) \\ 2J_2^S(0) & J_1^S(\omega_D) & -J_1^S(\omega_D)-2J_2^S(0)-\frac{1}{3\tau_R}-i\omega_I \end{bmatrix}^{-1}\begin{bmatrix}1\\0\\1\end{bmatrix}$$

$$= \frac{2}{3}\frac{(J_1(\omega_D)+4J_2(0)+1/3\tau_R)^{-1}(1+i\omega_I)}{1+\omega_I^2\left[(J_1(\omega_D)+4J_2(0)+1/3\tau_R)^{-1}\right]^2} = \frac{1}{3}\left(\frac{2\tau_{c,1}}{1+\omega_I^2\tau_{c,1}^2}+\frac{2i\omega_I\tau_{c,1}}{1+\omega_I^2\tau_{c,1}^2}\right) \qquad (4.54)$$

The spectral densities $s_{\pm1,\pm1}^{LF}$ are equal to:

$$s_{1,1}^{LF} = s_{-1,-1}^{LF} = 2[1+i(\omega_I+\omega_D)]\frac{\tau_{c,2}}{1+(\omega_I+\omega_D)^2\tau_{c,2}^2},$$ where the correlation times $\tau_{c,i}$ have been defined already ($\tau_{c,i}^{-1} = (3\tau_R)^{-1}+R_{iS}$).

IV.6. Effects of non-zero averaged quadrupolar coupling on relaxation of dipolar spin - quadrupolar spin systems

We begin the considerations from the quadrupolar spin S relaxation at low magnetic field. As usual we shall describe the spin S within the basis $\{|m_S\rangle\}$ constructed from the vectors corresponding to the magnetic quantum numbers. The quadrupolar Hamiltonian $H_Q^{(P)}(S)$ (Eq.3.2) for $S=1$ leads to three energy levels of the quadrupolar spin: $E_1 = 2a_Q(1+\eta)$, $E_2 = 2a_Q(1-\eta)$ and $E_3 = -2a_Q$. The corresponding eigenfunctions $|\psi_i\rangle$ are formed by the following combinations of the Zeeman vectors $|m_S\rangle$ (let us remind the labeling: $|1\rangle \equiv |1\rangle, |2\rangle \equiv |0\rangle, |3\rangle \equiv |-1\rangle$): $|\psi_1^S\rangle = \frac{1}{\sqrt{2}}[|1\rangle + |3\rangle]$, $|\psi_2^S\rangle = \frac{1}{\sqrt{2}}[-|1\rangle + |3\rangle]$ and $|\psi_3^S\rangle = |2\rangle$. The fact that the levels of $m_S = \pm 1$ are mixed is very important for the relaxation processes as we shall see soon. Following the ordinary procedure we intend to set up the relaxation matrix for the quadrupolar spin. We assume, in full analogy to the electron spin and the zero field splitting, that the main relaxation mechanism for the S spin is provided by deviations of the total quadrupolar interaction $H_Q^{(P)}(t)$ from its averaged value $\langle H_Q^{(P)}(t)\rangle$. To keep very close analogy to the description of the ZFS interaction let us denote the principal axis system of the electric field gradient tensor as (P_S^Q). This means that, consequently, the averaged quadrupolar coupling should be referred to as the static quadrupolar interaction: $H_Q^{(P)} = H_Q^{S(P_S^Q)}$ (see Section III.2; the index S denotes 'static'). The spread of the quadrupolar interactions (varying stochastically in time in the (P_S^Q) frame) is represented by the Hamiltonian $H_Q^{T(P_S^Q)}(S)(t) = H_Q^{(P_S^Q)}(t) - \langle H_Q^{(P_S^Q)}(t)\rangle = H_Q^{(P_S^Q)}(t) - H_Q^{S(P_S^Q)}$. This model of the quadrupolar coupling is absolutely analogous to the model of the ZFS interaction;

it has been illustrated in Fig.4.6 (this figure should be compared with Fig.3.2). Following this line of analogy the Hamiltonian $H_Q^{T(P_S^Q)}(S)(t)$ can be expressed as:

$$H_Q^{T(P_S^Q)}(S)(t) = \frac{1}{2}\sqrt{\frac{3}{2}} \frac{a_Q^T}{S(2S+1)} \sum_{m=-2}^{2}(-1)^m \left(\sum_{k=-2}^{2} A_k^{2T(P_T^Q)} D_{k,-m}^2\left(\Omega_{P_T^Q P_S^Q}(t)\right)\right) T_m^2(S)$$

$$= \sum_{m=-2}^{2}(-1)^m \tilde{A}_{-m}^{2T(P_S^Q)} T_m^2(S)$$
(4.55)

where $A_0^{(P_T^Q)} = 1$, $A_{\pm 1}^{(P_T^Q)} = 0$, $A_{\pm 2}^{(P_T^Q)} = \eta_T/\sqrt{6}$; the quantities a_Q^T and η_T are the parameters characterizing the transient quadrupolar coupling.

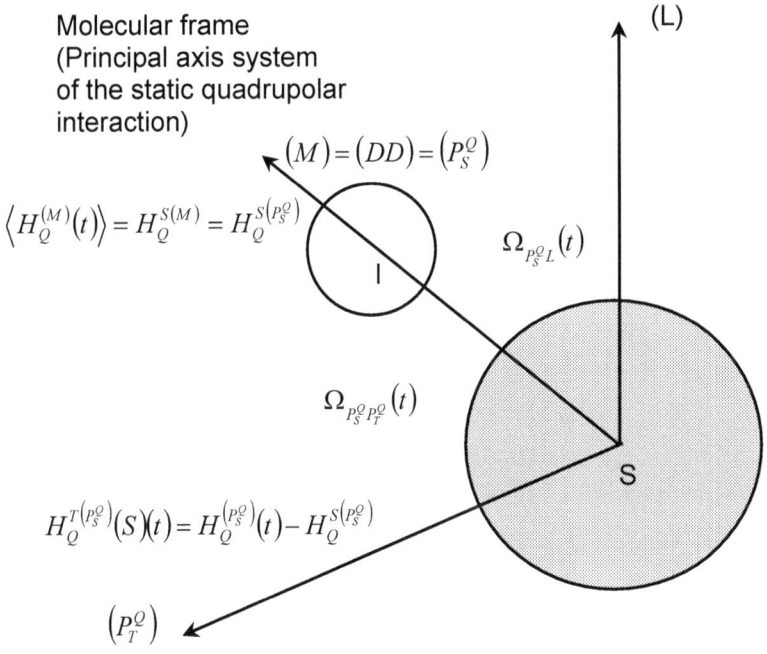

Fig.4.6

Decomposition of the quadrupolar interaction into the static and transient parts in full analogy to the model of the ZFS illustrated in Fig.3.2.

One should be aware that the WBR formula for the individual relaxation matrix elements (Eq.2.14) has been derived within the eigenbasis of the main Hamiltonian (H_0) and therefore it is correct only in this representation. To calculate properly the specific quadrupolar spectral densities $J_{\alpha\alpha'\beta\beta'}(\omega)$ one needs the matrix representation of the perturbing interaction $H_Q^{T(p_S^Q)}(S)(t)$ in the eigenbasis $\{|\psi_i^S\rangle\}$ of the main Hamiltonian of the spin S (i.e. $H_0^{S(p_S^Q)}$). The required matrix can be obtained by employing the relationship $|\psi_i\rangle = \sum_{n=1}^{3} c_{in}|n\rangle$ between the basis $\{|\psi_i^S\rangle\}$ and the Zeeman one $\{|n\rangle = |m_S\rangle\}$. The particular elements of the fluctuating quadrupolar coupling Hamiltonian, $H_Q^T(S)(t)$, calculated from the relation:

$$\langle \psi_i^S | H_Q^{T(p_S^Q)}(S) | \psi_j^S \rangle = \sum_{m=-2}^{2}(-1)^m \tilde{A}_{-m}^{2T(p_S^Q)}(t) \left[\sum_{n,n'=1}^{3} c_{in'}^* c_{jn} \langle n' | T_m^2(S) | n \rangle \right] \quad (4.56)$$

are provided explicitly below:

$$[H_Q^{T(p_S^Q)}(S)] =$$

$$\begin{bmatrix} & |\psi_1^S\rangle & |\psi_2^S\rangle & |\psi_3^S\rangle \\ \frac{1}{\sqrt{6}}\tilde{A}_0^{2T} + \frac{1}{2}\tilde{A}_2^{2T} + \frac{1}{2}\tilde{A}_{-2}^{2T} & \frac{1}{2}\tilde{A}_{-2}^{2T} - \frac{1}{2}\tilde{A}_2^{2T} & \frac{1}{2}\tilde{A}_{-1}^{2T} + \frac{1}{2}\tilde{A}_1^{2T} \\ \frac{1}{2}\tilde{A}_{-2}^{2T} - \frac{1}{2}\tilde{A}_2^{2T} & \frac{1}{\sqrt{6}}\tilde{A}_0^{2T} - \frac{1}{2}\tilde{A}_2^{2T} - \frac{1}{2}\tilde{A}_{-2}^{2T} & -\frac{1}{2}\tilde{A}_{-1}^{2T} + \frac{1}{2}\tilde{A}_1^{2T} \\ -\frac{1}{2}\tilde{A}_{-1}^{2T} - \frac{1}{2}\tilde{A}_1^{2T} & \frac{1}{2}\tilde{A}_1^{2T} - \frac{1}{2}\tilde{A}_{-1}^{2T} & -\frac{2}{\sqrt{6}}\tilde{A}_0^{2T} \end{bmatrix} \quad (4.57)$$

where we have simplified the labeling: $\tilde{A}_{-m}^{2T(p_S^Q)} \equiv \tilde{A}_{-m}^{2T}$. Using this matrix representation one can derive the relaxation rates using directly, once again, the recipe given by Eq.2.14. Explicit calculations bring one to the conclusion that the populations of the three eigenstates evolve in time according to the set of equation:

$$\frac{d}{dt}\begin{bmatrix} \rho_{11}^S \\ \rho_{22}^S \\ \rho_{33}^S \end{bmatrix} =$$

$$\begin{bmatrix} -J_1^S(\omega_Q+\omega_\eta)-J_2^S(2\omega_\eta) & J_2^S(2\omega_\eta) & J_1^S(\omega_Q+\omega_\eta) \\ J_2^S(2\omega_\eta) & -J_1^S(\omega_Q-\omega_\eta)-J_2^S(2\omega_\eta) & J_1^S(\omega_Q-\omega_\eta) \\ J_1^S(\omega_Q+\omega_\eta) & J_1^S(\omega_Q-\omega_\eta) & -J_1^S(\omega_Q+\omega_\eta)-J_1^S(\omega_Q-\omega_\eta) \end{bmatrix} \quad (4.58)$$

$$\times \begin{bmatrix} \rho_{11}^S \\ \rho_{22}^S \\ \rho_{33}^S \end{bmatrix}$$

The quadrupolar spin spectral densities are defined as:

$$J_m^S(\omega) \equiv J^Q(\omega) =$$

$$\int_0^\infty \left\langle \left(\tilde{A}_{-m}^{2T(P_S^Q)}(\tau)\right)^* \left(\tilde{A}_{-m}^{2T(P_S^Q)}(0)\right) \right\rangle \exp(-i\omega\tau) d\tau = \frac{1}{5}\left(\frac{a_Q^T}{S(2S-1)}\right)^2 \frac{(3+\eta_T^2)}{8} \frac{\tau_Q}{1+\omega^2\tau_Q^2} \quad (4.59)$$

where the correlation time τ_Q is the 'quadrupolar counterpart' of the distortional correlation time τ_D; it characterizes the fluctuations of the electric field gradient tensor. The frequencies ω_Q and ω_η are equal to the energies $4a_Q$ and $2a_Q\eta$, respectively, expressed in the angular frequency units. Writing down the relaxation rates we have made use of the fact that $J_m(\omega) = J_m(-\omega) = J_{-m}(\omega) = J_{-m}(-\omega)$. In the same way we calculate the relaxation rates $R_{\alpha\beta\alpha\beta}^S$ associated with the coherences $\rho_{\alpha\beta}^S$:

$$\frac{d\rho_{12}^S(t)}{dt} = -i\omega_\eta \rho_{12}^S(t)$$
$$-\left[2J_2^S(0)+J_2^S(\omega_\eta)+\frac{1}{2}J_1^S(\omega_Q-\omega_\eta)+\frac{1}{2}J_1^S(\omega_Q+\omega_\eta)\right]\rho_{12}^S(t) \quad (4.60a)$$

$$\frac{d\rho_{23}^S(t)}{dt} = -i(\omega_Q-\omega_\eta)\rho_{23}^S(t)$$
$$-\left[\frac{3}{2}J_0^S(0)+\frac{1}{2}J_2^S(0)+\frac{1}{2}J_2^S(\omega_\eta)+J_1^S(\omega_Q-\omega_\eta)+\frac{1}{2}J_1^S(\omega_Q+\omega_\eta)\right]\rho_{23}^S(t) \quad (4.60b)$$

$$\frac{d\rho_{13}^S(t)}{dt} = -i(\omega_Q + \omega_\eta)\rho_{13}^S(t)$$
$$-\left[\frac{3}{2}J_0^S(0) + \frac{1}{2}J_2^S(0) + \frac{1}{2}J_2^S(\omega_\eta) + \frac{1}{2}J_1^S(\omega_Q - \omega_\eta) + J_1^S(\omega_Q + \omega_\eta)\right]\rho_{13}^S(t) \quad (4.60c)$$

Looking at the equations, it is apparent that one can distinguish three single exponential relaxation processes described by the relaxation rates $R_{2S,1}$, $R_{2S,2}$ and $R_{2S,3}$, respectively. It is very interesting to consider, in the present case, the time evolution of the quantity $\langle S_z(t)\rangle$. The S_z operator can be expanded into the Liouville basis formed from the Zeeman functions, according to Eq.4.29a. Nevertheless, to set up the differential equation linking the derivative $\frac{d\langle S_z(t)\rangle}{dt}$ to $\langle S_z(t)\rangle$ one has to become aware of an appropriate representation of the S_z operator. Since the relaxation matrix has been calculated in the eigenbasis of the quadrupolar Hamiltonian $\{\psi_i\}$, one has to represent the S_z operator in the Liouville space, $\{|\psi_i\rangle\langle\psi_j|\}$. The inverse relation between the vectors $\{|m_S\rangle\}$ and $\{|\psi_i\rangle\}$, particularly: $|1\rangle = \frac{1}{\sqrt{2}}[|\psi_1\rangle - |\psi_2\rangle]$ and $|3\rangle = \frac{1}{\sqrt{2}}[|\psi_1\rangle + |\psi_2\rangle]$, leads to the expression: $S_z = |1\rangle\langle 1| - |3\rangle\langle 3| = -|\psi_1\rangle\langle\psi_2| - |\psi_2\rangle\langle\psi_1|$. This means that in fact the expectation value of the S_z operator, $\langle S_z(t)\rangle$, evolves in time as a single exponential process, however it is characterized by the already derived time constant $R_{2S,1}^{-1}$ considered as a spin-spin relaxation time. This result is a consequence of the mixing by the quadrupolar Hamiltonian the states with the different magnetic quantum numbers. Since the quadrupolar eigenstates combine the levels of $m_S = \pm 1$, the S_z operator is not associated any more with the populations of these Zeeman states.

Now we turn attention to the spin-lattice relaxation of the dipolar spin $I = \frac{1}{2}$. To evaluate the spin-lattice relaxation rate R_{1I} in terms of the spectral densities $s_{q,q}^{LF}$ one can modify in a straightforward manner Eq.4.53 by replacing the operators related to the zero field splitting by the corresponding operators associated with the quadrupolar interaction:

$$s_{q,q}^{LF}(-\omega_I) =$$
$$\int_0^\infty Tr_S \left\{ S_q^{1(P_S^Q)+} \exp\left[-\left(i\hat{L}_Q^{S(P_S^Q)}(S) + \hat{R}_S\right)\tau\right] S_q^{1(P_{SQ})} \rho_S^{eq} \right\} \exp\left(-i\omega_I\tau + \frac{\tau}{3\tau_R}\right) d\tau \quad (4.61)$$

$$= \frac{1}{2S+1} \left[S_q^{1(P_S^Q)}\right]^+ \left[\hat{M}_{LF}\right]^{-1} \left[S_q^{1(P_S^Q)}\right]$$

The operator $\hat{M}_{LF}$ includes now the Liouville superoperator $\hat{L}_Q^{S(P_S^Q)}$ generated by the static quadrupolar Hamiltonian and the relaxation operator $\hat{R}_S$:

$\hat{M}_{LF} = i\hat{L}_Q^{S(P_S^Q)}(S) + \hat{R}_S + \left(\frac{1}{3\tau_R} + i\omega_I\right)\hat{1}$. In the context of the discussion above it is particularly easy to write a closed form expression for the spectral density $s_{0,0}^{LF}$. The S_z operator projects out the element of the matrix $\left[\hat{M}_{LF}\right]^{-1}$ associated with the coherences ρ_{12}^S and ρ_{21}:

$$s_{0,0}^{LF}(-\omega_I) = \frac{1}{3} \frac{2\tau_{c,1}}{1 + (\omega_I + \omega_\eta)^2 \tau_{c,1}^2} \quad (4.62)$$

where the correlation time $\tau_{c,1}$ is defined as $\tau_{c,1}^{-1} = R_{S2,1} + 1/3\tau_R$. To evaluate the spectral densities $s_{\pm 1,\pm 1}^{LF}$ one has to derive the appropriate linear combinations of the Liouville space vectors $|\psi_i\rangle\langle\psi_j|$ representing the operators S_1 and S_{-1}, particularly:

$$S_1 = |2\rangle\langle 3| + |1\rangle\langle 2| = \frac{1}{\sqrt{2}}[|\psi_1\rangle\langle\psi_3| + |\psi_3\rangle\langle\psi_1| - |\psi_2\rangle\langle\psi_3| + |\psi_3\rangle\langle\psi_2|] \quad (4.63)$$

The resultant spectral densities $s_{\pm 1,\pm 1}^{LF}$ contain terms associated with the quadrupolar relaxation rate $R_{2S,2}$ as well as $R_{2S,3}$:

$$s_{1,1}^{LF}(-\omega_I) = s_{-1,-1}^{LF}(-\omega_I) = \frac{1}{3}\left[\frac{\tau_{c,2}}{1 + [\omega_I + (\omega_Q - \omega_\eta)]^2 \tau_{c,2}^2} + \frac{\tau_{c,3}}{1 + [\omega_I + (\omega_Q + \omega_\eta)]^2 \tau_{c,3}^2}\right] \quad (4.64)$$

The effective correlation times $\tau_{c,i}$ ($i=2,3$) are defined as $\tau_{c,i}^{-1} = R_{S2,i} + 1/3\tau_R$. Finally, the low field spin-lattice relaxation rate R_{1I} of the spin I due to its dipolar coupling to the quadrupolar spin of $S=1$ is given by the expression:

$$R_{1I} = \frac{4}{9}\left(\frac{\mu_0}{4\pi}\frac{\gamma_I \gamma_S \hbar}{r_{IS}^3}\right)^2 \times$$

$$\left\{\frac{4\tau_{c,1}}{1+(\omega_I+\omega_\eta)^2\tau_{c,1}^2} + \frac{\tau_{c,2}}{1+[\omega_I+(\omega_Q-\omega_\eta)]^2\tau_{c,2}^2} + \frac{\tau_{c,3}}{1+[\omega_I+(\omega_Q+\omega_\eta)]^2\tau_{c,3}^2}\right\}$$

(4.65)

Making use of the analogy between the quadrupolar coupling for a nuclear spin of $S=1$ and the zero field splitting for an electron spin of $S=1$ one can obtain directly from this equation an appropriate expression for the spin – lattice relaxation rate of a nuclear spin $I=\frac{1}{2}$ coupled to an electron spin characterized by a rhombic static ZFS. Comparing the energy level structures of the electron and nuclear spins (Fig.4.7) one can conclude that the quadrupolar spin transition frequencies $2\omega_\eta$, $\omega_Q - \omega_\eta$ and $\omega_Q + \omega_\eta$ correspond to the transition frequencies of the electron spin $2\omega_E$, $\omega_D - \omega_E$ and $\omega_D + \omega_E$ ($2E_S, D_S - E_S$ and $D_S + E_S$ expressed in angular frequency units); $\omega_Q = \frac{3}{4}\frac{a_Q}{S(2S-1)}$, $\omega_\eta = \frac{1}{4}\frac{a_Q \eta}{S(2S-1)}$.

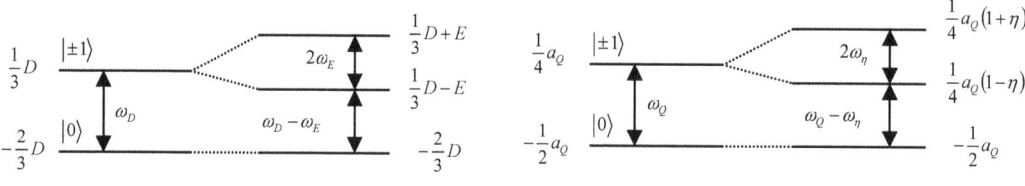

a)   b)

Fig.4.7
a) *Energy level splitting due to axial and rhombic ZFS for the electron spin quantum number $S=1$*
b) *Energy level splitting due to quadrupolar coupling for the nuclear spin quantum number $S=1$*

Some additional modifications require also the spectral densities determining the relaxation rates of the S spin. Despite the adjustment of the frequencies for the spectral densities of Eq.4.58 and Eqs.4.60a,b,c one has to modify the quadrupolar spin spectral densities expression of Eq.4.59 replacing the correlation time τ_Q by the distortional correlation time and the amplitude of the fluctuating part of the quadrupolar interaction by the transient ZFS (according to the relationship: $D \to \frac{3}{4}\frac{a_Q}{S(2S-1)}$ and $E \to \frac{1}{4}\eta\frac{a_Q}{S(2S-1)}$, see Section III.1). Thus, the explicit expression for the dipolar relaxation of a nuclear spin $I=\frac{1}{2}$ coupled to an electron spin $S=1$ characterized by a non-axial static ZFS takes the form [25]:

$$R_{1I} = \frac{4}{9}\left(\frac{\mu_0}{4\pi}\frac{\gamma_I \gamma_S \hbar}{r_{IS}^3}\right)^2 \times \tag{4.66}$$

$$\left\{\frac{4\tau_{c,1}}{1+(\omega_I+2\omega_E)^2 \tau_{c,1}^2} + \frac{\tau_{c,2}}{1+[\omega_I+(\omega_D-\omega_E)]^2 \tau_{c,2}^2} + \frac{\tau_{c,3}}{1+[\omega_I+(\omega_D+\omega_E)]^2 \tau_{c,3}^2}\right\}$$

where the electron spin correlation times are defined as: $\tau_{c,i} = R_{S2,i} + 1/3\tau_R$ ($i=1,2,3$). The relaxation rates $R_{S2,i}$ are given by the analogous combinations of the spectral densities:

$$R_{S2,1} = 2J_2^S(0) + J_2^S(2\omega_E) + \frac{1}{2}J_1^S(\omega_D - \omega_E) + \frac{1}{2}J_1^S(\omega_D + \omega_E) \tag{4.67a}$$

$$R_{S2,2} = \frac{3}{2}J_0^S(0) + \frac{1}{2}J_2^S(0) + \frac{1}{2}J_2^S(2\omega_E) + J_1^S(\omega_D - \omega_E) + \frac{1}{2}J_1^S(\omega_D + \omega_E) \tag{4.67b}$$

$$R_{S2,3} = \frac{3}{2}J_0^S(0) + \frac{1}{2}J_2^S(0) + \frac{1}{2}J_2^S(2\omega_E) + \frac{1}{2}J_1^S(\omega_D - \omega_E) + J_1^S(\omega_D + \omega_E) \tag{4.67c}$$

independently of the model applied to describe the fluctuations of the ZFS tensor. Obviously, following the line of analogies between the electron and quadrupolar spins, one can obtain from Eq.4.48 an expression for the spin-lattice relaxation rate of the I spin coupled to a quadrupolar spin $S=1$, which exhibit an axially symmetrical quadrupolar coupling ($\eta = 0$).

Closer inspection of the results of Eqs.4.67a,b,c leads to the conclusion that they converge to the formulation of Eqs.4.41a,b if the rhombicity of the static ZFS is set to zero, as expected. The spin-spin relaxation rates $R_{S2,2}$ and $R_{S2,3}$

become equal $R_{S2,2} = R_{S2,3} = R_{S2}$, while the relaxation rate $R_{S2,1}$ converges to the spin-lattice relaxation rate R_{S1}. Realizing that the relaxation rates $R_{S2,1}$ and R_{S1} describe time evolution of the expectation value of the S_z operator, $\langle S_z(t) \rangle$, the result becomes consistent.

IV.7. Validity conditions of the presented treatments

Finishing this chapter some caution must be exercised regarding the validity conditions of the perturbation approach applied to the considered spin system. The Redfield relaxation theory has the two essential limitations discussed in Chapter II. The perturbing Hamiltonian $H_1(t)$ has to fulfill the condition: $\omega_1 \tau_c \ll 1$, where τ_c is the correlation time for the fluctuating interaction and ω_1 is its amplitude expressed in the angular frequency units. This condition guaranties that the timescale of the relaxation is much slower than the fluctuations of the Hamiltonian $H_1(t)$ causing the relaxation. The main Hamiltonian H_0 must obey the relation $\omega_0 \gg \omega_1 \cdot \omega_1 \tau_c$. Under this condition the oscillating terms $(\omega_{\alpha\alpha'} - \omega_{\beta\beta'})t$ resulted from the transformation of the density operator to the interaction representation (generated by the Hamiltonian H_0) are effectively averaged out, unless they are not equal zero.

If the conditions are not fulfilled one cannot define explicitly time-independent relaxation rates. Apparently, we have assumed in this chapter than the S spin fulfills the two requirements. At this moment we would like to formulate them explicitly in terms of the interactions relevant for the considered spin systems. Let us start from the electron spin case. In the absence of the static ZFS, principally the electron spin Zeeman coupling acts as the main Hamiltonian determining the energy level structure, while the fluctuating part of the ZFS tensor causes the transitions between the energy levels. To apply the Redfield relaxation theory to the electron spin, treating the Hamiltonians $H_Z(S)$ and $H_{ZFS}^T(S)$ as the main and perturbing interactions H_0 and $H_1(t)$, respectively, one must check whether the relations $\omega_{ZFS}^T \tau_D \ll 1$ and $\omega_S \gg \omega_{ZFS}^T \cdot \omega_{ZFS}^T \tau_D$ are fulfilled (generally ω is the amplitude of the corresponding interaction H in the angular frequency units). The first relation (the Redfield condition) is satisfied if the fluctuations of the ZFS tensor are fast enough. The second relation indicates that if the amplitude of the transient ZFS is large the description of the electron spin relaxation presented in Sections IV.3 and IV.5 can break down. Analogous conditions would be required for the quadrupolar spin to assume that its Zeeman coupling can be treated as the main Hamiltonian, while the Hamiltonian generated by fluctuations of the electric field gradient tensor $H_Q^T(S)$ provides the perturbation $H_1(t)$, i.e.: $\omega_Q^T \tau_Q \ll 1$ and $\omega_I \gg \omega_Q^T \cdot \omega_Q^T \tau_Q$. In fact we have not considered this case explicitly; however the discussed analogies between the quadrupolar and electron spins make the statement clear. At the low field limit the relation between the static and fluctuating parts of the ZFS tensor for the electron spin is crucial. The static ZFS can be treated as the main Hamiltonian if it obeys the condition $\omega_{ZFS}^S \gg \omega_{ZFS}^T \cdot \omega_{ZFS}^T \tau_D$. At the same time the transient ZFS provides the relaxation mechanism, and therefore must obviously fulfill the Redfield condition $\omega_{ZFS}^T \tau_D \ll 1$. However if the amplitude of the transient ZFS is significantly larger than the amplitude of the static part of the ZFS interaction $\omega_{ZFS}^T \gg \omega_{ZFS}^S$, the secular approximation condition $\omega_{ZFS}^S \gg \omega_{ZFS}^T \cdot \omega_{ZFS}^T \tau_D$ can break down even though the correlation time τ_D is short enough to fulfill still the relation $\omega_{ZFS}^T \tau_D \ll 1$.

Now, we turn attention to the dipolar relaxation mechanism. In the simplest case, both the participating spins exhibit Zeeman energy structures (*i.e.* the Zeeman interactions provide the main Hamiltonian, H_0). The dipole-dipole interactions must satisfy the Redfield condition $\omega_{DD} \tau_c \ll 1$. This limitation is definitely not trivial, and the Redfield theory is not always applicable to spins coupled by dipole-dipole interactions. In section III.1 we have assumed that the orientation of the dipole-dipole axis fluctuates in time due to rotation, that $\tau_c = \tau_R$. For example, the rotational motion of relatively large molecules can be too slow to fulfill this condition. The spins are quantized in the laboratory frame as long as the relation $\frac{\omega_I}{\omega_{DD}} \gg \omega_{DD}\tau_c$ is true. Otherwise the role of the Zeeman coupling as the main Hamiltonian becomes questionable. This condition becomes relevant at low field. For slowly rotating nuclear spin – electron spin systems, the paramagnetic relaxation becomes the most important, effective source of modulations of the $I-S$ dipole-dipole coupling. Therefore, to be allowed to apply the perturbation approach to describe the nuclear spin relaxation, one must be sure that the electron spin relaxation is much faster than the nuclear spin relaxation caused by it. The desirable relation between R_I and R_S: $R_I \ll R_S$ is guaranteed by the condition $\omega_{DD} R_S^{-1} \ll 1$. Since the quadrupolar spin relaxation is usually significantly slower than the electron spin relaxation this condition must be treated with special caution. It is very important to realize that even though the Redfield theory is in principle applicable to the I spin (because of fast rotational motion, exchange dynamics, etc., averaging out the dipole-dipole interaction) it can happen that it does not apply to the electron or quadrupolar spin. On the contrary, in many cases the electron spin relaxation is very rapid (it is on the same timescale as the motion causing the relaxation). The derivations leading to the form of the spectral density $K_{1,1}^{DD}$ given by Eq.4.19 are based on the assumption that the I spin relaxation is within the Redfield limit, but the S spin relaxation does not need to be. Nevertheless, the further evaluations leading to the closed form expressions require obviously an explicit definition of the relaxation for the S spin.

References:

109. R. Shankar, Principles of Quantum Mechanics, *Plenum, New York* (1980)
110. A. Messiah, Quantum Mechanic, *North Holland, Amsterdam* (1962)
111. A. Abragam, The principles of nuclear magnetism, *Oxford University Press, Oxford* (1961)
112. R. Kimmich, NMR - Tomography, Diffusometry, Relaxometry, *Springer – Verlag Berlin* (1997)
113. J. Kowalewski, L. Mäler, Nuclear spin relaxation in liquids: theory, experiments and applications, Series in Chemical Physics, *Taylor & Francis Group* (2006)
114. D. Canet, Relaxation mechanisms: magnetization modes, in Encyclopedia of Nuclear Magnetic Resonance, D.M. Grant, R.K. Harris, (Eds.) *Wiley, Chichester* (1996) 4046-4053
115. P.-O. Westlund, Nuclear paramagnetic spin relaxation theory: Paramagnetic spin probes in homogenous and microheterogenous solutions, in Dynamics of solutions and fluid mixtures by NMR, J. J. Delpuech (Ed.), *Wiley, Chichester* (1995) 173-229
116. A. Kumar, R.C.R. Grace, P.K. Madhu, Cross-correlations in NMR, *Prog. Nucl. Magn. Reson. Spectr.* **37** (2000) 191-319
117. C.P. Slichter, Principles of magnetic resonance, *Springer-Verlag, Berlin* (1990)
118. R.R. Ernst, G. Bodenhausen, A, Wokaun, Principles of nuclear magnetic resonance in one and two dimensions, *Clarendon Press, Oxford* (1994)
119. N. Bloembergen, E.M. Purcell, R.V. Pound, Relaxation effects in nuclear magnetic resonance absorption, *Phys. Rev.* **73** (1948) 679-712
120. I. Solomon, Relaxation processes in a system of two spins, *Phys. Rev.* **99** (1955) 559-565
121. I. Solomon, N. Bloembergen, Nuclear magnetic interactions in the HF molecule, *J. Chem. Phys.* **25** (1956) 261-266
122. N. Bloembergen, Proton relaxation times in paramagnetic solutions, *J. Chem. Phys.* **27** (1957) 572-573

123. N. Bloembergen, L.O. Morgan, Proton relaxation times in paramagnetic solutions: Effects of electron spin relaxation, *J. Chem. Phys.* **34** (1961) 842-850

124. I. Bertini, C. Luchinat, G. Parigi, Solution NMR of paramagnetic molecules, *Elsevier, Amsterdam* (2001)

125. N. Benetis, J. Kowalewski, L. Nordenskiöld, H. Wennerström, P.-O. Westlund, Nuclear spin relaxation in paramagnetic systems. The slow motion problem for electron spin relaxation, *Mol. Phys.* **48** (1983) 329-346

126. J. Kowalewski, L. Nordenskiöld, N. Benetis, P.-O. Westlund, Theory of nuclear spin relaxation in paramagnetic systems, *Progr. NMR. Spectr.* **17** (1985) 141 – 185

127. J. Kowalewski, Paramagnetic relaxation in solution, in Encyclopedia of Nuclear Magnetic Resonance, *Wiley, Chichester* (1996) 141-185

128. J. Kowalewski, D. Kruk, G. Parigi, NMR relaxation in solution of paramagnetic complexes: Recent theoretical progress for S >= 1, *Advances in Inorganic Chemistry* **57** (2005) 41-104

129. A. R. Edmunds, Angular momentum in quantum mechanics, *Princeton University Press, Princeton* (1974)

130. D.A. Varshalovich, A.N. Moskalev, V.K. Khersonkii, Quantum theory of angular momentum, *Word Scientific Publishing, Singapore* (1988)

131. D. M. Brink, G.R. Satchler, Angular momentum, *Clarendon Press, Oxford* (1979)

132. B. L. Silver, Irreducible tensor methods, *Academic Press, New York* (1976)

133. T. Nilsson, J. Kowalewski, Low-field theory of nuclear spin relaxation in paramagnetic low – symmetry complexes for electron spin systems of S=1,3/2, 2, 5/2, 3 and 7/ 2, *Mol. Phys.* **98** (2000) 1617-1638, Erratum: *Mol. Phys.* **99** (2001) 369-370

134. I. Bertini, J. Kowalewski, C. Luchinat, T. Nilsson, G. Parigi, Nuclear spin relaxation in paramagnetic complexes of S=1: electron spin relaxation effects, *J. Chem. Phys.* **111** (1999) 5795-5807

135. D. Kruk, T. Nilsson, J. Kowalewski, Nuclear spin relaxation in paramagnetic systems with zero-field splitting and arbitrary electron spin, *Phys. Chem. Chem. Phys.* **3** (2001) 4907-4917

136. P.-O. Westlund, Nuclear paramagnetic spin relaxation theory: Paramagnetic spin probes in homogenous and microheterogenous solutions, in Dynamics of solutions and fluid mixtures by NMR, J. J. Delpuech (Ed.), *Wiley, Chichester* (1995) 173-229

137. M. Rotenberg, R. Bivins, N. Metropolis, J.K. Wooten, The 3-J and 6-J symbols, *Technology Press, Cambridge* (1959)

138. R. Sharp, Characteristic properties of the nuclear magnetic resonance – paramagnetic relaxation enhancement arising from integer and half-integer electron spins, *J. Chem. Phys.* **98** (1993) 2507-2515

139. R. Sharp, Nuclear spin relaxation due to paramagnetic species in solution: effect of anisotropy in the ZFS tensor, J. Chem. Phys. 98 (1993) 6092-6101

140. P.-O. Westlund, A low field paramagnetic nuclear spin relaxation theory, *J. Chem. Phys.* **108** (1998) 4945-4953

141. S.M. Abernathy, J. C. Miller, L.L. Lohr, R. Sharp, Nuclear magnetic resonance – paramagnetic relaxation enhancement: influence of spatial quantization of the electron spin when the zero-field splitting energy is larger than the Zeeman energy, J. Chem. Phys. 109 (1998) 4035-4046

142. R. Sharp, S.M. Abernathy, L.L. Lohr, Paramagnetically induced nuclear magnetic resonance relaxation in solutions containing S>1 ions, a molecular-frame theoretical and physical model, *J. Chem. Phys.* **107** (1997) 7620-7629

143. R. Sharp, L. Lohr, J. Miller, Paramagnetic NMR relaxation enhancement: recent advances in theory, *Prog. Nucl. Magn. Reson.* Spectr. **38** (2001) 115-158

CHAPTER V

Electron and quadrupolar spin relaxation: unified treatment

In the previous chapter we have discussed relaxation processes of electron and quadrupolar spins of the spin quantum number $S=1$ in the two limiting cases of low and high magnetic fields. Let us remind to the reader that the high field limit means that the Zeeman interaction is much stronger than the static zero field splitting (for an electron spin), or than the static quadrupolar interaction (for a quadrupolar spin): $\omega_S \gg \omega_{ZFS}^S$ or $\omega_S \gg \omega_Q^S$, respectively, while the low field limit is characterized by the opposite relations: $\omega_{ZFS}^S \gg \omega_S$ or $\omega_Q^S \gg \omega_S$. In this chapter I aim for a careful and detailed discussion of field dependent relaxation processes of electron and quadrupolar spins for an arbitrary spin quantum number $S \geq 1$, not restricting myself only to the limiting conditions regarding the applied magnetic field. I shall consider this subject pointing out close analogies between slowly rotating molecular systems in solution and solid state systems.

If an electron spin of $S \geq 1$ is carried by a slowly rotating low-symmetry molecule, one must expect a static ZFS, which has a profound effect on the energy level fine structure of the electron spin and in consequence on its relaxation. This is of course particularly important in the low field limit. In full analogy to the electron spin one can consider slowly rotating molecules carrying a quadrupolar spin, which exhibits a non-zero, averaged (static) quadrupolar interaction. In this case, the orientation of the principal axis system of the quadrupolar coupling (referred to as the molecular frame) is modulated by the molecular tumbling, exactly as the orientation of the static ZFS tensor for an electron spin (see Fig.3.2). Under the assumption of slow rotation the molecular and laboratory frames can be considered as fixed with respect to each other. This very close analogy can be however a little bit boring for the reader. It is much more interesting to consider solid state systems containing quadrupolar spins and characterized by a non-zero electric field gradient tensor resulting from a low symmetry environment.

This section is concerned with the manner in which the static ZFS (or the static quadrupolar coupling) affects the electron (or the quadrupolar) spin relaxation. The example of the electron spin relaxation at low field for $S=1$, discussed in Section IV.5, announces complex relaxation processes for higher spin quantum numbers, caused by the presence of a strong static ZFS interaction. We shall see that indeed the static ZFS influences the electron spin relaxation more markedly for higher spin quantum numbers than for $S=1$. The analogous statement can be of course addressed to quadrupolar spin systems and the static quadrupolar interaction. In the high field regime, where the Zeeman interaction dominates, we shall draw the attention to multiexponential relaxation processes of the electron and quadrupolar spins. I shall present a complete approach to the field dependent electron and quadrupolar spin relaxation, appropriate for such spin systems for which the motional conditions permit one to treat the static ZFS tensor (or the electric field gradient tensor) as fixed with respect to the laboratory frame.
This approach is based on the Redfield relaxation theory, which can be applied only under certain conditions, discussed in detail in Chapter II and summarized in Section IV.7. For clarity of the further considerations, I wish to specify them explicitly at this stage. First of all the condition referred to as the Redfield limit must be fulfilled. Since as explained already the relaxation mechanism is provided by the transient ZFS or by the transient quadrupolar interaction, the conditions $\omega_{ZFS}^{T}\tau_{D}<<1$ or $\omega_{Q}^{T}\tau_{Q}<<1$, (the quantities τ_{D} and τ_{Q} denote the characteristic correlation times introduced in Section III.2) have to be satisfied, for the electron

and quadrupolar spin system, respectively. The next condition (the secular approximation condition) concerns the relation between the main Hamiltonian H_0 and the perturbing interaction $H_1(t)$ ($H_1 = H_{ZFS}^T$ or $H_1 = H_Q^T$):

$\frac{\omega_0}{\omega_{ZFS}^T} \gg \omega_{ZFS}^T \tau_D$ or $\frac{\omega_0}{\omega_Q^T} \gg \omega_Q^T \tau_Q$. Since we aim for a description of relaxation processes, which is valid for an arbitrary magnetic field (particularly at low magnetic field, where the contribution of the effect of the Zeeman coupling is negligible), we have to consider carefully the relation between the static and transient terms of the ZFS or of the quadrupolar coupling, respectively. In the low field limit one can treat the static interactions as the unperturbed ones (and therefore determining the energy level structure of the spin under interest) if the relationships $\frac{\omega_{ZFS}^S}{\omega_{ZFS}^T} \gg \omega_{ZFS}^T \tau_D$ or $\frac{\omega_Q^S}{\omega_Q^T} \gg \omega_Q^T \tau_Q$ are satisfied. When the magnetic field increases the Zeeman coupling starts to play a noticeable role and supports the static ZFS to fulfill the condition: $\frac{\omega_{ZFS}^S + \omega_S}{\omega_{ZFS}^T} \gg \omega_{ZFS}^T \tau_D$ (and analogously: $\frac{\omega_Q^S + \omega_S}{\omega_Q^T} \gg \omega_Q^T \tau_Q$). It is very important to be aware that the static ZFS and the Zeeman coupling can contribute together to the energy level structure of the considered spins only if their principal axis systems do not fluctuate in time with respect to each other. Finally, the Zeeman coupling becomes strong enough, to fulfill the relation $\frac{\omega_S}{\omega_{ZFS}^T} \gg \omega_{ZFS}^T \tau_D$ by itself. Fig. 5.1 shows a pictorial view of the discussed conditions. The treatment, which we shall present below, can be applied to an arbitrary magnetic field and any spin quantum number and therefore it can give us a deep insight into various aspects of the rather complex relaxation processes.

Before we begin with this general approach, it may be useful and interesting for the reader to consider field dependences of the relaxation rates for the simplest case of $S=1$ and an axially symmetric static ZFS (for an electron spin) or an axial quadrupolar coupling (for a quadrupolar spin). To simplify the problem even more we assume that the direction of the applied magnetic field coincides with the static ZFS (or the static electric field gradient) axis. This case is somewhat unrealistic; however it has educational advantages and captures the essence of the field dependent relaxation in the presence of other interactions contributing to the energy level structure of the spin systems under interest.

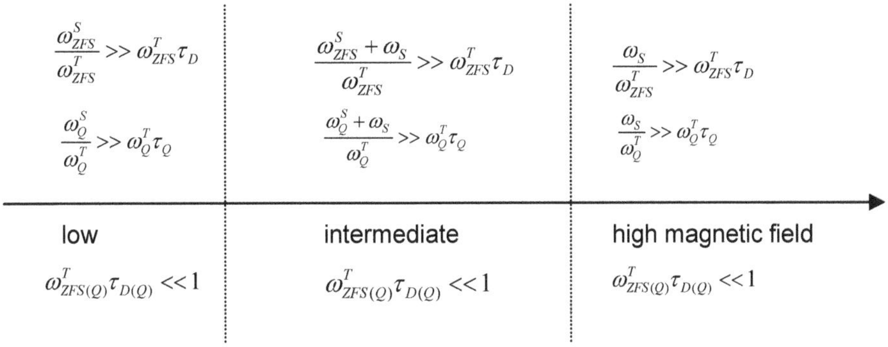

$\frac{\omega_{ZFS}^S}{\omega_{ZFS}^T} \gg \omega_{ZFS}^T \tau_D$	$\frac{\omega_{ZFS}^S + \omega_S}{\omega_{ZFS}^T} \gg \omega_{ZFS}^T \tau_D$	$\frac{\omega_S}{\omega_{ZFS}^T} \gg \omega_{ZFS}^T \tau_D$
$\frac{\omega_Q^S}{\omega_Q^T} \gg \omega_Q^T \tau_Q$	$\frac{\omega_Q^S + \omega_S}{\omega_Q^T} \gg \omega_Q^T \tau_Q$	$\frac{\omega_S}{\omega_Q^T} \gg \omega_Q^T \tau_Q$
low	intermediate	high magnetic field
$\omega_{ZFS(Q)}^T \tau_{D(Q)} \ll 1$	$\omega_{ZFS(Q)}^T \tau_{D(Q)} \ll 1$	$\omega_{ZFS(Q)}^T \tau_{D(Q)} \ll 1$

Fig.5.1
Validity conditions of the Redfield relaxation theory applied to an electron spin in the case of slow molecular tumbling.

V.1. Field dependent electron spin relaxation for axially symmetric ZFS and the spin quantum number S=1: the simplest example

The educational advantages of this example come from the fact that one can set up appropriate expressions for the individual relaxation matrix elements modifying straightforwardly Eq.4.40. It is so the assumed geometry that the molecular axis is taken to be parallel to the Zeeman axis (this is relaxed in the following sections). Taking into account that the Zeeman coupling alters the energy level structure of the electron spin in the following way: $E_1 = \frac{D_S}{3} + \omega_S$, $E_2 = -\frac{2D_S}{3}$ and $E_3 = \frac{D_S}{3} - \omega_S$ (I remind the reader that the energy levels E_1, E_2, E_3 correspond to the eigenstates $|m_S\rangle$ of $m_S = 1, 0, -1$, respectively), the population block of the relaxation matrix takes the form:

$$\frac{d}{dt}\begin{bmatrix} \rho_{11}^S \\ \rho_{22}^S \\ \rho_{33}^S \end{bmatrix} = \tag{5.1}$$

$$\begin{bmatrix} -J_1^S(\omega_D+\omega_S)-2J_2^S(2\omega_S) & J_1^S(\omega_D+\omega_S) & 2J_2^S(2\omega_S) \\ J_1^S(\omega_D+\omega_S) & -J_1^S(\omega_D+\omega_S)-J_1^S(\omega_D-\omega_S) & J_1^S(\omega_D-\omega_S) \\ 2J_2^S(2\omega_S) & J_1^S(\omega_D-\omega_S) & -J_1^S(\omega_D-\omega_S)-2J_2^S(2\omega_S) \end{bmatrix} \times \begin{bmatrix} \rho_{11}^S \\ \rho_{22}^S \\ \rho_{33}^S \end{bmatrix}$$

The relaxation matrix elements converge to the low field or high field expressions (Eq.4.40 and Eq.4.13), if $\omega_D \gg \omega_S$ or $\omega_S \gg \omega_D$, respectively. This is due to the fact that in the low field regime the spectral densities become obviously independent of the frequency ω_S: $J^S(\omega_D \pm \omega_S) \cong J^S(\omega_D)$ and $J^S(2\omega_S) \cong J^S(0)$, while at the high field the contribution of the static ZFS can be omitted: $J^S(\omega_D \pm \omega_S) \cong J^S(\omega_S)$. This guarantees that the expression for the electron spin-lattice relaxation rate R_{1S}, which results from the matrix of Eq.5.1 and describes the relaxation process for an arbitrary magnetic field, converges to the limiting formulas, presented in Chapter IV (Eq.4.41a and Eq.4.17a). In the intermediate range of the magnetic field (when the frequencies ω_D and ω_S are comparable) the spectral densities $J^S(\omega_D + \omega_S)$ and $J^S(\omega_D - \omega_S)$ can differ significantly. Actually, only if the characteristic correlation time τ_c is short enough the frequencies ω at which the spectral density are taken do not matter, since $\omega\tau_c \ll 1$. This condition is referred to in the literature as the extreme narrowing condition [1, 2]. Generally, the time evolution of the $\langle S_z \rangle$ quantity follows the equation:

$$\frac{d\langle S_z \rangle}{dt} = \frac{d(\rho_{11}^S - \rho_{33}^S)}{dt} =$$
$$-\rho_{11}^S \left[J_1^S(\omega_D+\omega_S) + 4J_2^S(2\omega_S) \right] + \rho_{33}^S \left[J_1^S(\omega_D-\omega_S) + 4J_2^S(2\omega_S) \right] \tag{5.2}$$
$$+\rho_{22}^S \left[J_1^S(\omega_D+\omega_S) - J_1^S(\omega_D-\omega_S) \right]$$

It is apparent from this expression that one cannot formulate a proper description of the spin-lattice relaxation processes of an electron (quadrupolar) spin $S = 1$, valid for an arbitrary magnetic field, in terms of only one time constant. The fact that we limit ourselves to static ZFS tensors of axial symmetry, which implies that the main Hamiltonian is diagonal in the Zeeman basis, does not help. Performing the calculations we have assumed in addition that the principal axis of the ZFS tensor and the direction of the magnetic field coincide. This assumption simplifies the calculations. However, it does not change the essentially different physical pictures of the relaxation processes occurring within the static ZFS energy level structure at low field and within the Zeeman energy levels in the high field limit. The spin-lattice relaxation is a consequence of transitions between the eigenstates of the spin system. At low field limit the eigenstates corresponding to $m_S = 1$ and $m_S = -1$ are degenerate, while in the high field they are well separated. Therefore it is not surprising that relaxation processes resulting from the transition dynamics have a different character. In the intermediate range of the magnetic field the low and high field relaxation features compete, leading to a complex relaxation dynamics. It is worthwhile to notice, that according to Eq.5.2, despite the different coefficients associated with the populations ρ_{11}^S and ρ_{33}^S, the $\langle S_z \rangle$ quantity is coupled also to the population ρ_{22}^S. This coupling is not present in the limiting cases of the low and high magnetic fields, since then $J^S(\omega_D + \omega_S) \cong J^S(\omega_D - \omega_S)$ (one of the frequencies, ω_D or ω_S, is negligible).

One can evaluate also the spin-spin relaxation rates taking into account the transition frequencies resulting from the superposition of the Zeeman coupling and the static ZFS interactions:

$$R_{1212}^S = \frac{3}{2} J_0^S(0) + J_1^S(\omega_D + \omega_S) + \frac{1}{2} J_1^S(\omega_D - \omega_S) + J_2^S(2\omega_S) \quad (5.3a)$$

$$R_{2323}^S = \frac{3}{2} J_0^S(0) + J_1^S(\omega_D - \omega_S) + \frac{1}{2} J_1^S(\omega_D + \omega_S) + J_2^S(2\omega_S) \quad (5.3b)$$

If the two interactions are of comparable magnitudes, the spin-spin relaxation cannot be described by a one time constant, either.

The example discussed in this section introduces us to the more extensive treatment of the static ZFS effects on the electron spin relaxation, which I shall present below.

V.2. Theory of electron spin relaxation for slowly rotating systems of low symmetry

As explained already, in the case of slowly rotating molecular systems the electron spin energy level structure is determined by a Hamiltonian $H_0(S)$ obtained as a combination of the electron spin Zeeman coupling and the static ZFS:

$$H_0(S) = H_Z(S) + H_{ZFS}^S(S) \tag{5.4}$$

To determine the energy levels one needs to express both the components of the main Hamiltonian in the same reference frame. Taking into account the forms of the static ZFS Hamiltonian and the Zeeman Hamiltonian (Eq.3.2 and Eq.3.8) it can be somewhat more convenient for the computations to represent the main Hamiltonian in the principal axis system of the static ZFS (P_S), that $H_0^{(P_S)}(S) = H_Z^{(P_S)}(S) + H_{ZFS}^{S(P_S)}(S)$. The $H_{ZFS}^{S(P_S)}(S)$ Hamiltonian is given by Eq.3.9 while the expression for $H_Z^{(P_S)}(S)$ can be obtained by applying the transformation rule (Eq.3.3) to the laboratory form of the Zeeman Hamiltonian (Eq.3.8):

$$H_Z^{(P_S)}(S) = \omega_S \sum_{m=-1}^{1} (-1)^m S_m^1 D_{0,-m}^1(\Omega_{P_S L}) \tag{5.5}$$

The electron spin energy levels depend on the orientation of the complex with respect to the external magnetic field. The orientation dependence is encoded in the Wigner rotation matrices $D_{0,-m}^1(\Omega_{P_S L})$. Since we consider slowly rotating molecules, the set of Euler angles can be treated as time independent. Using the formula of Eq.5.5 one can write down the explicit expression for the Zeeman Hamiltonian in the (P_S) frame:

$$\begin{aligned}H_Z^{(P_S)} &= S_z \omega_S \cos\beta_{P_S L} - \frac{1}{2}(S_+ + S_-)\omega_S \sin\beta_{P_S L} \cos\gamma_{P_S L} \\ &\quad - \frac{1}{2}i(S_+ - S_-)\omega_S \sin\beta_{P_S L} \sin\gamma_{P_S L}\end{aligned} \tag{5.6}$$

If the static ZFS exhibits an axial symmetry, the eigenvalues and eigenvectors of the Hamiltonian $H_0^{(P_S)}(S)$ are independent of the $\gamma_{P_S L}$ angle [3 - 5]. This permits one to put $\gamma_{P_S L} = 0$ and to consider the electron spin relaxation problem as dependent only on the β_{P_SL} angle. The electron spin relaxation has been calculated in this way in those works. One can also perform the calculations starting from the representation of the main Hamiltonian $H_0(S)$ in the laboratory frame. The final result is the same. Nevertheless when one is particularly interested in the low field electron or quadrupolar spin relaxation, it is highly advisable to express the main interaction (containing, in this case, actually only the static ZFS or the quadrupolar coupling) in the (P_S) (or (P_S^Q)) frame. In this chapter I prefer to deal with the laboratory frame representation of the contributing interactions. The reason for this is that performing the calculations in the laboratory representation one can relatively straightforwardly refer to some well known expressions (for example the SBM formula), which in fact correspond to limiting cases of the more general treatment. Actually, the small inconvenience of choosing the laboratory axis system as the reference frame concerns only the procedure of setting up the matrix representation of the main Hamiltonian, $H_0^{(L)}(S) = H_Z^{(L)}(S) + H_{ZFS}^{S(L)}(S)$, where the laboratory form of the static ZFS is given by Eq.3.10. The transformation of the second rank tensor operators, representing the static ZFS, generates more elements than in the case of the first rank tensors, describing the Zeeman Hamiltonian. If one deals with the static ZFS in the laboratory frame, the dependence of the energy level structure on the relative orientation of the laboratory and molecular frames is encoded in the second order Wigner rotation matrices $D_{k,-m}^2(\Omega_{P_SL})$, contained in the quantities $\widetilde{V}_{-m}^{2S(L)}$.

Let us consider as an example an electron spin of $S = \frac{7}{2}$. To set up the matrix form of the Hamiltonian $H_0^{(L)}(S)$ we chose the Zeeman basis formed by the $2S+1$ vectors $\{|m_S\rangle\}$. Obviously, this basis is not an eigenbasis for this Hamiltonian. However, since we intend to obtain in the next step the eigenvectors of the main Hamiltonian in terms of the Zeeman functions $|m_S\rangle$, it is appropriate to start from the Zeeman representation. Labeling the Zeeman states as follows:

$|1\rangle = \left|\frac{7}{2}\right\rangle$, $|2\rangle = \left|\frac{5}{2}\right\rangle$, $|3\rangle = \left|\frac{3}{2}\right\rangle$, $|4\rangle = \left|\frac{1}{2}\right\rangle$, $|5\rangle = \left|-\frac{1}{2}\right\rangle$, $|6\rangle = \left|-\frac{3}{2}\right\rangle$, $|7\rangle = \left|-\frac{5}{2}\right\rangle$ and

$|8\rangle = \left|-\frac{7}{2}\right\rangle$, we can write down the matrix elements of the Hamiltonian $H_0^{(L)}(S)$:

$$\langle 1|H_0^{(L)}(\Omega_{P_SL})|1\rangle = \frac{21}{\sqrt{6}}\widetilde{V}_0^{2S(L)}(\Omega_{P_SL})+\frac{7}{2}\omega_S, \quad \langle 7|H_0^{(L)}(\Omega_{P_SL})|7\rangle = \frac{21}{\sqrt{6}}\widetilde{V}_0^{2S(L)}(\Omega_{P_SL})-\frac{7}{2}\omega_S \quad (5.7a)$$

$$\langle 2|H_0^{(L)}(\Omega_{P_SL})|2\rangle = \frac{3}{\sqrt{6}}\widetilde{V}_0^{2S(L)}(\Omega_{P_SL})+\frac{5}{2}\omega_S, \quad \langle 2|H_0^{(L)}(\Omega_{P_SL})|2\rangle = \frac{3}{\sqrt{6}}\widetilde{V}_0^{2S(L)}(\Omega_{P_SL})-\frac{5}{2}\omega_S \quad (5.7b)$$

$$\langle 3|H_0^{(L)}(\Omega_{P_SL})|3\rangle = -\frac{9}{\sqrt{6}}\widetilde{V}_0^{2S(L)}(\Omega_{P_SL})+\frac{3}{2}\omega_S, \quad \langle 6|H_0^{(L)}(\Omega_{P_SL})|6\rangle = -\frac{9}{\sqrt{6}}\widetilde{V}_0^{2S(L)}(\Omega_{P_SL})-\frac{3}{2}\omega_S \quad (5.7c)$$

$$\langle 4|H_0^{(L)}(\Omega_{P_SL})|4\rangle = -\frac{15}{\sqrt{6}}\widetilde{V}_0^{2S(L)}(\Omega_{P_SL})+\frac{1}{2}\omega_S, \quad \langle 5|H_0^{(L)}(\Omega_{P_SL})|5\rangle = -\frac{15}{\sqrt{6}}\widetilde{V}_0^{2S(L)}(\Omega_{P_SL})-\frac{1}{2}\omega_S \quad (5.7d)$$

$$\langle 1|H_0^{(L)}(\Omega_{P_SL})|2\rangle = -\langle 7|H_0^{(L)}(\Omega_{P_SL})|8\rangle = 3\sqrt{7}\widetilde{V}_{-1}^{2S(L)}(\Omega_{P_SL}) \quad (5.7e)$$

$$\langle 1|H_0^{(L)}(\Omega_{P_SL})|3\rangle = \langle 6|H_0^{(L)}(\Omega_{P_SL})|8\rangle = \sqrt{21}\widetilde{V}_{-2}^{2S(L)}(\Omega_{P_SL}) \quad (5.7f)$$

$$\langle 2|H_0^{(L)}(\Omega_{P_SL})|3\rangle = -\langle 6|H_0^{(L)}(\Omega_{P_SL})|7\rangle = 4\sqrt{3}\widetilde{V}_{-1}^{2S(L)}(\Omega_{P_SL}) \quad (5.7g)$$

$$\langle 2|H_0^{(L)}(\Omega_{P_SL})|4\rangle = \langle 5|H_0^{(L)}(\Omega_{P_SL})|7\rangle = 3\sqrt{5}\widetilde{V}_{-2}^{2S(L)}(\Omega_{P_SL}) \quad (5.7h)$$

$$\langle 3|H_0^{(L)}(\Omega_{P_SL})|4\rangle = -\langle 5|H_0^{(L)}(\Omega_{P_SL})|6\rangle = \sqrt{15}\widetilde{V}_{-1}^{2S(L)}(\Omega_{P_SL}) \quad (5.7i)$$

$$\langle 3|H_0^{(L)}(\Omega_{P_SL})|5\rangle = \langle 4|H_0^{(L)}(\Omega_{P_SL})|6\rangle = 2\sqrt{15}\widetilde{V}_{-2}^{2S(L)}(\Omega_{P_SL}) \quad (5.7j)$$

$$\langle 4|H_0^{(L)}(\Omega_{P_SL})|5\rangle = 0 \quad (5.7k)$$

Replacing the static ZFS parameters D_S and E_S in the functions $\widetilde{V}_{-m}^{2S(L)}$ by the quadrupolar parameters, according to the relationships $D_S \to \frac{3}{4}\frac{a_Q^S}{S(2S-1)}$ and $E_S \to \frac{1}{4}\eta_S\frac{a_Q^S}{S(2S-1)}$ (the index 'S' associated with the quantities a_Q^S and η_S referrers to the static quadrupolar interaction) one can obtain the matrix representation of the main Hamiltonian $H_0^{(L)}(S) = H_Z^{(L)}(S) + H_Q^{S(L)}(S)$ for a quadrupolar spin of $S = \frac{7}{2}$ in the Zeeman basis $\{|m_S\rangle\}$. The matrix elements of Eqs.5.7a-k can be adapted to the case of the quadrupolar spin by replacing the quantities $\widetilde{V}_{-m}^{2S(L)}$ by their quadrupolar counterparts $\widetilde{A}_{-m}^{2S(L)}$, defined as:

$$\tilde{A}_{-m}^{2S(L)}(t) = \sum_{k=-2}^{2} \tilde{A}_{k}^{2S(P_S)} D_{k,-m}^2 (\Omega_{P_S L}), \quad \text{where} \quad \tilde{A}_0^{2S(P_S)} = \frac{1}{2}\sqrt{\frac{3}{2}} \frac{a_Q^S}{S(2S+1)}, \quad \tilde{A}_{\pm 1}^{2S(P_S)} = 0,$$

$$\tilde{A}_{\pm 2}^{2S(P_S)} = \frac{a_Q^S}{S(2S+1)} \eta_S$$ (see Section III.1). Obviously, in the same manner one can set up the matrix representation of the main Hamiltonian $H_0^{(L)}(S)$ for other spin quantum numbers.

Before proceeding to the evaluation of the individual relaxation rates it is of interest to see some examples of the energy level structures for an electron as well a quadrupolar spin. The energy levels for the spin quantum number $S = \frac{7}{2}$ for two orientations of the molecular frame relative to the laboratory frame: $\Omega_{P_S L} = (0,0,0)$ and $\Omega_{P_S L} = (0, 90°, 0)$ are presented. Fig.5.2a,b shows the electron spin energy levels for an axial ZFS tensor, while Fig.5.3a,b shows the energy level structures for the quadrupolar spin of lanthanum (^{139}La) depending on the orientation; the quadrupolar asymmetry parameter has been set to zero.

Starting from the proper form of the main Hamiltonian for any given magnetic field and any orientation of the molecule we can obtain, making use of the Redfield formula of Eq.2.14, the electron spin relaxation rates in terms of spectral densities at the transition frequencies at that particular field and orientation. The way of the evaluations of the relaxation rates is the same for all spin quantum numbers. As it has been pointed out in Chapters II and IV, to describe relaxation processes of an arbitrary spin system it is necessary to deal with its eigenlevels and eigenfunctions; the formula of Eq.2.14 has been obtained in just this way. The eigenfunctions $\{\psi_\alpha^S\}$ for a given spin system can be expressed in terms of the Zeeman basis functions, $\{m_S\} \equiv \{n\}$ by the relation:

$$|\psi_\alpha^S\rangle = \sum_{n=1}^{2S+1} c_{\alpha n}(\Omega_{P_S L}) |n\rangle \tag{5.8}$$

a)

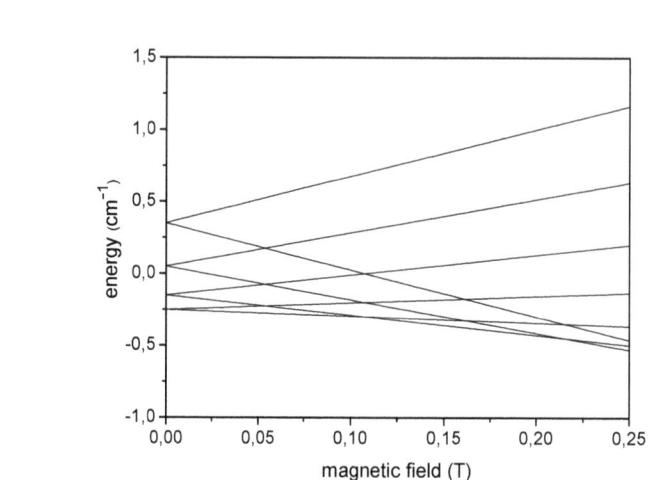

b)

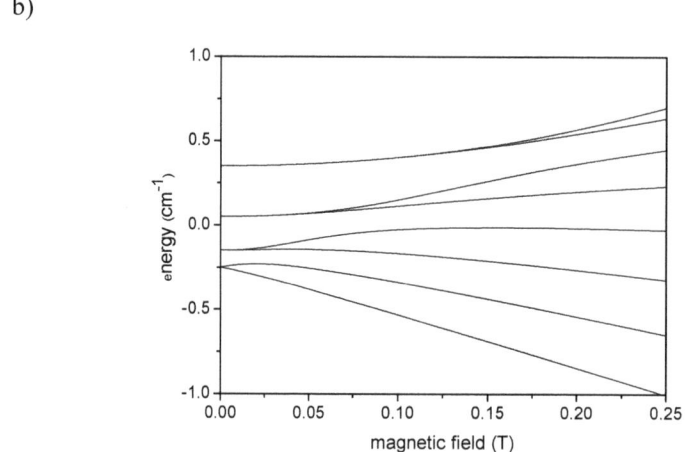

Fig.5.2

Energy level structure of the electron spin of $S = \frac{7}{2}$ versus the magnetic field. The static ZFS is axially symmetrical; $D_S = 0.05 cm^{-1}$. It has been assumed that the magnetic field is applied along the principal axis of the static ZFS, i.e. $\Omega_{P_S L} = 0$ a) and perpendicular to the principal axis of the static ZFS, i.e. $\Omega_{P_S L} = 90°$ b).

a)

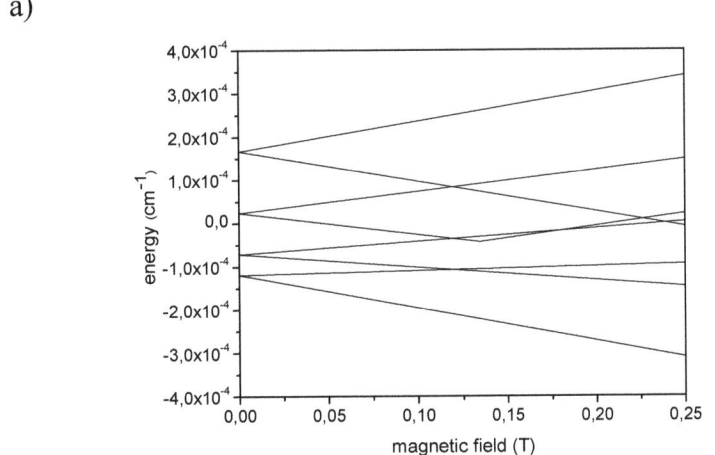

b)

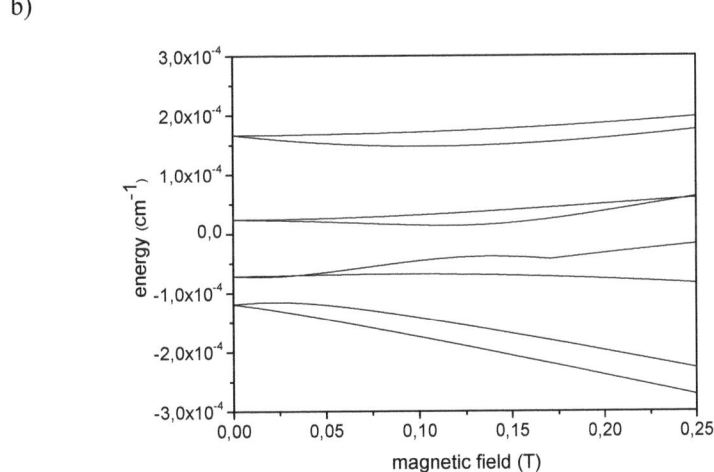

Fig.5.3

Energy level structure of the quadrupole spin of lanthanum ($S = \frac{7}{2}$) versus the magnetic field. The electric field gradient tensor is axially symmetrical; $a_Q^s = 20 MHz = 6.(6)*10^{-4} cm^{-1}$. It has been assumed that the magnetic field is applied along the principal axis of the electric field gradient tensor, i.e. $\Omega_{P_S^Q L} = 0$ a) and perpendicular to the principal axis system of the electric field gradient tensor, i.e. $\Omega_{P_S^Q L} = 90° $ b).

The specific, angular dependent coefficients $c_{\alpha n}(\Omega_{P_S L})$ are provided by a diagonalization of the matrix representation of the Hamiltonian $H_S^{0(L)}$ or $H_S^{0(P_S)}$ if for some reasons the molecular reference frame is preferable (for example, in the low field limit). Therefore, the corresponding Liouville basis $\{|\psi_\alpha^S\rangle\langle\psi_{\alpha'}^S|\}$ generated by the set of eigenfunctions of the main Hamiltonian and appropriate for applying of the perturbation approach is related to the Liouville basis constructed from the Zeeman eigenstates $\{|n\rangle\langle n'|\}$ as [3-8]:

$$\{|\psi_\alpha^S\rangle\langle\psi_{\alpha'}^S|\} = \left\{\sum_{n,n'=1}^{2S+1} c_{nr}^*(\Omega_{P_S L}) c_{\alpha'n'}(\Omega_{P_S L}) |n\rangle\langle n'|\right\} \tag{5.9}$$

The electron spin relaxation rates $R_{\alpha\beta\alpha\beta}^S$, $R_{\alpha\alpha\beta\beta}^S$ and $R_{\alpha\alpha\alpha\alpha}^S$ are related to the spectral densities $\Im_{\alpha\alpha'\beta\beta'}^S(\omega)$ taken at frequencies corresponding to differences between the energy levels, according to Eq.2.14. The computational procedure is now more complicated than in the limiting cases presented in Chapter IV; however the approach is basically the same. According to Eq.2.13 the spectral densities $\Im_{\alpha\alpha'\beta\beta'}^S(\omega)$ are determined by the corresponding matrix elements of the perturbing Hamiltonian H_1:

$$\Im_{\alpha\alpha'\beta\beta'}^S(\omega) = \int_0^\infty \langle\langle\psi_\alpha^S|H_1(\tau)|\psi_{\alpha'}^S\rangle\langle\psi_\beta^S|H_1(0)|\psi_{\beta'}^S\rangle\rangle \exp(-i\omega\tau)d\tau \tag{5.10}$$

The external brackets denote the averaging over the spin ensemble. The perturbing Hamiltonian H_1 corresponds now to the transient ZFS (or the transient quadrupolar coupling) Hamiltonian, i.e. $H_1(t) = H_{ZFS}^T(t)$ or $H_1(t) = H_Q^T(t)$. Although it should be understood from the derivations presented in this book so far, it is of primary importance at this stage to become aware of the fundamental aspect of the perturbation description of relaxation, which comes from its physical nature: relaxation processes occurring in a arbitrary spin system are determined by transition probabilities between its eigenstates $|\psi_\alpha^S\rangle$ and, in turn, the transition probabilities are determined by the corresponding elements of the perturbing interaction, $\langle\psi_\alpha^S|H_1|\psi_{\alpha'}^S\rangle$, taken between these eigenstates. Eq.5.10 reflects this fundamental aspect and one has to apply it, even though the calculations become quite cumbersome if the eigenbasis $\{|\psi_\alpha^S\rangle\}$ is not formed by the 'pure' Zeeman eigenstates. Using the relationship between the eigenvectors $|\psi_\alpha^S\rangle$ and the Zeeman vectors $|r\rangle = |m_S\rangle$ (Eq.5.8) one can express the Hamiltonian matrix elements $\langle\psi_\alpha^S|H_1|\psi_{\alpha'}^S\rangle$ as: $\langle\psi_\alpha^S|H_1|\psi_{\alpha'}^S\rangle = \sum_{n,n'=1}^{2S+1} c_{\alpha n}^* c_{\alpha' n'} \langle n|H_1|n'\rangle$. In consequence, the spectral densities $J_{\alpha\alpha'\beta\beta'}^S(\omega)$ can be obtained from the expression:

$$\mathfrak{J}_{\alpha\alpha'\beta\beta'}^S(\omega) = \sum_{n,n'} c_{\alpha n}^* c_{\alpha' n'} c_{\beta n'}^* c_{\beta' n'} \int_0^\infty \langle\langle n|H_1(\tau)|n'\rangle\langle n'|H_1(0)|n\rangle\rangle \exp(-i\omega\tau)d\tau \quad (5.11)$$

Assuming the ordinary representation of the Hamiltonian $H_1(t)$ in terms of the second rank tensor components $T_m^2(S)$ and the associated time dependent functions $\widetilde{V}_{-m}^2(t)$, $H_1(t) = \sum_{m=-2}^{2}(-1)^m T_m^2(S)\widetilde{V}_{-m}^2(t)$, the spectral densities $\mathfrak{J}_{\alpha\alpha'\beta\beta'}^S(\omega)$ can be related explicitly to the quantities $J_m^S(\omega) = \int_0^\infty \langle\widetilde{V}_m^{2*}(\tau)\widetilde{V}_m^2(0)\rangle \exp(-i\omega\tau)d\tau$ by employing the expression:

$$\int_0^\infty \langle n|H_1(\tau)|n'\rangle\langle n'|H_1(0)|n\rangle \exp(-i\omega\tau)d\tau = \sum_{m=-2}^{2}|\langle n|T_m^2(S)|n'\rangle|^2 J_m^S(\omega) \quad (5.12)$$

Then the explicit expressions for the individual relaxation rates $R_{\alpha\beta\alpha\beta}^S$ and $R_{\alpha\alpha\beta\beta}^S$ in terms of the spectral densities $J_m^S(\omega) = J^S(\omega)$ (for the motional models presented so far the spectral densities are independent of the order of the tensor components, m) take the form [3, 5-8]:

$$R_{\alpha\beta\alpha\beta}^S(\Omega_{P_SL}) = 2\xi_{\alpha\beta}(\Omega_{P_SL})J^S(0) - \sum_{\gamma=1}^{2S+1}\xi_{\gamma\alpha}(\Omega_{P_SL})J^S(\omega_{\gamma\alpha}) - \sum_{\gamma=1}^{2S+1}\xi_{\gamma\beta}(\Omega_{P_SL})J^S(\omega_{\gamma\beta}) \quad (5.13a)$$

$$R_{\alpha\alpha\beta\beta}^S(\Omega_{P_SL}) = \xi_{\alpha\beta}(\Omega_{P_SL})J^S(\omega_{rs}) \quad (5.13b)$$

Theory of evolution and relaxation of multi-spin systems

The orientation dependent coefficients $\xi_{\alpha\beta}(\Omega_{P_S L})$ result from the relationship between the representations of the tensor components $T_m^2(S)$ in the Zeeman basis $\{|n\rangle\}$ and the eigenbasis $\{|\psi_\alpha^S\rangle\}$ (Eq.5.8) and are defined as:

$$\xi_{\alpha\beta}(\Omega_{P_S L}) = \sum_{m=-2}^{2} \left(\sum_{n,n'}^{2S+1} \left[|c_{\alpha n}^* c_{\beta n'}|^2 (\Omega_{P_S L}) \right] |\langle n|T_m^2(S)|n'\rangle|^2 \right) \quad (5.14)$$

The matrix elements $\langle n|T_m^2(S)|n'\rangle$ are non-zero only if $\Delta m_S = m$, where Δm_S is the difference between the magnetic quantum numbers associated with the Zeeman eigenstates $|n\rangle$ and $|n'\rangle$, i.e. $\Delta m_S = m_{S|n'\rangle} - m_{S|n\rangle}$. If one assumes the pseudorotational model for the transient ZFS (and in analogy for the fluctuating part of the quadrupolar interaction) the spectral density $J^S(\omega)$ has the form of Eq.3.17 (or Eq.3.18 when the rotational correlation time is very long). The remaining relaxation rates $R_{\alpha\alpha\alpha\alpha}^S(\Omega_{P_S L})$ are given as a sum over the elements $R_{\alpha\alpha\beta\beta}^S(\Omega_{P_S L})$: $R_{\alpha\alpha\alpha\alpha}^S(\Omega_{P_S L}) = -\sum_{\beta\neq\alpha} R_{\alpha\alpha\beta\beta}^S(\Omega_{P_S L})$ and therefore one does not need to calculate them independently. The coefficients $\xi_{\alpha\beta}(\Omega_{P_S L})$ for the spin quantum number $S = \frac{7}{2}$ have been given in Appendix to [5], while the coefficients for other spin quantum numbers ($S = 1, 3/2, 2, 5/2, 3$) have been collected in Supplementary Material to this paper. However, for the completeness of our description and the convenience of the reader we present below the expression for $S = \frac{7}{2}$ (omitting the explicit dependence on the molecular orientation):

$$\xi_{\alpha\beta} = -\frac{3}{4}\begin{Bmatrix}7c^*_{\alpha,-7/2}c_{\beta,-7/2} + c^*_{\alpha,-5/2}c_{\beta,-5/2} - 3c^*_{\alpha,-3/2}c_{\beta,-3/2} - 5c^*_{\alpha,-1/2}c_{\beta,-1/2} \\ -5c^*_{\alpha,1/2}c_{\beta,1/2} - 3c^*_{\alpha,3/2}c_{\beta,3/2} + c^*_{\alpha,5/2}c_{\beta,5/2} + 7c^*_{\alpha,7/2}c_{\beta,7/2}\end{Bmatrix}^2$$

$$-\frac{1}{2}\begin{Bmatrix}3\sqrt{7}c^*_{\alpha,-7/2}c_{\beta,-5/2} + 4\sqrt{3}c^*_{\alpha,-5/2}c_{\beta,-3/2} + \sqrt{15}c^*_{\alpha,-3/2}c_{\beta,-1/2} \\ -\sqrt{15}c^*_{\alpha,1/2}c_{\beta,3/2} - 4\sqrt{3}c^*_{\alpha,3/2}c_{\beta,5/2} - 3\sqrt{7}c^*_{\alpha,5/2}c_{\beta,7/2}\end{Bmatrix}^2$$

$$+\frac{1}{2}\begin{Bmatrix}3\sqrt{7}c^*_{\alpha,-5/2}c_{\beta,-7/2} + 4\sqrt{3}c^*_{\alpha,-3/2}c_{\beta,-5/2} + \sqrt{15}c^*_{\alpha,-1/2}c_{\beta,-3/2} \\ -\sqrt{15}c^*_{\alpha,3/2}c_{\beta,1/2} - 4\sqrt{3}c^*_{\alpha,5/2}c_{\beta,3/2} - 3\sqrt{7}c^*_{\alpha,7/2}c_{\beta,5/2}\end{Bmatrix}^2 \qquad (5.15)$$

$$-\frac{1}{2}\begin{Bmatrix}\sqrt{21}c^*_{\alpha,-7/2}c_{\beta,-3/2} + 3\sqrt{5}c^*_{\alpha,-5/2}c_{\beta,-1/2} + 2\sqrt{15}c^*_{\alpha,-3/2}c_{\beta,1/2} \\ +2\sqrt{15}c^*_{\alpha,-1/2}c_{\beta,3/2} + 3\sqrt{5}c^*_{\alpha,1/2}c_{\beta,5/2} + \sqrt{21}c^*_{\alpha,3/2}c_{\beta,7/2}\end{Bmatrix}^2$$

$$+\frac{1}{2}\begin{Bmatrix}\sqrt{21}c^*_{\alpha,-3/2}c_{\beta,-7/2} + 3\sqrt{5}c^*_{\alpha,-1/2}c_{\beta,-5/2} + 2\sqrt{15}c^*_{\alpha,1/2}c_{\beta,-3/2} \\ +2\sqrt{15}c^*_{\alpha,3/2}c_{\beta,-1/2} + 3\sqrt{5}c^*_{\alpha,5/2}c_{\beta,1/2} + \sqrt{21}c^*_{\alpha,7/2}c_{\beta,3/2}\end{Bmatrix}^2$$

In the case of slowly rotating molecules or polycrystalline materials the relaxation rates result as a sum of responses from the different molecular orientations, thus one needs in the last step of the consideration to perform an averaging over the Ω_{P_SL} angle.

In the next sections I shall use the presented theory to discuss some aspects of field dependent relaxation processes for different spin quantum numbers. Unfortunately, in the intermediate region of the magnetic field, where the transition from the low to high field regime occurs, one cannot formulate analytical expressions for the relaxation rates. The fact that the quantization axis changes from the principal axis system of the static ZFS tensor (the electric field gradient tensor) to the laboratory frame leads to some interesting features of the relaxation profiles. It is particularly interesting to discuss the present treatment in a context of the Bloembergen-Morgan (BM) theory [9], which has been readily presented in Section IV.3. As has been pointed out in this section and demonstrated by examples, the BM approach is a high field theory and applying it at low field in the presence of the static ZFS or the static quadrupolar coupling contradicts to fundamental assumptions of the perturbation theory. However, even at high magnetic field a main restriction of the BM theory is that it ignores multiexponential relaxation effects. The effects are especially pronounced under non-extreme narrowing conditions, *i.e.* if the assumption that product of the transition frequencies ω_S of the considered spin (determined by the Zeeman coupling) and the correlation time $\tau_{D(Q)}$ is much less than unity: $\omega_S \tau_{D(Q)} << 1$ breaks down. More detailed discussion of such effects will occupy the remainder of the next sections.

It is very interesting and highly recommended to turn attention to other works dealing with the electron spin relaxation [10-21].

V.3. Field dependent electron (quadrupolar) spin relaxation processes for spin quantum number 3/2

In this chapter we shall work out in detail the electron spin relaxation for the spin quantum number $S = \frac{3}{2}$. Half-integer spin systems are significantly different from the integer spin systems, because of the Kramers degeneracy [4, 17-27] of some of their energy levels at low magnetic field. This feature of the energy level structure of half-integer spins applies both in the axial and in the rhombic case of the static ZFS. The Kramers degeneracy can only be lifted if the magnetic field is applied, which implies that even at very low magnetic fields the Kramers doublet $\left\{ \left| m_S = \frac{1}{2} \right\rangle, \left| m_S = -\frac{1}{2} \right\rangle \right\}$ is split by the frequency ω_S, like in the high field limit. This effect is particularly important for the nuclear spin relaxation at low field, as we shall see in the next chapter.

To elucidate better the effects of the Kramers degeneracy let us start with the low-field behavior of the electron spin relaxation. To avoid more cumbersome calculations we shall limit ourselves to a system of an axial static ZFS.

We begin, as usual, from the Zeeman basis of the spin of interest. In the case of $S = \frac{3}{2}$ it contains four vectors, labeled as follows: $|1\rangle = \left|\frac{3}{2}\right\rangle$, $|2\rangle = \left|\frac{1}{2}\right\rangle$, $|3\rangle = \left|-\frac{1}{2}\right\rangle$, $|4\rangle = \left|-\frac{3}{2}\right\rangle$. Next, we quote the main Hamiltonian $H_0(S)$ and write down its matrix in the Zeeman basis. Since, at this instant we aim for a discussion of the relaxation processes at low field, when $\omega_D >> \omega_S$, we consider the main

Hamiltonian in the (P_S) frame. For an axially symmetrical ZFS tensor, the Hamiltonian matrix is diagonal, with the elements: $\langle 1|H_{ZFS}^{(P_S)}|1\rangle = \langle 4|H_{ZFS}^{(P_S)}|4\rangle = \omega_D$ and $\langle 2|H_{ZFS}^{(P_S)}|2\rangle = \langle 3|H_{ZFS}^{(P_S)}|3\rangle = -\omega_D$. Fig.5.4 shows the transition frequency in the low field regime, i.e. under the assumption $\omega_D \gg \omega_S$.

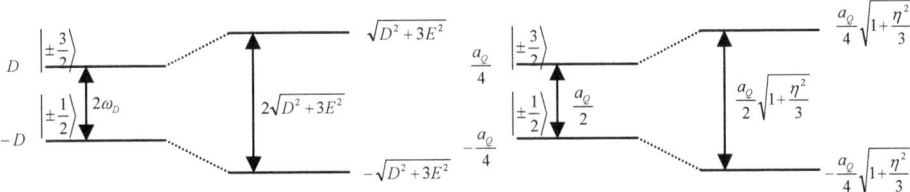

Fig.5.4

Energy level structure for electron and qudrupolar spins of $S = \dfrac{3}{2}$ when there is no Zeeman coupling.

The further computational procedure has been described in detail in Chapter IV. Following this procedure one can obtain the relaxation matrix elements $R^S_{\alpha\alpha'\beta\beta'}$ in terms of the electron spin spectral densities $J^S(\omega)$. To perform the calculations one needs to express the perturbing Hamiltonian, i.e. the transient ZFS in the eigenbasis of the main Hamiltonian. Fortunately, the main interaction is diagonal in the Zeeman basis $\{|m_S\rangle\}$ and this step can be achieved straightforwardly. For convenience we present below the matrix form of the transient ZFS Hamiltonian $H_{ZFS}^{T(P_S)}$:

$$H_{ZFS}^{T(P_S)}(S) = \begin{bmatrix} |1\rangle & |2\rangle & |3\rangle & |4\rangle \\ \frac{3}{\sqrt{6}} \widetilde{V}_0^{2T(P_S)} & \sqrt{3}\widetilde{V}_{-1}^{2T(P_S)} & \sqrt{3}\widetilde{V}_{-2}^{2T(P_S)} & 0 \\ -\sqrt{3}\widetilde{V}_1^{2T(P_S)} & -\frac{3}{\sqrt{6}} \widetilde{V}_0^{2T(P_S)} & 0 & \sqrt{3}\widetilde{V}_{-2}^{2T(P_S)} \\ \sqrt{3}\widetilde{V}_2^{2T(P_S)} & 0 & -\frac{3}{\sqrt{6}} \widetilde{V}_0^{2T(P_S)} & -\sqrt{3}\widetilde{V}_{-1}^{2T(P_S)} \\ 0 & \sqrt{3}\widetilde{V}_2^{2T(P_S)} & \sqrt{3}\widetilde{V}_1^{2T(P_S)} & \frac{3}{\sqrt{6}} \widetilde{V}_0^{2T(P_S)} \end{bmatrix} \quad (5.14)$$

As a result of the energy level structure consisting of the two doublets $\{|2\rangle, |3\rangle\}$ and $\{|1\rangle, |4\rangle\}$, the population part, $\hat{R}_1^S$, of the relaxation matrix takes the rather simple form:

$$\frac{d}{dt}\begin{bmatrix} \rho_{11}^S \\ \rho_{22}^S \\ \rho_{33}^S \\ \rho_{44}^S \end{bmatrix} = -\begin{bmatrix} -12J^S(2\omega_D) & 6J^S(2\omega_D) & 6J^S(2\omega_D) & 0 \\ 6J^S(2\omega_D) & -12J^S(2\omega_D) & 0 & 6J^S(2\omega_D) \\ 6J^S(2\omega_D) & 0 & -12J^S(2\omega_D) & 6J^S(2\omega_D) \\ 0 & 6J^S(2\omega_D) & 6J^S(2\omega_D) & -12J^S(2\omega_D) \end{bmatrix}\begin{bmatrix} \rho_{11}^S \\ \rho_{22}^S \\ \rho_{33}^S \\ \rho_{44}^S \end{bmatrix} \quad (5.15)$$

where I make use of the fact that $J_m^S(\omega) = J^S(\omega)$. One can find out from the relaxation matrix that at low field the electron spin-lattice relaxation rate is equal to $R_{1S} = \frac{12}{5}\Delta_T^2 \frac{\tau_D}{1+(2\omega_D\tau_D)^2}$ [4]. The spin-spin relaxation processes are more interesting. They are described by two time constants: $R_{2S,1} = R_{2323}^S = R_{1414}^S = \frac{12}{5}\Delta_T^2 \frac{\tau_D}{1+(2\omega_D\tau_D)^2}$, $R_{2S,2} = R_{1212}^S = R_{1313}^S = R_{2424}^S = R_{3434}^S = \frac{6}{5}\Delta_T^2\left[\tau_D + \frac{2\tau_D}{1+(2\omega_D\tau_D)^2}\right]$ [4]. The two distinct spin-spin relaxation rates are plotted in Fig.5.5 for several values of the distortional correlation time τ_D.

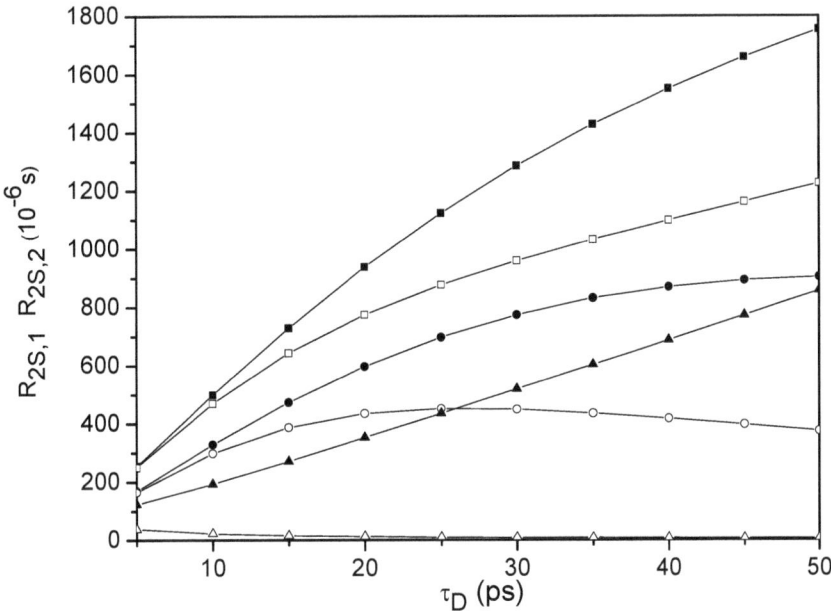

Fig.5.5

Electron spin – spin relaxation rates $R_{2S,1}$ and $R_{2S,2}$ at low field for the spin quantum number $S = \frac{3}{2}$ versus the distortional correlation time τ_D. The transient ZFS has been set to $\Delta_T = 0.02 cm^{-1}$. Open squares: $R_{2S,1}$, $D_s = 0.05 cm^{-1}$, solid squares $R_{2S,2}$, $D_s = 0.05 cm^{-1}$, open circles: $R_{2S,1}$, $D_s = 0.1 cm^{-1}$, solid circles $R_{2S,2}$, $D_s = 0.1 cm^{-1}$, open triangles: $R_{2S,1}$, $D_s = 1.0 cm^{-1}$, solid triangles $R_{2S,2}$, $D_s = 1.0 cm^{-1}$.

Looking at the energy level structure one can recognize that because of the degeneracy, in principle, we are not allowed to consider separately the spin-lattice and spin-spin relaxation. The secular approximation does not lead in this case to the decoupling of coherences of various orders. The time evolutions of the populations, represented by the derivatives $\frac{d\rho_{\alpha\alpha}^S}{dt}$ ($\alpha = 1,2,3,4$), are coupled to the coherences ρ_{14}^S, ρ_{41}^S and ρ_{23}^S, ρ_{32}^S, because $\omega_{14} = \omega_{23} = 0$. The degeneracy decreases the number of different transition frequencies within the energy level structure, leading in this way to some simplifications of the relaxation matrix elements (many spectral densities are taken at the same frequency), however this effect has a price. One has to consider carefully eventual additional couplings between coherences of different orders. Fortunately, if the relaxation mechanism is provided by the transient ZFS or the quadrupolar coupling, the form of the perturbing Hamiltonian and in consequence the orthogonality properties of the Wigner rotation matrices lead to vanishing of the relaxation matrix elements connecting the above mentioned coherences. The same situation takes place for the spin-spin relaxation. The pairs of coherences $\rho_{\alpha\alpha'}^S$ and $\rho_{\beta\beta'}^S$, which could be coupled because of the condition $\omega_{\alpha\alpha'} = \omega_{\beta\beta'}$ being fulfilled: ρ_{12}^S and ρ_{13}^S, ρ_{14}^S and ρ_{23}^S, and ρ_{24}^S and ρ_{34}^S remain separated due to the orthogonality of the Wigner matrices.

To gain more insight into the relaxation process in the Zeeman limit for higher spin quantum numbers, we examine now the electron spin relaxation for $S = \frac{3}{2}$ at high magnetic field. The individual relaxation rates of the population block of the relaxation matrix (Eq.5.15) can be easily adjusted to the Zeeman energy level structure, yielding:

$$\frac{d}{dt}\begin{bmatrix} \rho_{11}^S \\ \rho_{22}^S \\ \rho_{33}^S \\ \rho_{44}^S \end{bmatrix} = $$

$$-\begin{bmatrix} -6J^S(\omega_S)-6J^S(2\omega_S) & 6J^S(\omega_S) & 6J^S(2\omega_S) & 0 \\ 6J^S(\omega_S) & -6J^S(\omega_S)-6J^S(2\omega_S) & 0 & 6J^S(2\omega_S) \\ 6J^S(2\omega_S) & 0 & -6J^S(\omega_S)-6J^S(2\omega_S) & 6J^S(\omega_S) \\ 0 & 6J^S(2\omega_S) & 6J^S(\omega_S) & -6J^S(\omega_S)-6J^S(2\omega_S) \end{bmatrix} \times \begin{bmatrix} \rho_{11}^S \\ \rho_{22}^S \\ \rho_{33}^S \\ \rho_{44}^S \end{bmatrix}$$

(5.16)

The fundamental difference between the form of the spin-lattice relaxation matrices at low and high field limits is that in the last case two transition frequencies, ω_S and $2\omega_S$ are involved in the relaxation process, while at low field limit all relevant spectral densities are taken at the frequency $2\omega_D$. The expectation value of the S_Z operator is given as a sum of the population differences: $\langle S_Z \rangle = \frac{3}{2}(\rho_{11}^S - \rho_{44}^S) + \frac{1}{2}(\rho_{22}^S - \rho_{33}^S)$. Its time evolution, determined by the relaxation rates included in the above matrix, follows the equation:

$$\frac{d\langle S_Z \rangle}{dt} = -[8J^S(\omega_S) + 4J^S(2\omega_S)]\left\{\frac{3}{2}(\rho_{11}^S - \rho_{44}^S)\right\}$$
$$-[24J^S(2\omega_S) - 12J^S(\omega_S)]\left\{\frac{1}{2}(\rho_{22}^S - \rho_{33}^S)\right\} \quad (5.17)$$

The equation shows clearly that, in general, the spin-lattice relaxation of an electron (quadrupolar) spin of $S = \frac{3}{2}$ is not single exponential at high magnetic field. In this context one should appreciate the simplifications of the relaxation dynamics in the low field regime caused by the degeneracy of the electron spin energy levels. Actually, a single exponential spin-lattice relaxation process of the spin $S = \frac{3}{2}$ can be achieved under the extreme narrowing condition [1, 2] $\omega_S \tau_D \ll 1$. Then the spectral densities $J^S(\omega_S)$ and $J^S(2\omega_S)$ become equal: $J^S(\omega_S) \cong J^S(2\omega_S) \cong J^S(0)$, and the quantity $\langle S_Z \rangle$ follows the simplified equation:

$$\frac{d\langle S_Z \rangle}{dt} \cong -R_{1S}\langle S_Z \rangle = -12J^S(0)\langle S_Z \rangle \quad (5.18)$$

The expression reflects an obvious, but pleasing, consistency with the analogous low field limit result, where the spectral density is taken at the frequency $2\omega_D$; under the conditions $\omega_D \tau_D \ll 1$ one has $J^S(2\omega_D) \cong J^S(0)$. To describe the spin-lattice relaxation beyond the extreme narrowing regime, it is very useful to introduce an appropriate set of population combinations, representing some relevant observables. Besides the sum of the populations $E = \rho_{11}^S + \rho_{22}^S + \rho_{33}^S + \rho_{44}^S$, which remains unchanged (or in other words evolves in time with infinitely long time constant) and the already defined quantity $\langle S_z \rangle$, it is worthwhile to consider two more combinations representing the higher order alignments: $\langle 3S_z^2 - S(S+1) \rangle = 3(\rho_{11}^S - \rho_{22}^S - \rho_{33}^S + \rho_{44}^S)$ and $\langle S_z(5S_z^2 - 5S(S+1)+1) \rangle = \frac{3}{2}\rho_{11}^S - \frac{9}{2}\rho_{22}^S + \frac{9}{2}\rho_{33}^S - \frac{3}{2}\rho_{44}^S$ [28-30]. The population block of the relaxation matrix can be rewritten in terms of these four observables. I do not intend to develop intensively such a treatment in this book; however I shall refer to this approach sometimes in next chapters. The spin-spin relaxation at high field is also more complicated than in the low field limit; it is described by three relaxation rates, which in addition to this complication contain now the spectral densities at three frequencies: $0, \omega_S, 2\omega_S$:

$$R_{1414}^S = R_{2323}^S = \frac{6}{5}\Delta_T^2 \left[\frac{\tau_D}{1+(\omega_S\tau_D)^2} + \frac{\tau_D}{1+(2\omega_S\tau_D)^2} \right],$$

$$R_{1212}^S = R_{1313}^S = R_{3434}^S = \frac{6}{5}\Delta_T^2 \left[9\tau_D + \frac{\tau_D}{1+(\omega_S\tau_D)^2} + \frac{\tau_D}{1+(2\omega_S\tau_D)^2} \right] \text{ and}$$

$$R_{2424}^S = \frac{6}{5}\Delta_T^2 \left[9\tau_D + \frac{\tau_D}{1+(\omega_S\tau_D)^2} + \frac{3}{2}\frac{\tau_D}{1+(2\omega_S\tau_D)^2} \right].$$

The case of the spin quantum number $S = \frac{3}{2}$ is also very suitable for an examination of the effects of the rhombic terms in the static ZFS on the electron spin relaxation. So far we have discussed only the quadrupolar spin relaxation for $S = 1$ allowing for a non-zero quadrupolar anisotropy parameter. Choosing the spin quantum number $S = \frac{3}{2}$ we can profit from a relatively simply mathematical treatment (we will be able still to provide a fully analytical description) and discuss at the same time all relevant aspects of the rhombicity.

The rhombic component of the static ZFS alters the energy level splitting for the electron spin. At low magnetic field (when the Zeeman coupling is negligible) there are still two degenerate energy levels, however now they are equal to $\pm\sqrt{D_S^2 + 3E_S^2}$ (see Fig.5.4). The four eigenvectors $\{\psi_\alpha^S\}$ of the ZFS Hamiltonian are given as the following combinations of the Zeeman vectors $\{|m_S\rangle\}$:

$$|\psi_1\rangle = \frac{1}{\sqrt{1+\mu^2}}\left[\left|-\frac{1}{2}\right\rangle + \mu\left|\frac{3}{2}\right\rangle\right], \quad |\psi_2\rangle = \frac{1}{\sqrt{1+\nu^2}}\left[\left|\frac{1}{2}\right\rangle + \nu\left|-\frac{3}{2}\right\rangle\right], \quad |\psi_3\rangle = \frac{1}{\sqrt{1+\nu^2}}\left[\left|-\frac{1}{2}\right\rangle + \nu\left|\frac{3}{2}\right\rangle\right]$$

and $|\psi_4\rangle = \frac{1}{\sqrt{1+\mu^2}}\left[\left|\frac{1}{2}\right\rangle + \mu\left|-\frac{3}{2}\right\rangle\right]$, with the coefficients $\mu = \dfrac{\omega_D + \sqrt{\omega_D^2 + 3\omega_E^2}}{\sqrt{3}\omega_E}$ and

$\nu = \dfrac{\omega_D - \sqrt{\omega_D^2 + 3\omega_E^2}}{\sqrt{3}\omega_E}$. The fact, that the Zeeman vectors do not provide the eigenbasis for the rhombic ZFS tensor makes the calculations of the electron relaxation rates more cumbersome. As it has been pointed out many times in this book, the perturbation treatment requires to consider the perturbing Hamiltonian in the eigenbasis of the main interaction. In fact, the set of coefficients $\xi_{\alpha\beta}$ defined by Eq.5.12 comes from the required representation of the perturbing Hamiltonian and one can just use them. Nevertheless, it is much more revealing to start from the representation of the transient ZFS in the eigenbasis of the static ZFS, required by the perturbation theory for the considered case, and perform the calculations being aware of their physical meaning. Therefore, we shall proceed in this way instead of using the final recipe of Eq.5.12. To make possible for the reader a straightforward analysis of the forthcoming calculations, I collect below the matrix elements of the transient ZFS evaluated within the basis $\{|\psi_\alpha^S\rangle\}$:

$$\langle\psi_1^S|H_{ZFS}^{T(P_S)}|\psi_1^S\rangle = \frac{1}{1+\mu^2}\left[(\mu^2-1)\frac{3}{\sqrt{6}}\widetilde{V}_0^{2T(P_S)} + \mu\sqrt{3}\left(\widetilde{V}_{-2}^{2T(P_S)} + \widetilde{V}_2^{2T(P_S)}\right)\right] \quad (5.19a)$$

$$\langle\psi_1^S|H_{ZFS}^{T(P_S)}|\psi_2^S\rangle = \frac{1}{\sqrt{(1+\mu^2)(1+\nu^2)}}(\mu-\nu)\sqrt{3}\widetilde{V}_{-1}^{2T(P_S)} \quad (5.19b)$$

$$\langle\psi_1^S|H_{ZFS}^{T(P_S)}|\psi_3^S\rangle = \frac{1}{\sqrt{(1+\mu^2)(1+\nu^2)}}\left[(\mu\nu-1)\frac{3}{\sqrt{6}}\widetilde{V}_0^{2T(P_S)} + \mu\sqrt{3}\widetilde{V}_{-2}^{2T(P_S)} + \nu\sqrt{3}\widetilde{V}_2^{2T(P_S)}\right] \quad (5.19c)$$

$$\langle\psi_2^S|H_{ZFS}^{T(P_S)}|\psi_2^S\rangle = \frac{1}{1+\nu^2}\left[(\nu^2-1)\frac{3}{\sqrt{6}}\widetilde{V}_0^{2T(P_S)} + \nu\sqrt{3}\left(\widetilde{V}_{-2}^{2T(P_S)} + \widetilde{V}_2^{2T(P_S)}\right)\right] \quad (5.19d)$$

$$\langle\psi_2^S|H_{ZFS}^{T(P_S)}|\psi_3^S\rangle = \frac{1}{1+\nu^2}\sqrt{3}\nu\left(\widetilde{V}_{-1}^{2T(P_S)} + \widetilde{V}_1^{2T(P_S)}\right) \quad (5.19e)$$

$$\langle\psi_2^S|H_{ZFS}^{T(P_S)}|\psi_4^S\rangle = \frac{1}{\sqrt{(1+\mu^2)(1+\nu^2)}}\left[(1-\mu\nu)\frac{\sqrt{3}}{6}\tilde{V}_0^{2T(P_S)} + \mu\sqrt{3}\tilde{V}_{-2}^{2T(P_S)} + \nu\sqrt{3}\tilde{V}_2^{2T(P_S)}\right] \quad (5.19f)$$

$$\langle\psi_3^S|H_{ZFS}^{T(P_S)}|\psi_3^S\rangle = \frac{1}{1+\nu^2}\left[(\nu^2-1)\frac{3}{\sqrt{6}}V_0^{2T(P_S)} + \nu\sqrt{3}\left(V_{-2}^{2T(P_S)} + V_2^{2T(P_S)}\right)\right] \quad (5.19g)$$

$$\langle\psi_3^S|H_{ZFS}^{T(P_S)}|\psi_4^S\rangle = \frac{1}{\sqrt{(1+\mu^2)(1+\nu^2)}}(\mu-\nu)\sqrt{3}\tilde{V}_{-1}^{2T(P_S)} \quad (5.19h)$$

$$\langle\psi_4^S|H_{ZFS}^{T(P_S)}|\psi_4^S\rangle = \frac{1}{1+\mu^2}\left[(1-\mu^2)\frac{3}{\sqrt{6}}\tilde{V}_0^{2T(P_S)} + \mu\sqrt{3}\left(\tilde{V}_{-2}^{2T(P_S)} + \tilde{V}_2^{2T(P_S)}\right)\right] \quad (5.19i)$$

Utilizing the above expressions we can obtain the relaxation matrix elements directly from the formula of Eq.2.14. The orthogonality properties of the Wigner rotation matrices encoded into the functions $\tilde{V}_m^{2T(P_S)}$ simplify the calculations a lot. The coefficients $R_{\alpha\alpha\beta\beta}^S$ take the form:

$$R_{1122}^S = R_{3344}^S = \frac{1}{\omega_D^2 + 3\omega_E^2}\left(6\omega_D^2 + 27\omega_E^2\right)J_1^S\left(2\sqrt{\omega_D^2 + 3\omega_E^2}\right) \quad (5.20)$$

$$R_{1133}^S = R_{2244}^S = \frac{1}{\omega_D^2 + 3\omega_E^2}\left[\left(6\omega_D^2 + 9\omega_E^2\right)J_2^S\left(2\sqrt{\omega_D^2 + 3\omega_E^2}\right) + 9\omega_E^2 J_0^S\left(2\sqrt{\omega_D^2 + 3\omega_E^2}\right)\right] \quad (5.21)$$

In the equations the different orders of the electron spin spectral densities of $J_m^S(\omega)$ are distinguished to illustrate the effect of the corresponding terms $\tilde{V}_m^{2T(P_S)}$ contributing to the matrix elements of the perturbing Hamiltonian. Within the pseudorotational model of the transient ZFS the form of the spectral densities is independent of their order. The diagonal relaxation matrix element result, as usual, from the relation: $R_{\alpha\alpha\alpha\alpha}^S = -\sum_{\beta\neq\alpha}R_{\alpha\alpha\beta\beta}^S$. Apparently the relaxation rates of Eq.5.20 and Eq.5.21 converge to the term $6J^S(\omega_D)$ if the static ZFS exhibit an axial symmetry. The rhombic component of the static ZFS affects also the spin-spin relaxation processes. They are still characterized by two time constants; however they are influenced by the rhombicity:

$$R_{2S,1} = \frac{1}{\omega_D^2 + 3\omega_E^2}\times$$
$$\left[9\omega_E^2 J_0^S\left(2\sqrt{\omega_D^2 + 3\omega_E^2}\right) + 3\omega_E\sqrt{3\omega_D^2 + 9\omega_E^2}J_1^S\left(2\sqrt{\omega_D^2 + 3\omega_E^2}\right) + \left(12\omega_D^2 + 18\omega_E^2\right)J_2^S\left(2\sqrt{\omega_D^2 + 3\omega_E^2}\right)\right] \quad (5.22)$$

$$R_{2S,2} = \frac{1}{\omega_D^2 + 3\omega_E^2} \times$$

$$\left[6\omega_D^2 J_0^S(0) + 18\omega_E^2 J_0^S\left(2\sqrt{\omega_D^2 + 3\omega_E^2}\right) + 6\omega_D^2 J_1^S\left(2\sqrt{\omega_D^2 + 3\omega_E^2}\right) + 18\omega_E^2 J_2^S(0) + \right.$$
$$\left. \left(6\omega_D^2 + 18\omega_E^2\right) J_2^S\left(2\sqrt{\omega_D^2 + 3\omega_E^2}\right) \right] \quad (5.23)$$

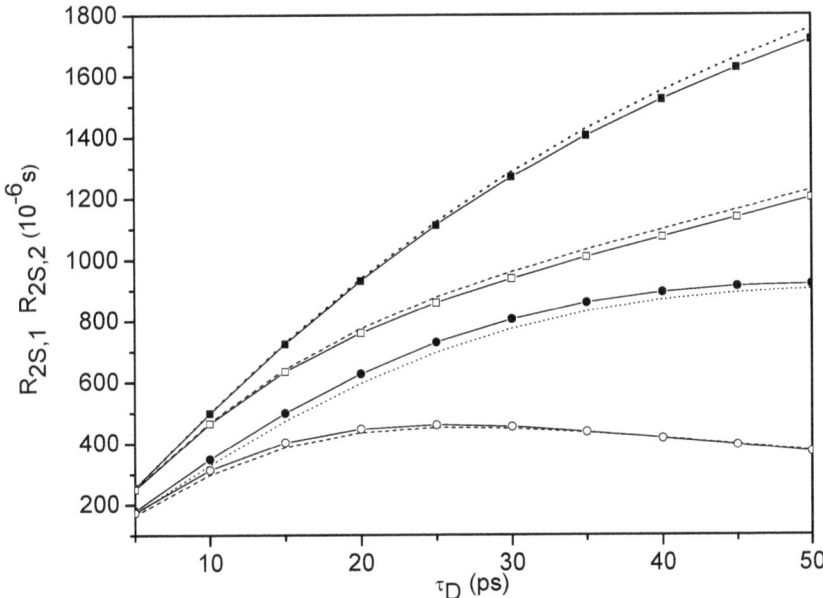

Fig.5.6

Effects of the rhombic component of the static ZFS ($E_S = \frac{D_S}{3}$) on the electron spin – spin relaxation rates $R_{2S,1}$ and $R_{2S,2}$ at low field for the spin quantum number $S = \frac{3}{2}$ ($\Delta_T = 0.02 cm^{-1}$). Open squares: $R_{2S,1}$, $D_S = 0.05 cm^{-1}$, solid squares $R_{2S,2}$, $D_S = 0.05 cm^{-1}$, open circles: $R_{2S,1}$, $D_S = 0.1 cm^{-1}$, solid circles $R_{2S,2}$, $D_S = 0.1 cm^{-1}$; dotted and dashed lines show the corresponding $R_{2S,1}$ and $R_{2S,2}$ relaxation rates (presented in Fig.5.5, already) for the axially symmetrical static ZFS.

Theory of evolution and relaxation of multi-spin systems

The spin-lattice relaxation rates have been obtained by a diagonalization of the population block of the relaxation matrix. Fig. 5.6 shows the effects of the rhombicity of the static ZFS on the low field electron spin relaxation rates. They are not very significant in terms of numbers (because of the rather short correlation time τ_D), but it is worthwhile to put attention on the physical picture of the relaxation processes when the static ZFS contains the rhombic components.

It is of primary importance to realize at this stage that the electron spin relaxation rates calculated in the low field limit refer to the (P_S) frame. Since we have assumed a slow molecular tumbling, the relative orientation of this axis system and the laboratory frame does not change in time. Therefore, the already calculated quantities describe well the relaxation processes also from the perspective of the laboratory frame. In other words, the correlation functions do not oscillate with respect to the laboratory frame. They decay with the same characteristic time constant with respect to the laboratory and molecular frames. The second important remark concerns the correspondence between the coherences $|\psi_\alpha^S\rangle\langle\psi_\beta^S|$ and some observables. Since the presence of the rhombic terms changes the eigenbasis of the static ZFS (the Zeeman states are no longer the eigenvectors of the ZFS tensor), one must adjust in an appropriate manner the linear combinations of coherences representing relevant observables, like for example $\langle S_z \rangle = \frac{3}{2}(\rho_{11}^S - \rho_{44}^S) + \frac{1}{2}(\rho_{22}^S - \rho_{33}^S)$. The coherences $\rho_{\alpha\beta}^S$ are defined in terms of the Zeeman vectors (for example $\rho_{11}^S \equiv \left|\frac{3}{2}\right\rangle\left\langle\frac{3}{2}\right|$). To express an operator $\langle \hat{O} \rangle$ in terms of the molecular frame coherences $|\psi_\alpha^S\rangle\langle\psi_\beta^S|$ one needs to employ the relationship between these quantities defined in the laboratory and in the molecular frames. This relationship is obviously determined by the expansion coefficients of the eigenvectors $|\psi_\alpha^S\rangle$ in the Zeeman basis: $|\psi_\alpha^S\rangle = \sum_{n=1}^{2S+1} c_{\alpha n}|n\rangle$. I shall come to this problem in the forthcoming chapters of this book.

V.4 Multiexponential relaxation processes for high spin quantum numbers at low and high magnetic fields

The preceding chapters have dealt mainly with spin systems of rather low quantum numbers, $S=1$ and $S=\frac{3}{2}$ in different regimes of the magnetic field. In this context, it is interesting for comparison purposes to turn now attention to high spin systems, like the case of $S=\frac{7}{2}$. Since we have developed in detail the appropriate mathematical apparatus, the treatment of higher spins is now quite straightforward.

The energy levels for an electron spin $S=\frac{7}{2}$ affected by the static ZFS can be obtained from a diagonalization of the Hamiltonian matrix with the elements given by Eq.5.7. They form at low magnetic field four doublets. For the axially symmetric static ZFS, the degenerated levels have the energies $7D, D, -3D$ and $-5D$ (the index 'S' denoting the static interactions has been here omitted), and correspond to the pairs of eigenstates $\left\{\left|\frac{7}{2}\right\rangle,\left|-\frac{7}{2}\right\rangle\right\}$, $\left\{\left|\frac{5}{2}\right\rangle,\left|-\frac{5}{2}\right\rangle\right\}$, $\left\{\left|\frac{3}{2}\right\rangle,\left|-\frac{3}{2}\right\rangle\right\}$ and $\left\{\left|\frac{1}{2}\right\rangle,\left|-\frac{1}{2}\right\rangle\right\}$, respectively. Since the form of the transient ZFS Hamiltonian in the molecular frame is formally analogous to the form of the static ZFS in the laboratory frame (Eq.3.12 and Eq.3.10), the matrix representation of the transient ZFS can be obtained directly from Eq.5.7 by omitting the Zeeman contribution and replacing the orientation dependent functions $\widetilde{V}_m^{2S(L)}(\Omega_{P_SL}(t))$ by their time dependent counterparts $\widetilde{V}_m^{2T(P_S)}(\Omega_{P_TP_S}(t))$. Then, we can derive the electron spin relaxation matrix elements at low field. For the axial symmetry case, they population block, $\hat{\hat{R}}_1^S$, of the relaxation matrix has the form:

$$\hat{\hat{R}}_1^S =$$

$$\begin{bmatrix} |1\rangle & |2\rangle & |3\rangle & |4\rangle & |5\rangle & |6\rangle & |7\rangle & |8\rangle \\ R_{1111} & 126J_1^S(6\omega_D) & 42J_2^S(10\omega_D) & 0 & 0 & 0 & 0 & 0 \\ 126J_1^S(6\omega_D) & R_{2222} & 96J_1^S(4\omega_D) & 90J_2^S(6\omega_D) & 0 & 0 & 0 & 0 \\ 42J_2^S(10\omega_D) & 96J_1^S(4\omega_D) & R_{3333} & 30J_1^S(2\omega_D) & 120J_2^S(2\omega_D) & 0 & 0 & 0 \\ 0 & 90J_2^S(6\omega_D) & 30J_1^S(2\omega_D) & R_{4444} & 0 & 60J_2^S(2\omega_D) & 0 & 0 \\ 0 & 0 & 120J_2^S(2\omega_D) & 0 & R_{5555} & 30J_1^S(2\omega_D) & 120J_2^S(6\omega_D) & 0 \\ 0 & 0 & 0 & 60J_2^S(2\omega_D) & 30J_1^S(2\omega_D) & R_{6666} & 96J_1^S(4\omega_D) & 42J_2^S(10\omega_D) \\ 0 & 0 & 0 & 0 & 120J_2^S(6\omega_D) & 96J_1^S(4\omega_D) & R_{7777} & 126J_1^S(6\omega_D) \\ 0 & 0 & 0 & 0 & 0 & 42J_2^S(10\omega_D) & 126J_1^S(6\omega_D) & R_{8888} \end{bmatrix}$$

(5.24)

with the diagonal elements $R_{\alpha\alpha\alpha\alpha}^S$ given by the sums of the associated coefficients $R_{\alpha\alpha\beta\beta}^S$: $R_{\alpha\alpha\alpha\alpha}^S = -\sum_{\beta \neq \alpha} R_{\alpha\alpha\beta\beta}^S$. Also in this case all spectral densities taken at the same frequencies are equal, $J_m^S(\omega) = J^S(\omega)$, however I decided to keep the index, m, to see the order of the transitions corresponding to the particular frequencies $2\omega_D$, $4\omega_D$, $6\omega_D$ and $10\omega_D$. Even though the degeneracy reduces the number of different transition frequencies, the relaxation dynamics at low field for the spin quantum number $S = \frac{7}{2}$ is quite complicated and obviously multiexponential. The spin-spin relaxation is complex as well. They are four different relaxation rates $R_{\alpha\beta\alpha\beta}^S$ [4] associated with the individual single-quantum coherences $\rho_{\alpha\beta}^S$:

$$R_{1212}^S = R_{5656}^S = \\ 54J_0^S(0) + 48J_1^S(4\omega_D) + 126J_1^S(10\omega_D) + 45J_2^S(6\omega_D) + 21J_2^S(10\omega_D)$$ (5.25a)

$$R_{2323}^S = R_{6767}^S = \\ 24J_0^S(0) + 15J_1^S(2\omega_D) + 96J_1^S(4\omega_D) + 63J_1^S(10\omega_D) + 60J_2^S(2\omega_D) \\ + 45J_2^S(6\omega_D) + 21J_2^S(2\omega_D)$$ (5.25b)

$$R_{3434}^S = R_{7878}^S = \\ 6J_0^S(0) + 30J_1^S(2\omega_D) + 48J_1^S(4\omega_D) + 120J_2^S(2\omega_D) + 45J_2^S(6\omega_D) \\ + 21J_2^S(10\omega_D)$$ (5.25c)

$$R_{4545}^S = 30J_1^S(2\omega_D) + 120J_2^S(2\omega_D) + 90J_2^S(6\omega_D)$$ (5.25d)

For higher magnetic field, if the Zeeman coupling is non-negligible compared to the static ZFS one is not able to provide the relaxation description in terms of 'purely' analytical expressions for the relaxation matrix elements. The difficulties are caused by the necessity of a numerical diagonalization of the matrix of the main Hamiltonian as it has been described in detail in Section V.2.

The Bloembergen-Morgan (BM) theory cannot be treated as a satisfactory approach also at the high field. As pointed out already, the idea of one longitudinal and transverse relaxation rate in the high field regime is suitable only under the extreme narrowing condition. It is appropriate in this context to formulate a complete description of the relaxation processes at high field and discuss carefully the oversimplification encoded in the BM approach even in the high field limit. Apparently, for the Zeeman energy level structure all single-quantum transitions correspond to the frequency ω_S, while all double-quantum transitions are characterized by the frequency $2\omega_S$. Taking into account that at low field the relaxation is affected by four transition frequencies, one can expect a much simpler relaxation picture in the high field, where only the two frequencies, ω_S and $2\omega_S$, are involved. Indeed, the spin-lattice relaxation part of the relaxation matrix has a simpler form:

$$\hat{\hat{R}}_1^S =$$

	$|1\rangle$	$|2\rangle$	$|3\rangle$	$|4\rangle$	$|5\rangle$	$|6\rangle$	$|7\rangle$	$|8\rangle$
	R_{1111}	$126J^S(\omega_S)$	$42J^S(2\omega_S)$	0	0	0	0	0
	$126J^S(\omega_S)$	R_{2222}	$96J^S(\omega_S)$	$90J^S(2\omega_S)$	0	0	0	0
	$42J^S(2\omega_S)$	$96J^S(\omega_S)$	R_{3333}	$30J^S(\omega_S)$	$120J^S(2\omega_S)$	0	0	0
	0	$90J^S(2\omega_S)$	$30J^S(\omega_S)$	R_{4444}	0	$60J^S(2\omega_S)$	0	0
	0	0	$120J^S(2\omega_S)$	0	R_{5555}	$30J^S(\omega_S)$	$120J^S(2\omega_S)$	0
	0	0	0	$60J_2^S(2\omega_S)$	$30J^S(\omega_S)$	R_{6666}	$96J^S(\omega_S)$	$42J^S(2\omega_S)$
	0	0	0	0	$120J^S(2\omega_S)$	$96J^S(\omega_S)$	R_{7777}	$126J^S(\omega_S)$
	0	0	0	0	0	$42J^S(2\omega_S)$	$126J^S(\omega_S)$	R_{8888}

(5.26)

Nevertheless, the spin-lattice relaxation does not become single-exponential, if the extreme narrowing condition, $\omega_S\tau_D \ll 1$, is not fulfilled.

One can carry out the same type of calculations for a rhombic static ZFS. However, from the computational point of view the evaluations are much more complicated. Since they do not introduce any new elements from the point of view of the methodology, I do not present them here.

Finishing this section, it is of some interest to formulate more precisely the assumption that the molecular reorientation is slow. From the electron spin point of view it means that the rotational correlation time is long relative to the electron spin relaxation times, therefore the electron spins sense a time independent energy level structure during its relaxation process. The answer to this question is more complicated from the perspective of the nuclear spin. I shall come back to this problem in Chapter IX.

All the expression provided in this chapter in terms of the ZFS parameters, D_S, E_S, Δ_T and τ_D have their quadrupolar counterparts, written in terms of the quantities $a_Q^S, \eta_S, \Delta_{T(Q)}$ and τ_Q. All the obtained results are valid for quadrupolar spins of the corresponding spin quantum numbers by making the identifications:

$$D_S \to \frac{3}{4}\frac{a_Q^S}{S(2S-1)} = \frac{3}{4}\frac{e^2qQ_S}{\hbar S(2S-1)}, \quad E_S \to \frac{1}{4}\eta_S \frac{a_Q^S}{S(2S-1)} = \frac{1}{4}\eta_S \frac{e^2qQ_S}{\hbar S(2S-1)}, \text{ (see Fig. 5.4)}$$

$$\Delta_T \to \Delta_{T(Q)} \left(\Delta_{T(Q)}^2 = \left(\frac{a_Q^T}{S(2S-1)}\right)^2 \frac{3+\eta_T^2}{8}\right) \text{ and } \tau_D \to \tau_Q. \text{ Let us set up, as an example,}$$

the population block of the relaxation matrix for a quadrupolar spin of $S = \frac{3}{2}$ characterized by an axial static quadrupolar coupling (described by the parameter a_Q^S) at low field limit, when the quadrupolar interaction dominates over the Zeeman interaction:

$$\frac{d}{dt}\begin{bmatrix} \rho_{11}^S \\ \rho_{22}^S \\ \rho_{33}^S \\ \rho_{44}^S \end{bmatrix} = -\begin{bmatrix} -12J^{S(Q)}(2\omega_Q) & 6J^{S(Q)}(2\omega_Q) & 6J^{S(Q)}(2\omega_Q) & 0 \\ 6J^{S(Q)}(2\omega_Q) & -12J^{S(Q)}(2\omega_Q) & 0 & 6J^{S(Q)}(2\omega_Q) \\ 6J^{S(Q)}(2\omega_Q) & 0 & -12J^{S(Q)}(2\omega_Q) & 6J^{S(Q)}(2\omega_Q) \\ 0 & 6J^{S(Q)}(2\omega_Q) & 6J^{S(Q)}(2\omega_Q) & -12J^{S(Q)}(2\omega_Q) \end{bmatrix} \times \begin{bmatrix} \rho_{11}^S \\ \rho_{22}^S \\ \rho_{33}^S \\ \rho_{44}^S \end{bmatrix}$$

(5.27)

Chapter V - Electron and quadrupole spin relaxation: unified treatment

The frequency ω_Q is defined as $\omega_Q = \frac{3}{4}\frac{a_Q^S}{S(2S-1)}$ expressed in the angular frequency units. The resulted quadrupolar spin-lattice relaxation rate is, in full analogy to the case of an electron spin, given as: $R_{1S} = 12 J^{S(Q)}(2\omega_Q) = \frac{12}{5}\Delta_{T(Q)}^2 \frac{\tau_Q}{1+(2\omega_Q \tau_Q)^2}$. The approach, directly adapted from systems consisting of large molecules slowly rotating in a solution and carrying an electron spin, can be obviously applied to analogous systems containing instead of an electron spin a quadrupolar spin. However, this treatment is also directly applicable to solid polycrystalline systems containing quadrupolar spins. Individual molecules carry quadrupolar spins which exhibit a non-zero averaged quadrupolar interaction. The principal axis system of the electric field gradient tensor at the position of the spin is fixed in the molecule and therefore it can be treated as a molecular frame. The orientation of individual molecules with respect to the laboratory frame does not change in time. Since the system covers differently oriented molecules, to describe the resulting relaxation it is necessary to average over the molecular orientations. Crystal systems also fulfill the basic requirement that the orientation of principal axis system of the electric field gradient is fixed. In addition the fact that all quadrupolar spins belonging to one crystal lattice occupy crystallographically equivalent positions simplifies the considerations a lot. It is enough to carry the calculations only for one orientation.

The descriptions presented so far are based on the assumption that the transient ZFS or the fluctuating quadrupolar coupling provides a predominant relaxation mechanism for the electron or the quadrupolar spin, respectively. The electron spin relaxation is usually very efficient and this assumption is well motivated. Nevertheless, considering systems containing quadrupolar and dipolar spins it is worthwhile to notice that there are in principle two relaxation pathways for the relaxation of the quadrupolar spins. The first one, already described, is provided by fluctuations of the quadrupolar interactions and therefore it is independent of the presence of other spins in the system, while the second relaxation channel is provided by the mutual $I - S$ dipole-dipole couplings. Some motional degrees of freedom of the dipolar spins, like for example jump diffusion in solid state crystal systems, lead to time fluctuations of the dipole-dipole interactions. Both coupled spins relax via this dipolar mechanism. Relaxation processes of a spin $I = \dfrac{1}{2}$ coupled to a quadrupolar spin are the subject of Chapter VII, while Chapter XI is devoted to the quadrupolar spins relaxation due to the $I - S$ dipole-dipole interactions.

References:

1. C.P. Slichter, Principles of magnetic resonance, *Springer-Verlag, Berlin* (1990)
2. J. Kowalewski, L. Mäler, Nuclear spin relaxation in liquids: theory, experiments and applications, Series in Chemical Physics, *Taylor & Francis Group* (2006)
3. I. Bertini, J. Kowalewski, C. Luchinat, T. Nilsson, G. Parigi, Nuclear spin relaxation in paramagnetic complexes of S=1: electron spin relaxation effects, *J. Chem. Phys.* **111** (1999) 5795-5807
4. T. Nilsson, J. Kowalewski, Low-field theory of nuclear spin relaxation in paramagnetic low – symmetry complexes for electron spin systems of S=1, 3/2, 2, 5/2, 3 and 7/2, *Mol. Phys.* **98** (2000) 1617-1638, Erratum: *Mol. Phys.* **99** (2001) 369-370
5. D. Kruk, T. Nilsson, J. Kowalewski, Nuclear spin relaxation in paramagnetic systems with zero-field splitting and arbitrary electron spin, *Phys. Chem. Chem. Phys.* **3** (2001) 4907-4917
6. D. Kruk, O. Lips, Field-dependent nuclear relaxation of spins ½ induced by dipole –dipole couplings to quadrupolar spins: LaF_3 crystals as an example, *J. Magn. Reson.* **179** (2006) 250 – 262
7. D. Kruk, O. Lips, Evolution of solid state systems containing mutually coupled dipolar and quadrupolar spins: a perturbation treatment, *Solid State Nucl. Magn. Reson.* **28 (2-4)** (2005) 180-192
8. D. Kruk, Field dependent electron and quadrupole spin relaxation: A unified treatment, *Mol. Phys. Rep.* – in press
9. N. Bloembergen, L.O. Morgan, Proton relaxation times in paramagnetic solutions: Effects of electron spin relaxation, *J. Chem. Phys.* **34** (1961) 842-850
10. M. Rubinstein, A. Baram, Z. Luz, Electronic and nuclear relaxation in solutions of transition metal ions with S=3/2 and S=5/2, *Mol. Phys.* **20** (1971) 67-80
11. R. Sharp, Characteristic properties of the nuclear magnetic resonance – paramagnetic relaxation enhancement arising from integer and half-integer electron spins, *J. Chem. Phys.* **98** (1993) 2507-2515

12. R. Sharp, Effect of zero-field splitting interactions on the paramagnetic relaxation enhancement of nuclear-spin relaxation rates in solution, *J. Chem. Phys.* **98** (1993) 912-921

13. R. Sharp, Characteristic properties of the nuclear – magnetic resonance paramagnetic relaxation enhancement arising from integer and half – integer electron spins, *J. Chem. Phys.* **98** (1993) 2507-2515

14. R. Sharp, Nuclear-spin relaxation due to paramagnetic species in solution – effect of anisotropy in the zero-field splitting tensor, *J. Chem. Phys.* **98** (1993) 6092-6101

15. P.-O. Westlund, A low-field paramagnetic nuclear spin relaxation theory, *J. Chem. Phys.* **108** (1998) 4945-4953

16. E. Stranberg, P.-O. Westlund, Paramagnetic proton nuclear spin relaxation theory of low-symmetry complexes for electron spin quantum number S=5/2, *J. Magn. Reson.* **137** (1999) 333-344

17. S. Rast, P.H. Fries, E. Belorizky, Static zero field splitting effects on the electronic relaxation of paramagnetic metal ion complexes in solution, *J. Chem. Phys.* **113** (2000) 8724-8735

18. A. Borel, F. Yearly, L. Helm. A. E. Merbach, Multiexponential electronic spin relaxation and Redield's limit in Gd (III) complexes in solution: consequences for O-17/H-1 NMR and EPR simultaneous analysis, *J. Am. Chem. Soc.* **123** (2001) 2637-2644

19. R. Sharp, L. Lohr, Thermal relaxation of electron spin motion in a thermal equilibrium ensemble: relation to paramagnetic nuclear magnetic resonance relaxation, *J. Chem. Phys.* **115** (2001) 5005-5014

20. R. Sharp, L. Lohr, J. Miller, Paramagnetic NMR relaxation enhancement: recent advances in theory, *Prog. Nucl. Magn. Reson. Spectr.* **38** (2001) 115-158

21. R. Sharp, Closed-form expressions for level-averaged electron spin relaxation times outside the Zeeman limit: application to paramagnetic NMR relaxation, *J. Magn. Reson.* **154** (2002) 269-279

22. J.S. Griffith, The Theory of Transition metal ions, *Cambridge University Press*, Cambridge (1961)

23. A. Abragam, B. Bleaney, Electron paramagnetic resonance of transition ions, *Clarendon Press, Oxford* (1970)
24. J. E. Wertz, J. R. Bolton, Electron Spin Resonance: Elementary theory and practical applications, *McGraw-Hill, New York* (1972)
25. C.P. Poole, A. H. Farach, Theory of magnetic resonance, *A Wiley-Interscience Publication, John Wiley & Sons* (1987)
26. L. Banci, I. Bertini, C. Luchinat, Nuclear and electron relaxation. The magnetic nucleus-unpared electron coupling in solution, *VCH, Weinheim* (1991)
27. I. Bertini, C. Luchinat, G. Parigi, Solution NMR of paramagnetic molecules, *Elsevier, Amsterdam* (2001)
28. D. Canet, Relaxation mechanisms: magnetization modes, in Encyclopedia of Nuclear Magnetic Resonance, D.M. Grant, R.K. Harris, (Eds.) *Wiley, Chichester* (1996) 4046-4053
29. A. Kumar, R.C.R. Grace, P.K. Madhu, Cross-correlations in NMR, *Prog. Nucl. Magn. Reson. Spectr.* **37** (2000) 191-319
30. L. Werbelow, Relaxation-Induced Transfer of Nuclear Spin Polarization as a Probe of Molecular Structure and Dynamics in Mobile Phases, in Nuclear Magnetic Resonance Probes of Molecular Dynamics, R. Tycko (Ed.) Kluwer *Academic Publishers* (1994) 223-263

CHAPTER VI

Nuclear spin relaxation in the presence of neighboring electron spins for slowly rotating systems

The previous chapter has emphasized the multiexponential electron spin relaxation affected by the static ZFS contributing to the electron spin energy level structure. This chapter is concerned with the nuclear spin relaxation caused by a dipole-dipole coupling to an electron spin relaxing in the complicated, multiexponential manner.

As discussed previously from the nuclear spin point of view the electron spin dynamics is an important component of the lattice dynamics. Actually, under the slow-rotation conditions, the electron spin dynamics is identical to the lattice dynamics. This fact that the nuclear spin relaxation fully results from the electron spin relaxation makes the theoretical treatment of the combined $I-S$ spin system very demanding. One cannot hope that some insufficiencies or oversimplifications encoded in the description of the electron spin dynamics will be masked by the rotational motion, which in the case of a faster molecular tumbling competes with the electron spin relaxation in modulations of the mutual $I-S$ dipole-dipole coupling. The present chapter begins with formulating a consistent description of the nuclear spin relaxation for slowly rotating nuclear spin – electron spin systems within the framework of the quantum-mechanical correlation function, introduced in Chapter II and employed to some illustrative calculations in Chapter IV. The earlier examples have been devoted to the limiting regimes of the magnetic field (the low and high field limits). Now, I shall generalize this treatment to an arbitrary magnetic field and an arbitrary electron spin quantum number. Therefore, in this chapter one has the opportunity to appreciate fully the advantages of this concept to treat the electron spin as a part of the composite lattice within the Liouville operator formalism. The remainder of this chapter is devoted to a discussion of the nuclear spin relaxation, depending on the relative strengths of the static ZFS and the Zeeman coupling, for different electron spin quantum numbers.

It has been explained in Chapter IV why and how the electron spin dynamics affects the nuclear spin relaxation at low field. The detailed calculations of the nuclear spin relaxation based on the concept of the composite lattice (including the electron spin subsystem) have been performed at low and high magnetic fields for the simplest case of the electron spin quantum number $S = 1$. The examples illustrate the effects of the electron spin energy level structure and its relaxation on the nuclear spin dynamics, as well as the computational procedure. In this chapter we shall derive for some selected cases appropriate, analytical (semi-analytical) expressions for the nuclear spin relaxation and confront them with the traditional SBM theory at the low as well as high field limits. Interesting conclusions, which one can obtain from the considerations of the limiting cases and more practical experience with the evaluation procedure, will turn out to be useful for the formulation of the more general description, not restricted to the limiting regimes of the magnetic field.

VI.1 Closed form description of nuclear spin-lattice relaxation at low and high magnetic fields for electron spin quantum numbers 3/2 and 7/2

In this section we shall examine in detail the effects of the individual electron spin relaxation rates on the nuclear spin relaxation in the low and high magnetic field regimes. Let use choose for this purpose two cases: the electron spin quantum number $S = \frac{3}{2}$ and the most complicated case of the highest spin quantum number $S = \frac{7}{2}$. The closed analytical expressions for the nuclear spin – lattice relaxation rate at low field for an arbitrary spin quantum number S, in the case of an axial static ZFS have been derived in [1]. I would like to put more emphasis on the evaluating of the quantum-mechanical spectral density within the Liouville formalism and discuss more examples in detail. It is of great importance to understand why several electron spin relaxation rates are relevant for the nuclear spin relaxation and why any simplified treatments are questionable. The advantage of the case of $S = \frac{3}{2}$ is the rather low quantum number and, in consequence, the relatively simple mathematical formulations, while the second example ($S = \frac{7}{2}$) requires dealing with a large set of the electron relaxation rates, and this provides a rich and interesting material for the discussion. I shall extend the considerations to a rhombic static ZFS and discuss the effects of the multiexponential electron spin relaxation also in the high field regime.

Talking, in the previous chapter, about the electron spin relaxation I have mentioned the Kramers degeneracy [2-4] for half-integer spin quantum numbers. It implies that, for half-integer spins, the electron Larmor frequency enters the low field expressions for the nuclear spin relaxation rate, while for integer spins one can neglect such terms altogether [1]. The further discussion will motivate this statement.

I shall base the low field calculations on the formula of Eq.4.49, beginning from the case of an axially symmetric static ZFS. Let us assume, for simplicity reasons, that the molecular (dipole-dipole) frame coincides with the principal axis system of the ZFS tensor, i.e.: $(M) = (DD) = (P_S)$. At low magnetic field, we do not need to consider the molecular orientation. In fact, we can assume that the magnetic field has been applied along the dipole-dipole axis, i.e. $\Omega_{DDL} = 0$. First of all, the molecular orientation is not important from the point of view of the electron spin. It has been pointed out in the previous chapter that since the influence of the Zeeman coupling on the electron spin energy level structure is negligible; the direction of the applied magnetic field with respect to the molecular frame does not matter. In consequence, the superoperator $\hat{\hat{M}}$ does not contain any terms depending on the molecular orientation. This statement is quite obvious, however discussing later some aspects of the Kramers degeneracy we will need to consider more carefully the effects of a small magnetic field which lifts this degeneracy, on the nuclear spin relaxation.

Let us begin from the electron spin quantum number $S = \frac{3}{2}$. In the previous chapter we have derived the relaxation matrix $\hat{\hat{R}}_S$ for the electron spin assuming that the static ZFS is axially symmetrical. It provides now a background for the evaluation of the spectral densities $s_{q,q}^{LF}(-\omega_I)$, defined by Eq.4.53. The quantities $s_{\pm 1, \pm 1}^{LF}(-\omega_I)$ involve the spin-spin part of the electron spin relaxation matrix, while the spectral density $s_{0,0}^{LF}(-\omega_I)$ is related to the spin-lattice electron spin relaxation. This statement is apparent if one looks at the form of the projection vectors: $[S_q^{1(DD)}]$. In Sections IV.4 and IV.5. the idea of the matrix representation of the operators S_q^1 has been presented and applied to the spin quantum number $S = 1$ at the low and high field limit. Now it is suitable to extend it to an arbitrary spin quantum number. Generally, the rank-one tensor operators S_q^1 can be written in the Zeeman basis $\{|S, m_S\rangle\}$ as [1, 5 - 8]:

$$S_0^1 = S_z = \sum_{m_S=-S}^{S} m_S |S, m_S\rangle\langle S, m_S| \tag{6.1a}$$

$$S_1^1 = \frac{1}{\sqrt{2}} S_+ = \frac{1}{\sqrt{2}} \sum_{m_S=-S}^{S-1} \sqrt{(S-m_S)(S+m_S+1)} |S, m_S+1\rangle\langle S, m_S| \tag{6.1b}$$

$$S_{-1}^1 = -\frac{1}{\sqrt{2}} S_- = -\frac{1}{\sqrt{2}} \sum_{m_S=-S+1}^{S} \sqrt{(S+m_S)(S-m_S+1)} |S, m_S-1\rangle\langle S, m_S| \tag{6.1c}$$

Eqs.4.29a-c provide the above formulas for the simplest case of the spin quantum number $S = 1$. In the general case the projection vectors contain $(2S+1)^2$ elements defined by Eq.6.1a-c. I wish to remind the reader at this moment that we can use directly these expressions, because the Zeeman basis $\{|m_S\rangle\}$ is the eigenbasis of the axially symmetric static ZFS Hamiltonian, i.e. the representations of the operators S_q^1 in the laboratory and in the (P_S) frames are the same. Thus, the vector $[S_q^1]$ for $S = \frac{3}{2}$ contains only three non-zero elements corresponding to the single quantum coherences of the electron spin $|m_S+1\rangle\langle m_S| \equiv \rho_{m_S+1,m_S}^S$. The elements are: $\sqrt{3}, 2, \sqrt{3}$ for the Liouville states $|1\rangle\langle 2| = \left|\frac{3}{2}\right\rangle\left\langle\frac{1}{2}\right|$, $|2\rangle\langle 3| = \left|\frac{1}{2}\right\rangle\left\langle-\frac{1}{2}\right|$ and $|3\rangle\langle 4| = \left|-\frac{1}{2}\right\rangle\left\langle-\frac{3}{2}\right|$, respectively. It implies that the spectral density $s_{1,1}^{LF}(-\omega_I)$ can be obtained by inverting only this part matrix representation of the operator $\hat{\hat{M}}$ which is associated with the single-quantum block of the relaxation matrix:

$$s_{1,1}^{LF}(-\omega_I) = \frac{1}{2S+1} \times$$

$$[\sqrt{3} \quad 2 \quad \sqrt{3}] \times \begin{bmatrix} i(2\omega_D + \omega_S + \omega_I) + R_{1212}^S & 0 & 0 \\ 0 & i(\omega_S + \omega_I) + R_{2323}^S & 0 \\ 0 & 0 & i(-2\omega_D + \omega_S + \omega_I) + R_{3434}^S \end{bmatrix}^{-1} \quad (6.2)$$

$$\times \begin{bmatrix} \sqrt{3} \\ 2 \\ \sqrt{3} \end{bmatrix}$$

The operator $\hat{\hat{M}}$ defined in Section IV.4 contains the rotational part. For slowly rotating systems it is negligible altogether and therefore it is not present in the expressions provided in this chapter. At this stage it is necessary to explain why the electron Larmor frequency ω_S has been included in the matrix $[\hat{\hat{M}}]$. So far, considering in Chapter IV the nuclear and electron spin relaxation for $S = 1$ and in Chapter IV the electron spin relaxation for $S = \frac{3}{2}$ and $\frac{7}{2}$, we neglected the Zeeman contribution to the electron spin energy level structure because of the relationship $\omega_D \gg \omega_S$. Generally, the electron spin relaxation is described in terms of the spectral densities $J^S(\omega) \propto \dfrac{\tau_D}{1+(\omega\tau_D)^2}$, while the nuclear spin relaxation is expressed as a combination of the quantities $J(\omega) \propto \dfrac{\tau_c}{1+(\omega\tau_c)^2}$. It is clear that calculating the spectral densities $J^S(\omega)$ or $J(\omega)$ at the frequency $\omega = m\omega_D \pm \omega_S$, $m \neq 0$, one can omit the term ω_S altogether. Also for the electron spin spectral densities $J^S(\omega)$, taken at the frequency ω_S at low field, one can write $J^S(\omega_S) \equiv J^S(0)$. It is due to the fact that the distortional correlation time τ_D is of the order of ps and therefore we can be sure that the term $(\omega_S \tau_D)^2$ is almost zero at low field. Nevertheless, one should notice that the correlation time τ_c for the nuclear spin is determined by the electron spin relaxation time which, according to the Redfield condition, occurs on much slower timescale than the motion causing the relaxation, i.e. $\tau_c \gg \tau_D$. Therefore, the product $\omega_S \tau_c$ can affect the nuclear spin spectral density $J(\omega_S)$, even though the magnetic field is low, so that $\omega_D \gg \omega_S$. At low magnetic field, the Kramers doublet $\left\{\left|\tfrac{1}{2}\right\rangle, \left|-\tfrac{1}{2}\right\rangle\right\}$ is split only by the frequency ω_S and, therefore, the spectral density $J(\omega_S)$ appears in the expression for the nuclear spin relaxation for half-integer spin. Setting up Eq.6.2 we have included the Zeeman contribution to the energy levels of the electron spin. Actually, it is relevant only for the element $i(\omega_I + \omega_S) + R^S_{2323}$, which does not contain any transition frequency of the order ω_D. The resultant spectral density $s^{LF}_{1,1}(-\omega_I)$ has the form:

$$\operatorname{Re}\left[s^{LF}_{\pm 1,\pm 1}(-\omega_I)\right] = \dfrac{\tau_{c,21}}{1+(\omega_S+\omega_I)^2 \tau_{c,21}^2} + \dfrac{3}{4}\dfrac{\tau_{c,22}}{1+(2\omega_D+\omega_I)^2 \tau_{c,22}^2} + \dfrac{3}{4}\dfrac{\tau_{c,22}}{1+(2\omega_D-\omega_I)^2 \tau_{c,22}^2} \qquad (6.3)$$

where the correlation times $\tau_{c,2i}$ are related to the electron spin relaxation rates:

$\tau_{c,21}^{-1} = R^S_{2323} = \dfrac{12}{5}\Delta_T^2 \dfrac{\tau_D}{1+(2\omega_D\tau_D)^2}$ and $\tau_{c,22}^{-1} = R^S_{1212} = \dfrac{6}{5}\Delta_T^2\left[\tau_D + \dfrac{\tau_D}{1+(2\omega_D\tau_D)^2}\right]$, obtained in

Section V.3. Since the relaxation rates $R^S_{1212} = R^S_{3434}$ are equal, the spectral densities

Theory of evolution and relaxation of multi-spin systems

$s^{LF}_{\pm 1,\pm 1}(-\omega_I)$ include only two correlation times $\tau_{c,2i}$, $i=1,2$. In the last two terms of Eq.6.3 containing the frequency $2\omega_D$ ω_S contribution has been omitted.

The spectral density $s^{LF}_{0,0}(-\omega_I)$, related to the spin-lattice relaxation of the electron spin, can be evaluated from the expression:

$$s^{LF}_{0,0}(-\omega_I) = \frac{1}{2S+1}\begin{bmatrix}\frac{3}{2} & \frac{1}{2} & -\frac{1}{2} & -\frac{3}{2}\end{bmatrix} \times \begin{bmatrix} i\omega_I + R_{1S} & -\frac{1}{2}R_{1S} & -\frac{1}{2}R_{1S} & 0 \\ -\frac{1}{2}R_{1S} & i\omega_I + R_{1S} & 0 & -\frac{1}{2}R_{1S} \\ -\frac{1}{2}R_{1S} & 0 & i\omega_I + R_{1S} & 0 \\ 0 & -\frac{1}{2}R_{1S} & 0 & i\omega_I + R_{1S} \end{bmatrix}^{-1} \times \begin{bmatrix} \frac{3}{2} \\ \frac{1}{2} \\ -\frac{1}{2} \\ -\frac{3}{2} \end{bmatrix} \quad (6.4)$$

where the spin-lattice relaxation rate $R_{1S} = \frac{12}{5}\Delta_T^2 \frac{\tau_D}{1+(2\omega_D\tau_D)^2}$ has been evaluated in Chapter V. The simple physical picture of the electron spin-lattice relaxation implies also a simple form of the spectral density $s^{LF}_{0,0}(-\omega_I)$:

$$s^{LF}_{0,0}(-\omega_I) = \frac{5}{4}\frac{\tau_{c,1}}{1+(\omega_I\tau_{c,1})^2} \quad (6.5)$$

which involves only one correlation time:

$$\tau_{c1}^{-1} = R_{1S} = \frac{12}{5}\Delta_T^2 \frac{\tau_D}{1+(2\omega_D\tau_D)^2} \quad (6.6)$$

Thus, the final expression for the nuclear spin-lattice relaxation rate R_{1I} at low magnetic field yields the form [1]:

$$R_{1I} = \frac{5}{6}\left(\frac{\mu_0}{4\pi}\frac{\gamma_I\gamma_S\hbar}{r_{IS}^3}\right)^2 \times$$

$$\left[4\frac{\tau_{c,1}}{1+(\omega_I\tau_{c,1})^2} + \frac{4}{5}\frac{\tau_{c,21}}{1+(\omega_I+\omega_S)^2\tau_{c,21}^2} + \frac{3}{5}\frac{\tau_{c,22}}{1+(\omega_I+2\omega_D)^2\tau_{c,22}^2} + \frac{3}{5}\frac{\tau_{c,22}}{1+(\omega_I-2\omega_D)^2\tau_{c,22}^2}\right] \quad (6.7)$$

In the previous chapter we have discussed the effects of the static ZFS rhombicity on the electron spin relaxation in the low field regime. We have seen that the symmetry of the static ZFS influences considerably the spin-lattice as well as spin-spin relaxation processes. The nuclear spin senses the rhombicity effects through the electron spin dynamics. However, the altered energy level structure of the electron spin influences the nuclear spin relaxation by changing the transition frequencies occurring in the dipolar spectral densities relevant for the nuclear spin relaxation. To follow these mechanisms and obtain an expression for the low field nuclear spin relaxation comprising the rhombicity effects, we need to modify the matrix form of the operator $\hat{\hat{M}}$. It must be expressed in the Liouville eigenbasis of the main Hamiltonian $\{|\psi_\alpha^S\rangle\langle\psi_\beta^S| \equiv |\alpha,\beta\rangle\}$, which is not formed any more by the Zeeman eigenstates. However, due to our previous considerations we are quite familiar with such problems.

Let us write down explicitly the matrix elements of the operator $\hat{\hat{M}}$ starting from the population part of the matrix $\left[\hat{\hat{M}}\right]$ denoted as $\left[\hat{\hat{M}}_0\right]$:

$$\left[\hat{\hat{M}}_0\right] = \begin{bmatrix} & |1,1) & |2,2) & |3,3) & |4,4) \\ i\omega_I - \widetilde{R}_{1,1}^S - \widetilde{R}_{1,2}^S & \widetilde{R}_{1,1}^S & \widetilde{R}_{1,2}^S & 0 \\ \widetilde{R}_{1,1}^S & i\omega_I - \widetilde{R}_{1,1}^S - \widetilde{R}_{1,2}^S & 0 & \widetilde{R}_{1,2}^S \\ \widetilde{R}_{1,2}^S & 0 & i\omega_I - \widetilde{R}_{1,1}^S - \widetilde{R}_{1,2}^S & \widetilde{R}_{1,1}^S \\ 0 & \widetilde{R}_{1,2}^S & \widetilde{R}_{1,1}^S & i\omega_I - \widetilde{R}_{1,1}^S - \widetilde{R}_{1,2}^S \end{bmatrix} \quad (6.8)$$

where the quantities $\widetilde{R}_{1,1}^S = R_{1122}^S = R_{2233}^S$ and $\widetilde{R}_{1,2}^S = R_{1133}^S = R_{2244}^S$ have been calculated in the previous chapter devoted to the electron spin relaxation; they are given by Eq.5.20 and Eq.5.21. Setting up the coherence parts of the operator matrix at low field for half-integer spins one should take into account a weak Zeeman coupling, lifting nevertheless the Kramers degeneracy. To capture the effects of the Zeeman term on the electron spin energy levels and consequently on the nuclear spin relaxation, we will deal with the Zeeman Hamiltonian of the simplest possible form $H_Z = \omega_S S_Z$, neglecting altogether its dependence on the molecular orientation. I have commented on this above, discussing the case of an axially symmetric static ZFS for $S = \frac{3}{2}$. For the purpose of accounting for the effects of the electron Zeeman interaction on the nuclear spin relaxation at low field one does not need to treat the Zeeman contribution in detail, including its orientation dependent form. The off-diagonal terms resulting from the general form of the Zeeman Hamiltonian of Eq.5.6 are negligible anyway, while the role of the diagonal contribution is reflected sufficiently well by setting $\beta = 0$. Nevertheless, one should be aware that for a larger static ZFS the non-diagonal terms cause very interesting low field dispersive features and the relaxation curves for $\beta = 0$ and $\beta = 90^0$ are pretty different.

As a result the coherence block of the $\hat{M}$ operator covers the elements:

$$(1,2|\hat{M}|1,2) = i\omega_I + R^S_{1212} \tag{6.9a}$$

$$(1,3|\hat{M}|1,3) = i\omega_I + i2\sqrt{\omega_D^2 + 3\omega_E^2} + R^S_{1313} \tag{6.9b}$$

$$(1,4|\hat{M}|1,4) = i\omega_I + i\omega_S/(1+v^2) + R^S_{1414} \tag{6.9c}$$

$$(2,3|\hat{M}|2,3) = i\omega_I + i\omega_S/(1+\mu^2) + R^S_{2323} \tag{6.9d}$$

$$(2,4|\hat{M}|2,4) = i\omega_I - i2\sqrt{\omega_D^2 + 3\omega_E^2} + R^S_{2424} \tag{6.9e}$$

$$(3,4|\hat{M}|3,4) = i\omega_I + R^S_{3434} \tag{6.9f}$$

where $\mu = \dfrac{\omega_D + \sqrt{\omega_D^2 + 3\omega_E^2}}{\sqrt{3}\omega_E}$ and $v = \dfrac{\omega_D - \sqrt{\omega_D^2 + 3\omega_E^2}}{\sqrt{3}\omega_E}$ (see Chapter V). The terms containing the electron spin Larmor frequency ω_S are obtained as the corresponding elements of the Zeeman Hamiltonian $\langle 1|\omega_S S_z|3\rangle$ and $\langle 2|\omega_S S_z|4\rangle$; the elements taken between other pairs of the eigenstates $|\psi_\alpha^S\rangle$ are equal to zero. To obtain closed form expressions for the dipolar spectral densities $s_{p,q}^{LF}(-\omega_I)$ we have to evaluate the projection vectors $[S_q^{1(DD)}]$. It means that we have to express the electron spin tensor operators S_z, S_+ and S_- in terms of the Liouville basis $\{|\psi_\alpha^S\rangle\langle\psi_\beta^S|\}$, constructed from the eigenstates $\{|\psi_\alpha^S\rangle\}$ of the main Hamiltonian (i.e. the rhombic static ZFS). In general, the eigenfunctions of the main Hamiltonian for the electron spin, H_0^S, are related to the Zeeman basis via the set of coefficients $c_{\alpha r}(\Omega_{DDL})$ (Eq.5.8). The vectors $[S_q^{1(DD)}(\Omega_{DDL})]$ contain the coefficients $a_{\alpha\beta}^q(\Omega_{DDL})$ (referred to as the projection vectors in the literature [1, 5, 9, 10]) resulting from the expansion of the electron spin tensor components S_q^1 into the Liouville basis $\{|\psi_\alpha^S\rangle\langle\psi_\beta^S|\}$. The projection vectors have been introduced in Section IV.4. Applying the inverse relation between the Zeeman states $|n\rangle = |S, m_s\rangle$ and the eigenfunctions $|\psi_\alpha^S\rangle$: $|S, m_S\rangle = \sum_\alpha c_{m_S\alpha}^{-1}(\Omega_{DDL})|\psi_\alpha^S\rangle$, one can obtain from Eq.6.1a-c the coefficients $a_{\alpha\beta}^q(\Omega_{DDL})$ in a general form:

$$a_{\alpha\beta}^0(\Omega_{DDL}) = \sum_{m_S=-S}^{S} m_S \left[\sum_{\alpha=1}^{2S+1} c_{m_S\alpha}^{-1*}(\Omega_{DDL}) \times \sum_{\beta=1}^{2S+1} c_{m_S\beta}^{-1}(\Omega_{DDL})\right] |\psi_\alpha^S\rangle\langle\psi_\beta^S| \qquad (6.10a)$$

$$a_{\alpha\beta}^1(\Omega_{DDL}) = \frac{1}{\sqrt{2}} \sum_{m_S=-S}^{S-1} \sqrt{(S-m_S)(S+m_S+1)} \left[\sum_{\alpha=1}^{2S+1} c_{m_S+1,\alpha}^{-1*}(\Omega_{DDL}) \times \sum_{\beta=1}^{2S+1} c_{m_S\beta}^{-1}(\Omega_{DDL})\right] |\psi_\alpha^S\rangle\langle\psi_\beta^S| \qquad (6.10b)$$

$$a_{\alpha\beta}^{-1}(\Omega_{DDL}) = -\frac{1}{\sqrt{2}} \sum_{m_S=-S+1}^{S} \sqrt{(S+m_S)(S-m_S+1)} \left[\sum_{\alpha=1}^{2S+1} c_{m_S-1,\alpha}^{-1*}(\Omega_{DDL}) \times \sum_{\beta=1}^{2S+1} c_{m_S\beta}^{-1}(\Omega_{DDL})\right] |\psi_\alpha^S\rangle\langle\psi_\beta^S| \qquad (6.10c)$$

The angular dependencies of the coefficients $[S_q^1] = [a^q(\Omega_{DDL})]$ result from the angular dependent form of the static Hamiltonian for the electron spin.

This does not concern the present case when we need just to use the explicit expressions for the eigenstates $\{|\psi_\alpha^S\rangle\}$ of the rhombic static ZFS Hamiltonian, given in Section V.3; for a convenience of the reader I write them down once again:

$$|\psi_1^S\rangle = \frac{1}{\sqrt{1+\mu^2}}\left[\left|-\frac{1}{2}\right\rangle + \mu\left|\frac{3}{2}\right\rangle\right],$$

$$|\psi_2^S\rangle = \frac{1}{\sqrt{1+\nu^2}}\left[\left|\frac{1}{2}\right\rangle + \nu\left|-\frac{3}{2}\right\rangle\right],$$

$$|\psi_3^S\rangle = \frac{1}{\sqrt{1+\nu^2}}\left[\left|-\frac{1}{2}\right\rangle + \nu\left|\frac{3}{2}\right\rangle\right] \text{ and}$$

$$|\psi_4^S\rangle = \frac{1}{\sqrt{1+\mu^2}}\left[\left|\frac{1}{2}\right\rangle + \mu\left|-\frac{3}{2}\right\rangle\right].$$

The calculations of the projection vectors are quite cumbersome; however their results elucidate the physical essence of the effects of the static ZFS symmetry on the nuclear spin relaxation. The expansion of the operator S_Z in the basis

$\{|\psi_\alpha^S\rangle\langle\psi_\beta^S|\}$ encompasses not only the four population elements $|\psi_\alpha^S\rangle\langle\psi_\alpha^S|$ ($\alpha = 1,2,3,4$), but also four other elements associated with the electron spin-spin relaxation: $|\psi_1^S\rangle\langle\psi_3^S|, |\psi_2^S\rangle\langle\psi_4^S|$ and $|\psi_3^S\rangle\langle\psi_1^S|, |\psi_4^S\rangle\langle\psi_2^S|$. The coefficients $a_{\alpha\alpha}^0$ are equal to: $a_{22}^0 = -a_{11}^0 = \dfrac{2\omega_D + \sqrt{\omega_D^2 + 3\omega_E^2}}{2\sqrt{\omega_D^2 + 3\omega_E^2}}$, $a_{33}^0 = -a_{44}^0 = \dfrac{-2\omega_D + \sqrt{\omega_D^2 + 3\omega_E^2}}{2\sqrt{\omega_D^2 + 3\omega_E^2}}$, while the coefficients $a_{\alpha\beta}^0$ are given as $a_{13}^0 = a_{31}^0 = -a_{24}^0 = -a_{42}^0 = \dfrac{-2\sqrt{3}\omega_E}{\sqrt{\omega_D^2 + 3\omega_E^2}}$. The practical result of the expansion is that one cannot obtain the dipolar spectral density $s_{0,0}^{LF}$ involving only the $\hat{M}_0$ part of the operator associated with the electron spin-lattice relaxation. This result has an interesting physical background. Employing the analogy between the ZFS and quadrupolar interaction one can conclude from Section IV.6 that in the case of $S = 1$ a rhombic static ZFS only the spin-spin electron spin relaxation is relevant for the nuclear spin relaxation [1]. It is due to the fact that if the Zeeman coupling is negligible, the energy level structure generated by a rhombic static ZFS does not lead to any permanent dipole moment for the spin quantum number $S = 1$. One can verify this statement calculating the expectation value of the S_z operator: $\langle S_z \rangle = \sum_{\alpha,\beta=1}^{3} \langle \psi_\alpha^S | S_z | \psi_\beta^S \rangle$ in the eigenbasis of the rhombic static ZFS Hamiltonian for $S = 1$ (this issue is discussed in Chapter VIII). For the spin quantum number $S = \dfrac{3}{2}$ the $\langle S_z \rangle$ quantity is not equal zero, that is the effect of the Kramers degeneracy, and therefore the spectral density $s_{0,0}^{ZFS}$ is still affected by the spin-lattice electron spin relaxation. Nevertheless, because of the rhombic symmetry of the static ZFS, the electron spin-spin relaxation becomes relevant for the spectral density $s_{0,0}^{ZFS}$ as well. The projection vectors for the $S_{\pm 1}$ operators contain also eight elements:

$a_{21}^1 = a_{43}^1 = \sqrt{3}\omega_E / \sqrt{2(\omega_D^2 + 3\omega_E^2)}$,

$a_{12}^1 = a_{34}^1 = \sqrt{3}\omega_D / \sqrt{2(\omega_D^2 + 3\omega_E^2)}$,

$$a_{14}^1 = -a_{32}^1 = \omega_E / \sqrt{2(\omega_D^2 + 3\omega_E^2)},$$

$$a_{23}^1 = \left(\omega_D + \sqrt{\omega_D^2 + 3\omega_E^2}\right) / \sqrt{2} \text{ and}$$

$$a_{41}^1 = \left(-\omega_D + \sqrt{\omega_D^2 + 3\omega_E^2}\right) / \sqrt{2}.$$

Knowing the coefficients one can evaluate relatively easily the $s_{1,1}^{LF}$ contribution to the nuclear spin relaxation [1]:

$$s_{1,1}^{LF} = \frac{1}{10} \frac{1}{\omega_D^2 + 3\omega_E^2} \left[\begin{array}{l} \left(\omega_D^2 + 6\omega_E^2 - \omega_D\sqrt{\omega_D^2 + 3\omega_E^2}\right) \dfrac{\tau_{c2,1}}{1 + \left[\omega_I + \omega_S/(1+\mu^2)\right]^2 \tau_{c2,1}^2} + \\ \left(\omega_D^2 + 6\omega_E^2 + \omega_D\sqrt{\omega_D^2 + 3\omega_E^2}\right) \dfrac{\tau_{c2,1}}{1 + \left[\omega_I + \omega_S/(1+\nu^2)\right]^2 \tau_{c2,1}^2} + \\ 3\left(\omega_D^2 + \omega_E^2\right) \dfrac{\tau_{c2,2}}{1 + \left(\omega_I + 2\sqrt{\omega_D^2 + 3\omega_E^2}\right)^2 \tau_{c2,2}^2} \end{array} \right] \quad (6.11)$$

where the correlation times are determined by the electron spin relaxation rates given by Eq.5.22 and Eq.5.23: $\tau_{c2,1} = R_{2S,1}^{-1}$ and $\tau_{c2,2} = R_{2S,2}^{-1}$. After some algebraic calculations one can obtain also the spectral density $s_{0,0}^{LF}$ [1]:

$$s_{0,0}^{LF} = \frac{1}{5} \frac{5\omega_D^2 + 3\omega_E^2}{\omega_D^2 + 3\omega_E^2} \left(\frac{\tau_{c1}}{1 + \omega_I^2 \tau_{c1}^2} + \frac{\tau_{c2,2}}{1 + \left(\omega_I + 2\sqrt{\omega_D^2 + 3\omega_E^2}\right)^2 \tau_{c2,2}^2} \right) \quad (6.12)$$

where the second term comes from the electron spin-spin relaxation. The correlation time τ_{c1} is defined as $\tau_{c1}^{-1} = 12 J_S\left(2\sqrt{\omega_D^2 + 3\omega_E^2}\right)$. Substituting the spectral densities $s_{q,q}^{LF}(-\omega_I)$ into the formula of Eq.4.49 one obtains the low field relaxation rate R_{1I} for the nuclear spin.

Now, let us turn attention to the high spin quantum number $S = \frac{7}{2}$ to gain more insight into the effects of the multiexponential electron spin relaxation on the nuclear spin relaxation. As pointed out already in Chapter IV, the nuclear spin-lattice relaxation results from joined transitions of the spins I and S. The $I - S$ dipole-dipole coupling allows for the zero, first and second order transitions. This issue is particularly easy to explain in the high field limit, when the S spin exhibits Zeeman energy levels corresponding to the eigenstates characterized by the magnetic spin quantum number m_S. The spin-lattice relaxation of the I spin is caused by a set of joined $I-S$ spin transitions described by the conditions: $\Delta m = \Delta m_I + \Delta m_S = 0,1,2$ and $\Delta m_I = 1$; it implies that $\Delta m_S = -1,0,1$. In fact, the indices of the spectral densities $s_{q,q}(-\omega_I)$ denote just the transition order: $1-q = \Delta m$. The problem is that for high spin quantum numbers S we have many transitions of the S spin to our disposal. There are $2S$ single quantum transitions, $\Delta m_S = \pm 1$, and $2S+1$ zero quantum transitions (populations) for which $\Delta m_S = 0$. In addition, they are characterized by different time constants (relaxation rates). To describe properly the I spin relaxation we have to select from the whole ensemble of the S spin transitions the appropriate ones and determine their contributions to this process. This selection takes place when we expand the tensor operators S_q^1 into the Liouville basis $\{|\psi_\alpha^S\rangle\langle\psi_\beta^S|\}$, calculating the projection coefficients $a_{\alpha\beta}^q$. They indicate the relevant coherences (populations) of the spin S, $|\psi_\alpha^S\rangle\langle\psi_\beta^S| \equiv \rho_{\alpha\beta}^S$ and determine their 'weight factors'. We have seen that the picture is more complicated if the Zeeman functions $\{|m_S\rangle\}$ do not form the eigenbasis for the spin S, like in the case of the rhombic static ZFS in the low field limit.

The already discussed case of $S = \frac{3}{2}$ explains in detail why some electron spin relaxation rates are relevant for the nuclear spin relaxation, while some others are not. Nevertheless, it is of some interest to follow this mechanism for the higher spin quantum number $S = \frac{7}{2}$, because of the more complicated electron spin relaxation. In the low field limit there are four single quantum transitions of the electron spin, characterized by the distinct relaxation rates. They have been evaluated in Chapter V, Eqs.5.25a-d. Looking at the representation of the S_q^1 tensor components in the Liouville basis constructed from the Zeeman eigenstates (Eq.6.1a-c) one can easily conclude that all the coherences will affect the nuclear spin relaxation. The spectral density $s_{1,1}^{LF}(-\omega_I)$ is given by the matrix product:

$$s_{11}^{ZFS}(-\omega_I) = \frac{1}{2} \times \begin{bmatrix} \sqrt{7} & 2\sqrt{3} & \sqrt{15} & 4 & \sqrt{15} & 2\sqrt{3} & \sqrt{7} \end{bmatrix} \times$$

$$\begin{bmatrix} i(\omega_I + 6\omega_D + \omega_S) + R_{1212}^S & 0 & 0 & 0 & 0 & 0 & 0 \\ 0 & i(\omega_I + 4\omega_D + \omega_S) + R_{2323}^S & 0 & 0 & 0 & 0 & 0 \\ 0 & 0 & i(\omega_I + 2\omega_D + \omega_S) + R_{3434}^S & 0 & 0 & 0 & 0 \\ 0 & 0 & 0 & i(\omega_I + \omega_S) + R_{4545}^S & 0 & 0 & 0 \\ 0 & 0 & 0 & 0 & i(\omega_I + 2\omega_D + \omega_S) + R_{5656}^S & 0 & 0 \\ 0 & 0 & 0 & 0 & 0 & i(\omega_I + 4\omega_D + \omega_S) + R_{6767}^S & 0 \\ 0 & 0 & 0 & 0 & 0 & 0 & i(\omega_I + 6\omega_D + \omega_S) + R_{7878}^S \end{bmatrix}^{-1}$$

$$\times \begin{bmatrix} \sqrt{7} \\ 2\sqrt{3} \\ \sqrt{15} \\ 4 \\ \sqrt{15} \\ 2\sqrt{3} \\ \sqrt{7} \end{bmatrix} \quad (6.13)$$

Setting up this expression we have used the labeling of the eigenstates introduced in Chapter V, i.e.: $|1\rangle \equiv \left|m_S = \frac{7}{2}\right\rangle$, ..., $|8\rangle \equiv \left|m_S = -\frac{7}{2}\right\rangle$. The transition frequencies result from the energy levels structure consisting of the four doublets $7D, D, -3D$ and $-5D$, corresponding to the pairs of eigenstates $\left\{\left|\frac{7}{2}\right\rangle, \left|-\frac{7}{2}\right\rangle\right\}$, $\left\{\left|\frac{5}{2}\right\rangle, \left|-\frac{5}{2}\right\rangle\right\}$, $\left\{\left|\frac{3}{2}\right\rangle, \left|-\frac{3}{2}\right\rangle\right\}$ and $\left\{\left|\frac{1}{2}\right\rangle, \left|-\frac{1}{2}\right\rangle\right\}$, respectively, as it has been explained in the same chapter. The electron spin Larmor frequency ω_S appears in the expression due to the effect of the low (but non-zero) magnetic field on the electron spin energy level structure. It can be important only for the fourth diagonal element, which corresponds to the single quantum transition of the electron spin between the Kramers doublet $\left\{\left|\frac{1}{2}\right\rangle, \left|-\frac{1}{2}\right\rangle\right\}$, if the extreme narrowing condition $\omega_S R_{4545}^{-1} \ll 1$ is not fulfilled (the other transitions contain the ω_D frequency and $\omega_S \ll \omega_D$). This issue has been discussed above for the case of $S = \frac{3}{2}$.

The expression of Eq.6.13 can be adapted in a straightforward manner to the opposite limit of the high magnetic field, when the static ZFS is negligible. At the high field limit all the single quantum transitions have the same frequency ω_S; therefore, it is enough in practice to omit the ω_D terms in the matrix. The electron spin relaxation rates can be also deduced from Eqs.5.25a-d, if one realizes that the m-th order spectral densities correspond to the frequencies $m\omega_S$, that:

$$R^S_{1212} = R^S_{5656} = 54 J^S_0(0) + 174 J^S_1(\omega_S) + 66 J^S_2(2\omega_S) \tag{6.14a}$$

$$R^S_{2323} = R^S_{6767} = 24 J^S_0(0) + 174 J^S_1(\omega_S) + 126 J^S_2(2\omega_S) \tag{6.14b}$$

$$R^S_{3434} = R^S_{7878} = 6 J^S_0(0) + 78 J^S_1(\omega_S) + 186 J^S_2(2\omega_S) \tag{6.14c}$$

$$R^S_{4545} = 30 J^S_1(\omega_S) + 210 J^S_2(2\omega_S) \tag{6.14d}$$

The spectral densities $s_{0,0}(-\omega_I)$ can be evaluated by employing the spin-lattice parts of the relaxation matrix, given by Eq.5.14 and Eq.5.26 at the low and high field, respectively. However, the calculations are rather cumbersome and do not provide enough new information to be reported here.

Finishing this section I would like to point out that the limiting approaches presented so far are very valuable and informative. Especially, low-field expressions provide a tool to investigate the properties of the nuclear spin relaxation, affected by the static ZFS. By developing an analytical low field theory of the nuclear spin relaxation one can explain in more detail various effects which cannot be fully understood using a numerical treatment.

In the next section I shall present a description of the spin-lattice relaxation of the I spin valid for an arbitrary magnetic field and an arbitrary spin quantum number S.

VI.2. Theory of field dependent spin-lattice relaxation of nuclear spins 1/2 coupled to an electron spin: the general formulation

I shall begin the considerations, from the general form of the spectral density, Eq.4.19, taking full advantage of the Redfield description of the electron spin relaxation, formulated in the previous chapter. To make use of the perturbation treatment of the electron spin relaxation and incorporate it in a straightforward manner into the description of the nuclear spin relaxation, we need to express the electron spin tensor operators $S_q^{1(L)}$ in the same reference frame, which we have used for evaluating the electron spin relaxation. I shall present here the theory of the nuclear spin relaxation referring to the molecular frame (the principal axis system of the static ZFS) as well as to the laboratory frame. It is interesting and useful from the conceptual point of view to develop the two descriptions in a parallel manner. In addition, this parallel treatment gives us some flexibility; we can use the preferable one in the limiting cases of the low and high field limits.

Expressing the tensor operators $S_q^{1(L)}$ in the molecular frame and contracting the Wigner rotation matrices according to the angular momentum theory, one obtains Eq.4.49. The equation has been derived step by step in Section IV in the context of the nuclear spin relaxation at the low field limit for the electron spin quantum number $S = 1$. It is of primary importance to realize that the formulations of Eq.4.19 and of Eq.4.49 are general to the same extent. The first one gives the expression for the dipolar spectral density, $K_{1,1}^{DD}(\omega_I)$, if the main Hamiltonian for the electron spin H_0^S has been written in the laboratory frame, $H_0^S \equiv H_0^{S(L)}$ while the second one provides the analogous formulation appropriate for the case when the Hamiltonian H_0^S has been represented in the molecular frame, $H_0^S \equiv H_0^{S(P_S)}$ (we assume that $(P_S) \equiv (DD)$). The expressions can be just treated as the laboratory and molecular frame counterparts. Both of them provide an appropriate starting point for the calculations of the field dependent nuclear spin relaxation. Of course, if one is particularly interested in the low or high field theory of the nuclear spin relaxation, it is highly preferable to use the formulation dedicated to the principal axis system of the static ZFS or to the laboratory frame, respectively.

Under the assumption of slow molecular tumbling, the orientation of the molecular axis system with respect to the laboratory frame is fixed. The calculations need to be carried out for all possible molecular orientations represented by the angle Ω_{DDL}. The product of the two Wigner matrices, $D_{m,1}^{1*}(\Omega_{DDL})D_{k,1}^{1}(\Omega_{DDL})$, which appear in Eq.4.49 plays the role of a weight factor for the spectral density $(K_{1,1}^{DD}(-\omega_I))(\Omega_{DDL})$, calculated separately for a given molecular orientation [5, 9, 11]:

$$(K_{1,1}^{DD}(-\omega_I))(\Omega_{DDL}) = \frac{10}{3}\left(\frac{\mu_0}{4\pi}\frac{\gamma_I\gamma_S\hbar}{r_{IS}^3}\right)^2 \sum_{p,q=-1}^{1}\begin{pmatrix}2 & 1 & 1\\1-q & q & -1\end{pmatrix}^2\begin{pmatrix}2 & 1 & 1\\1-p & p & -1\end{pmatrix}^2 \times$$

$$\sum_{m,k=-1}^{1}\begin{pmatrix}2 & 1 & 1\\0 & m & -m\end{pmatrix}\begin{pmatrix}2 & 1 & 1\\0 & k & -k\end{pmatrix}D_{m,1}^{1*}(\Omega_{DDL})D_{k,1}^{1}(\Omega_{DDL}) \times \quad (6.15)$$

$$\int_0^\infty Tr_S\left\{S_m^{1(DD)+}\left[\exp\left(-i\hat{L}_S^{(DD)}(\Omega_{DDL})\tau\right)S_k^{1(DD)}\right]\rho_S^{eq}\right\}\exp(-i\omega_I\tau)d\tau$$

The final spectral density $K_{1,1}^{DD}(-\omega_I)$ has to be evaluated as a result of the averaging over the angle Ω_{DDL}: $K_{1,1}^{DD}(\omega_I) = \left\langle(K_{1,1}^{DD}(-\omega_I))(\Omega_{DDL})\right\rangle$. Since the lattice dynamics for the considered systems is equivalent to the electron spin dynamics, the lattice Liouvilian $\hat{L}_L$ of Eq. 4.20 has been replaced in the above equation by the angle-specific operator $\hat{L}_S^{(DD)}(\Omega_{DDL})$. It includes the main interactions for the electron spin (the static ZFS and the Zeeman interaction) and the relaxation superoperator $\hat{R}_S(\Omega_{DDL})$:

$$\hat{L}_S^{(DD)}(\Omega_{DDL}) = \hat{L}_{ZFS}^{S(DD)} + \hat{L}_Z^{(DD)}(\Omega_{DDL}) + \hat{R}_S(\Omega_{DDL}) \quad (6.16)$$

Looking at Eq.6.1 it is apparent that the dipolar spectral density $K_{1,1}^{DD}(-\omega_I)$ can be written in terms of the orientation dependent electron spin spectral densities,

$\left(s_{m,k}^{(DD)}(-\omega_I)\right)\left(\Omega_{DDL}\right) \equiv \left(s_{m,k}^{(P_S)}(-\omega_I)\right)\left(\Omega_{P_SL}\right)$ (under the assumption that $(P_S) \equiv (DD)$)
[5]:

$$\left(s_{m,k}^{(DD)}(-\omega_I)\right)\left(\Omega_{DDL}\right) = \int_0^\infty Tr_S\left\{S_m^{1(DD)+}\left[\exp\left(-i\hat{L}_S^{(DD)}(\Omega_{ML})\tau\right)S_k^{1(DD)}\right]\rho_S^{eq}\right\}\exp(-i\omega_I\tau)d\tau \quad (6.17)$$

weighted by an appropriate factors $f_{mk}(\Omega_{DDL})$:

$$K_{1,1}^{DD}(-\omega_I) = \left(\frac{\mu_0}{4\pi}\frac{\gamma_I\gamma_S\hbar}{r_{IS}^3}\right)^2 \sum_{m,k=-1}^{1}\left\langle f_{mk}(\Omega_{DDL})\left(s_{m,k}^{(DD)}(-\omega_I)(\Omega_{DDL})\right)\right\rangle$$

$$= \left(\frac{\mu_0}{4\pi}\frac{\gamma_I\gamma_S\hbar}{r_{IS}^3}\right)^2 \sum_{m,k=-1}^{1} s_{m,k}^{(DD)}(-\omega_I) \quad (6.18)$$

The function $f_{mk}(\Omega_{DDL})$ encompasses the angular dependencies of the dipolar spectral densities, $\left(K_{1,1}^{DD}(-\omega_I)\right)(\Omega_{DDL})$, resulting from the orientation of the dipole-dipole axis (which coincides with the principal axis of the static ZFS) with respect to the laboratory and the representation of the electron spin tensor operators in the molecular frame, which is required to keep the consistence with the treatment of the electron spin relaxation [5]:

$$f_{mk}(\Omega_{DDL}) = \frac{10}{3}\sum_{p,q=-1}^{1}\begin{pmatrix}2 & 1 & 1\\ 1-q & q & -1\end{pmatrix}^2\begin{pmatrix}2 & 1 & 1\\ 1-p & p & -1\end{pmatrix}^2\begin{pmatrix}2 & 1 & 1\\ 0 & m & -m\end{pmatrix}\begin{pmatrix}2 & 1 & 1\\ 0 & k & -k\end{pmatrix}D_{m,1}^{1*}(\Omega_{DDL})D_{k,1}^{1}(\Omega_{DDL}) \quad (6.19)$$

Finally, we proceed to calculate the electron spin spectral densities, $s_{m,k}^{(DD)}(-\omega_I)$. Following the procedure demonstrated in Chapter IV, let us introduce the orientation specific operator $\hat{M}(\Omega_{DDL})$ including besides the terms associated with the electron spin Liouville operator, $\hat{L}_S$, the term containing the frequency ω_I of the nuclear spin transition:

$$\hat{M}^{(DD)}(\Omega_{DDL}) = -i\hat{L}_S^{(DD)}(\Omega_{DDL}) - i\omega_I\hat{1} \quad (6.20)$$

We know from Section IV.4 that the spectral densities can be written very compactly using this operator, leading to a straightforward evaluation of the Fourier transform. Details of the evaluations have been presented for several cases in Chapter IV as well as in the first section of the present chapter and, therefore, they do not require further comments. The essential difference between the two operators is that the present one is angle dependent, while the operator of Eq.4.32 does not contain any angular dependence. Actually, the explanation is quite obvious. In the low field limit, we do not need to bother how the applied magnetic field is oriented with respect to the molecular frame, because the contribution of the Zeeman coupling to the electron spin energy level structure is negligible anyway, independently of the orientation. In a general case, the explicit, closed form of the quantities $s_{m,k}^{(DD)}(-\omega_I)$ can be evaluated as:

$$s_{m,k}^{(DD)}(-\omega_I) = \left\langle \int_0^\infty f_{mk}(\Omega_{DDL}) Tr_S \left\{ S_m^{1(DD)+} \left[\exp\left(-i\hat{\hat{L}}_S^{(DD)}(\Omega_{DDL})\tau\right) S_k^{1(DD)} \right] \rho_S^{eq} \right\} \exp(-i\omega_I\tau) d\tau \right\rangle =$$

$$\frac{1}{2S+1} \left\langle \int_0^\infty f_{mk}(\Omega_{DDL}) Tr_S \left\{ S_m^{1(DD)+} \left[\exp\left(\hat{M}^{(DD)}(\Omega_{DDL})\tau\right) S_k^{1(DD)} \right] \right\} d\tau \right\rangle = \qquad (6.21)$$

$$\frac{1}{2S+1} \left\langle f_{mk}(\Omega_{DDL}) [S_m^{1(DD)+}(\Omega_{DDL})] \left[\hat{M}^{(DD)}(\Omega_{DDL})\right]^{-1} [S_k^{1(DD)}(\Omega_{DDL})] \right\rangle =$$

$$\frac{1}{2S+1} \left\langle f_{mk}(\Omega_{DDL}) \sum_{\mu\nu} [S_m^{1(DD)+}(\Omega_{DDL})]_\mu \left[\hat{M}^{(DD)}(\Omega_{DDL})^{-1}\right]_{\mu\nu} [S_k^{1(DD)}(\Omega_{DDL})]_\nu \right\rangle$$

The calculations are a generalized version of the treatment dedicated to the low field limit (see Section IV.4). The matrix form $\left[\hat{M}^{(DD)}(\Omega_{DDL})\right]$ of the operator $\hat{M}^{(DD)}(\Omega_{DDL})$ can be obtained, in analogy to Eq.4.28, from the electron spin relaxation matrix $\hat{\hat{R}}_S(\Omega_{DDL})$ including the orientation dependent relaxation rates, $R^S_{\alpha\alpha'\beta\beta'}(\Omega_{DDL})$, by adding to the diagonal the appropriate transition frequencies and the ω_I term:

$$\left(\psi_{\alpha'}^S, \psi_{\beta'}^S \left| \hat{M}(\Omega_{DDL}) \right| \psi_\alpha^S, \psi_\beta^S \right) = i\left(\omega_{\alpha\beta}(\Omega_{DDL}) + \omega_I\right)\delta_{\alpha'\alpha}\delta_{\beta'\beta} + R^S_{\alpha'\beta'\alpha\beta}(\Omega_{DDL})\bigg|_{\omega_{\alpha\alpha'}=\omega_{\beta\beta'}} \qquad (6.22)$$

As it has been discussed in details in the previous chapter, devoted to the electron spin relaxation, the appropriate Liouville space for the electron spin $\{|\psi_\alpha^S\rangle\langle\psi_\beta^S|\} \equiv \{|\psi_\alpha^S, \psi_\beta^S\rangle\}$ is generated by the set of eigenfunctions of the main Hamiltonian, $\{|\psi_\alpha^S\rangle\}$.

The electron spin spectral densities, s_{mk}, determine the nuclear spin-lattice

relaxation rate [5]:
$$R_{1I} = 2\operatorname{Re}\{K_{1,1}^{DD}(-\omega_I)\} =$$
$$2\left(\frac{\mu_0}{4\pi}\frac{\gamma_I\gamma_S\hbar}{r_{IS}^3}\right)^2 \frac{1}{2S+1}\{s_{0,0}^{(DD)} + s_{1,1}^{(DD)} + s_{-1,-1}^{(DD)} + 2s_{1,-1}^{(DD)} + 2s_{1,0}^{(DD)} + 2s_{0,-1}^{(DD)}\}$$
(6.23)

It has been demonstrated in Chapter IV.4 that the spectral density $s_{0,0}$ is associated with the spin-lattice electron spin relaxation, while the spectral densities $s_{1,1}$ and $s_{-1,-1}$ contain the spin-spin relaxation matrix elements. However, Eq.6.23 contains also the spectral densities of $m \neq k$ mixing up coherences of different orders, which makes impossible a clear separation on the effects of the longitudinal and transverse electron spin relaxation on the nuclear spin relaxation. At low and high magnetic fields, the mixed spectral densities disappear, because of the form of the projection vectors in these regimes. In these limiting magnetic field regimes, in principle, one is able to obtain closed analytical expressions for the nuclear spin relaxation, as it has been shown above. Nevertheless, for high spin quantum numbers and a rhombic static ZFS the calculations are quite cumbersome; some examples have already been given.

Let us now formulate the alternative expression for the nuclear spin relaxation, based on the laboratory representation of the main Hamiltonian for the electron spin. It can be quite easily handled, by starting once again from the dipolar spectral density of Eq.4.19 and referring to the main steps of the above presented treatment. The electron spin tensor components S_q^1 present in Eq.4.19 are already expressed in the laboratory frame and therefore we do not need to perform any further operation on them. Therefore, one can write down, in a straightforward manner, the laboratory counterpart of Eq.6.18:

$$K_{1,1}^{DD}(-\omega_I) =$$
$$\left(\frac{\mu_0}{4\pi}\frac{\gamma_I\gamma_S\hbar}{r_{IS}^3}\right)^2 \sum_{p,q=-1}^{1}\langle h_{pq}(\Omega_{DDL})(s_{pq}^{(L)}(-\omega_I)(\Omega_{DDL}))\rangle = \left(\frac{\mu_0}{4\pi}\frac{\gamma_I\gamma_S\hbar}{r_{IS}^3}\right)^2 \sum_{p,q=-1}^{1} s_{pq}^{(L)}(-\omega_I)$$
(6.24)

where
$$h_{pq}(\Omega_{DDL}) = 30\begin{pmatrix}2 & 1 & 1\\ 1-q & q & -1\end{pmatrix}\begin{pmatrix}2 & 1 & 1\\ 1-p & p & -1\end{pmatrix}D_{0,1-q}^{2*}(\Omega_{DDL})D_{0,1-p}^{2}(\Omega_{DDL})$$
(6.25)

The spectral densities, $s_{p,q}^{(L)}(-\omega_I)$, are defined in analogy to the molecular frame functions, $s_{m,k}^{(DD)}(-\omega_I)$, as a result of the averaging over molecular orientation of the orientation specific spectral densities $s_{p,q}^{(L)}(-\omega_I)(\Omega_{DDL})$, taken with the appropriate weight factors $h_{pq}(\Omega_{DDL})$:

$$s_{p,q}^{(L)}(-\omega_I) = \langle h_{pq}(\Omega_{DDL})(s_{p,q}^{(L)}(-\omega_I)(\Omega_{DDL})) \rangle =$$
$$\left\langle \int_0^\infty h_{pq}(\Omega_{DDL}) Tr_S \left\{ S_p^{1(L)+} \left[\exp\left(-i\hat{L}_S^{(L)}(\Omega_{DDL})\tau\right) S_q^{1(L)} \right] \rho_S^{eq} \right\} \exp(-i\omega_I \tau) d\tau \right\rangle \quad (6.26)$$

The electron spin Liouvilian $\hat{L}_S^{(L)}(\Omega_{DDL})$ contains the Zeeman Liouville operator, the static ZFS operator, both defined in the laboratory frame, and the relaxation operator:

$$\hat{L}_S^{(L)}(\Omega_{DDL}) = \hat{L}_{ZFS}^{S(L)}(\Omega_{DDL}) + \hat{L}_Z^{(L)} + \hat{R}_S(\Omega_{DDL}) \quad (6.27)$$

The fact that the previous calculations (based on the molecular frame representation of the electron spin Hamiltonian) and the present ones are being developed in the same Liouville basis $\{|\psi_\alpha^S\rangle\langle\psi_\beta^S|\}$ makes the next steps quite straightforward. The averaged spectral densities, $s_{pq}^{(L)}(-\omega_I)$, can be calculated directly from the expression:

$$s_{pq}^{(L)}(-\omega_I) = \frac{1}{2S+1} \left\langle h_{pq}(\Omega_{DDL}) \sum_{\mu\nu} \left[S_p^{1(L)+}(\Omega_{DDL})\right]_\mu^* \left[\hat{M}^{(L)}(\Omega_{DDL})^{-1}\right]_{\mu\nu} \left[S_q^{1(L)1}(\Omega_{DDL})\right]_\nu \right\rangle \quad (6.28)$$

Figs.6.2-6.3 show examples of proton and carbon relaxation profiles for the electron spin quantum numbers $S = 3$, and $S = \frac{7}{2}$, respectively. The evaluations have been performed applying the presented method for the static ZFS of the axial symmetry. The results have been compared with the predictions of SBM theory [12-15]. More examples one can find in [5].

In the next section I shall put more effort to explain some practical aspects of the evaluations described so far in a quite general manner.

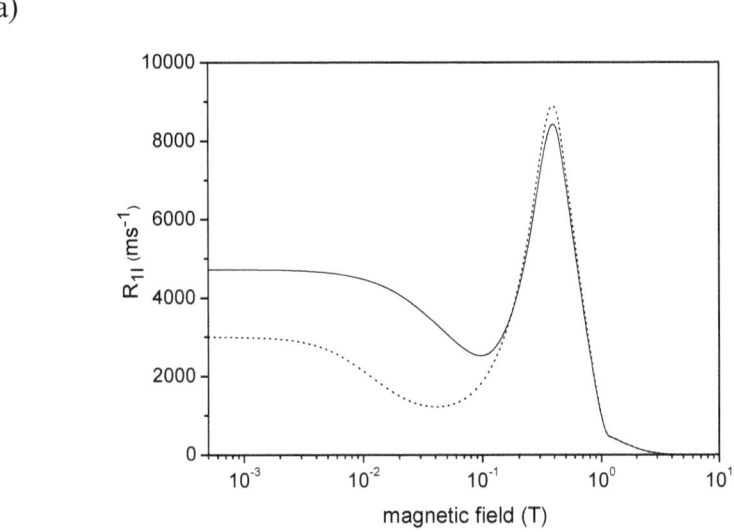

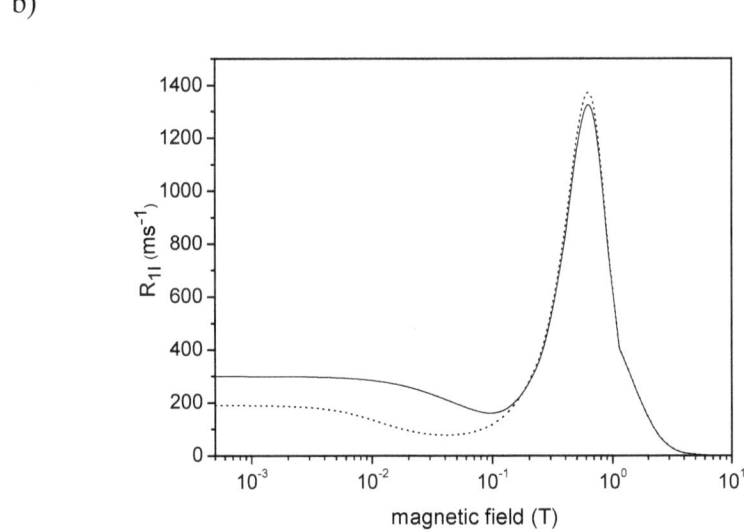

Fig.6.1

Proton a) and carbon b) spin-lattice relaxation profiles for the electron spin quantum number $S = 3$ for slowly rotating molecular systems. Solid line: the present theory, $D_S = 0.05 cm^{-1}$, $D_T = 0.02 cm^{-1}$, $\tau_D = 30 ps$, $r_{IS} = 300 pm$; dotted line: the corresponding SBM profile.

a)

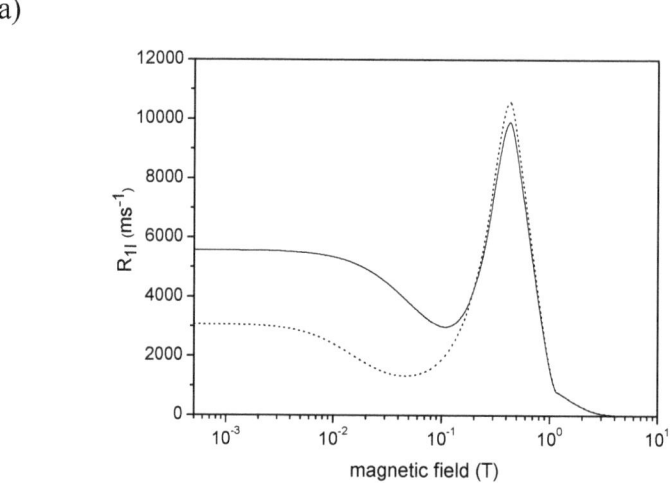

b)

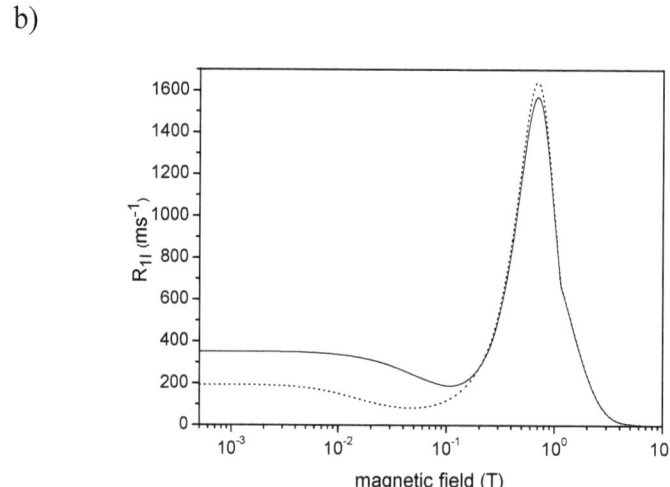

Fig.6.2

Proton a) and carbon b) spin-lattice relaxation profiles for the electron spin quantum number $S = \frac{7}{2}$ for slowly rotating molecular systems. Solid line: the present theory, $D_S = 0.05 cm^{-1}$, $D_T = 0.02 cm^{-1}$, $\tau_D = 30 ps$, $r_{IS} = 300 pm$; dotted line: the corresponding SBM profile.

VI.3. Nuclear spin relaxation in intermediate magnetic field range – examples of the semi-analytical treatment

The first section of this chapter introduced the basis of the perturbation treatment dedicated to spin-lattice relaxation of a dipolar spin $I = \frac{1}{2}$ caused by its interaction with an electron spin S. The section provided us with a set of equations derived under the fundamental assumption that the static ZFS does not change its orientation with respect to the laboratory frame. Even though the presented approach is valid for an arbitrary magnetic field, I applied it so far to some selected spin systems (which I consider as noteworthy and explicatory) in the limiting, high and low field regimes. In this section I shall illustrate how the described procedure works in an intermediate range of the magnetic field, using one specific example. If one wishes to deal with as simple as possible formulas, one must choose as the example the simplest possible case of the electron spin quantum number $S = 1$. In fact, the simplest example supported by the limiting cases examined in the previous section should fully illustrate how to apply the general approach to an arbitrary spin system in the whole magnetic field range.

The essential physical element of the discussed treatment is the hierarchy of events in the spin system of interest. One should consider this aspect carefully, being aware of the fact that the S spin possesses, in fact, two relaxation channels: it can relax by the ZFS mechanism and due to the I-S dipole-dipole coupling; Chapter XI is devoted to this issue. If the first relaxation mechanism dominates (*i.e.* the second one is negligible compared to it), the dynamics of the spin S becomes independent of the presence of the spin I and therefore the S spin can be treated separately. The next very important aspect concerns the electron as well as the nuclear spins. First, one must be conscious regarding the restrictions of the Redfield relaxation theory when applied to the electron spin subsystems. This subject has been justified in the previous chapter. The electron spin relaxation contributes to the modulations of the I-S dipole-dipole causing the I spin relaxation. Actually, for slowly rotating systems, the relaxation component of these modulations competes only with chemical exchange processes (we shall come later to the problem of the exchange motion). Therefore one must be very careful considering the validity regimes of the perturbation theory for the spin I. The correlation time resulting from the S spin relaxation and the exchange, $\tau_c^{-1} = R_S + \tau_M^{-1}$, must be short enough to fulfill the Redfield condition: $\omega_{DD}^{IS} \tau_c \ll 1$. Under this condition we can express the spin-lattice relaxation rate in terms of the quantum mechanical spectral density, $K_{1,1}^{DD}(-\omega_I)$, and appreciate the fact that the spin S is within the Redfield limit, if we intend to calculate it explicitly. It is very important to notice that, apart from the cases of an axially symmetric static ZFS at low field and the high magnetic field region, one may not attribute the quantity $s_{0,0}$ to the spin-lattice electron spin relaxation only, at the same time as the other spectral densities $s_{\pm 1,\pm 1}$ do not contain only the spin-spin electron spin relaxation. They must be evaluated in the basis of the $(2S+1)^2$ matrix representation of the operator $\hat{M}$ covering the populations as well as all multi-quantum coherences available for the spin S. The elements of the matrix $[\hat{M}]$, relevant for the nuclear spin relaxation, are selected by the non-zero expansion coefficients of the operators S_z and $S_\pm$. In the case of $S=1$, the full matrix $[\hat{M}]$ covers the following elements:

$[\hat{M}] =$ (6.29)

$$\begin{bmatrix} |\psi_1^S\rangle\langle\psi_1^S| & |\psi_2^S\rangle\langle\psi_2^S| & |\psi_3^S\rangle\langle\psi_3^S| & |\psi_1^S\rangle\langle\psi_2^S| & |\psi_2^S\rangle\langle\psi_3^S| & |\psi_1^S\rangle\langle\psi_3^S| & |\psi_2^S\rangle\langle\psi_1^S| & |\psi_3^S\rangle\langle\psi_2^S| & |\psi_3^S\rangle\langle\psi_1^S| \\ i\omega_I + R_{1111}^S & R_{1122}^S & R_{1133}^S & 0 & 0 & 0 & 0 & 0 & 0 \\ R_{1122}^S & i\omega_I + R_{2222}^S & R_{2233}^S & 0 & 0 & 0 & 0 & 0 & 0 \\ R_{1133}^S & R_{2233}^S & i\omega_I + R_{3333}^S & 0 & 0 & 0 & 0 & 0 & 0 \\ 0 & 0 & 0 & i(\omega_I+\omega_{12}) + R_{1212}^S & 0 & 0 & 0 & 0 & 0 \\ 0 & 0 & 0 & 0 & i(\omega_I+\omega_{23}) + R_{2323}^S & 0 & 0 & 0 & 0 \\ 0 & 0 & 0 & 0 & 0 & i(\omega_I+\omega_{13}) + R_{1313}^S & 0 & 0 & 0 \\ 0 & 0 & 0 & 0 & 0 & 0 & i(\omega_I-\omega_{12}) + R_{1212}^S & 0 & 0 \\ 0 & 0 & 0 & 0 & 0 & 0 & 0 & i(\omega_I-\omega_{23}) + R_{2323}^S & 0 \\ 0 & 0 & 0 & 0 & 0 & 0 & 0 & 0 & i(\omega_I-\omega_{13}) + R_{1313}^S \end{bmatrix}$$

The relaxation matrix element, $R^S_{\alpha\beta\alpha'\beta'}(\Omega_{ML})$, as well as the transition frequencies, $\omega_{\alpha\beta}(\Omega_{ML}) = E_\alpha(\Omega_{ML}) - E_\beta(\Omega_{ML})$, depend on the molecular orientation; I have not denoted this explicitly in the above expression. Setting up the part of the matrix $\left[\hat{M}\right]$ related to the spin-spin electron spin relaxation one must be aware of the fact that depending on the amplitude of the magnetic field and the molecular orientation, the secular approximation condition can be broken for some of the coherences. Therefore, the diagonal form of the coherence block should be treated as a simplifying assumption. The problem is that we are not able to follow in detail the energy level structure for every molecular orientation to figure out when some of the energy levels cross. It is possible only in some special cases like for example in the high field limit, when we know that the pairs of coherences $|\psi_1^S\rangle\langle\psi_2^S|, |\psi_2^S\rangle\langle\psi_3^S|$ and $|\psi_2^S\rangle\langle\psi_1^S|, |\psi_3^S\rangle\langle\psi_2^S|$ are coupled, however this knowledge does not help very much. Therefore, instead of searching for energy level crossings (this task is in practice unmanageable) it is best to use just the diagonal form of the coherence block of the matrix $\left[\hat{M}\right]$. Such a treatment is fully satisfactorily. One can prove this by testing its predictions against the 'general model', described in detail in Chapter X [5]. At this stage, I would like to convince the reader that this simplification is completely acceptable.

The problem of nuclear spin relaxation in the presence of an electron spin (Paramagnetic Relaxation Enhancement, PRE) is a challenging subject for theoretical studies. Several very interesting and theoretically demanding treatments are present in the literature. Finishing this chapter let me refer to some of them [16-23].

References:

1. T. Nilsson, J. Kowalewski, Low-field theory of nuclear spin relaxation in paramagnetic low – symmetry complexes for electron spin systems of S=1, 3/2, 2, 5/2, 3 and 7/ 2, *Mol. Phys.* **98** (2000) 1617-1638, Erratum: *Mol. Phys.* **99** (2001) 369-370
2. J.S. Griffith, The Theory of Transition metal ions, *Cambridge University Press*, Cambridge (1961)
3. A. Abragam, B. Bleaney, Electron paramagnetic resonance of transition ions, *Clarendon Press, Oxford* (1970)
4. J. E. Wertz, J. R. Bolton, Electron Spin Resonance: Elementary theory and practical applications, *McGraw-Hill, New York* (1972)
5. D. Kruk, T. Nilsson, J. Kowalewski, Nuclear spin relaxation in paramagnetic systems with zero-field splitting and arbitrary electron spin, *Phys. Chem. Chem. Phys.* **3** (2001) 4907-4917
6. A. Kumar, R.C.R. Grace, P.K. Madhu, Cross-correlations in NMR, *Prog. Nucl. Magn. Reson. Spectr.* **37** (2000) 191-319
7. U. Fano, Description of states in quantum mechanics by density matrix and operator techniques, *Rev. Mod. Phys.* **29** (1957) 74-93
8. L. Werbelow, Relaxation-Induced Transfer of Nuclear Spin Polarization as a Probe of Molecular Structure and Dynamics in Mobile Phases, in Nuclear Magnetic Resonance Probes of Molecular Dynamics, R. Tycko (Ed.) *Kluwer Academic Publishers* (1994) 223-263
9. I. Bertini, J. Kowalewski, C. Luchinat, T. Nilsson, G. Parigi, Nuclear spin relaxation in paramagnetic complexes of S=1: electron spin relaxation effects, *J. Chem. Phys.* **111** (1999) 5795-5807
10. D. Kruk, O. Lips, Field-dependent nuclear relaxation of spins ½ induced by dipole –dipole couplings to quadrupolar spins: LaF_3 crystals as an example, *J. Magn. Reson.* **179** (2006) 250 – 262
11. I. Bertini, O. Galas, C. Luchinat, G. Parigi, A computer program for the calculation of paramagnetic enhancements of nuclear – relaxation rates in slowly rotating systems, *J. Magn. Reson. A* **113** (1995) 151-158
12. I. Solomon, Relaxation processes in a system of two spins, *Phys. Rev.* **99** (1955) 559-565

13. I. Solomon, N. Bloembergen, Nuclear magnetic interactions in the HF molecule, *J. Chem. Phys.* **25** (1956) 261-266
14. N. Bloembergen, Proton relaxation times in paramagnetic solutions, *J. Chem. Phys.* **27** (1957) 572-573
15. N. Bloembergen, L.O. Morgan, Proton relaxation times in paramagnetic solutions: Effects of electron spin relaxation, *J. Chem. Phys.* **34** (1961) 842-850
16. I. Bertini, C. Luchinat, S. Aime, NMR of paramagnetic substances, *Elsevier, Amsterdam* (1996)
17. L. Banci, I. Bertini, C. Luchinat, Nuclear and Electron Relaxation, *VCH, Weinheim* (1991)
18. S. Rast, P.H. Fries, E. Belorizky, Static zero field splitting effects on the electronic relaxation of paramagnetic metal ion complexes in solution, *J. Chem. Phys.* **113** (2000) 8724-8735
19. A. Borel, F. Yearly, L. Helm. A. E. Merbach, Multiexponential electronic spin relaxation and Redield's limit in Gd (III) complexes in solution: consequences for O-17/H-1 NMR and EPR simultaneous analysis, *J. Am. Chem. Soc.* **123** (2001) 2637-2644
20. R. Sharp, L. Lohr, Thermal relaxation of electron spin motion in a thermal equilibrium ensemble: relation to paramagnetic nuclear magnetic resonance relaxation, *J. Chem. Phys.* **115** (2001) 5005-5014
21. R. Sharp, L. Lohr, J. Miller, Paramagnetic NMR relaxation enhancement: recent advances in theory, *Prog. Nucl. Magn. Reson. Spectr.* **38** (2001) 115-158
22. R. Sharp, Closed-form expressions for level-averaged electron spin relaxation times outside the Zeeman limit: application to paramagnetic NMR relaxation, *J. Magn. Reson.* **154** (2002) 269-279
23. N. Schaefle, R. Sharp, Four theoretical approaches for the analysis of NMR paramagnetic relaxation, *J. Magn. Reson.* **176** (2005) 160-170

CHAPTER VII

Dipolar spin relaxation in solid state systems containing quadrupolar spins

The present chapter will examine the spin - lattice nuclear relaxation of spins $I = \frac{1}{2}$ caused by dipole-dipole couplings to quadrupolar spins $S \geq 1$ (a second nuclear spin), which exhibit a non-zero averaged (permanent) nuclear electric quadrupole interaction (as opposed to the permanent ZFS interaction). It is assumed that the dipolar as well as quadrupolar spins fulfill the conditions of the Redfield relaxation theory. I shall formulate, following the line of analogies between the quadrupolar interaction and the zero field splitting, a 'quadrupolar' counterpart of the paramagnetic relaxation enhancement theory dedicated to solid state systems.

VII.1. Spin-lattice relaxation of spins 1/2 due to dipole-dipole couplings to quadrupolar spins in crystals

This section deals with an assembly of equivalent spins $I = \frac{1}{2}$ and equivalent spins S, which occupy well defined position in a crystal lattice. Let us for a while restrict the considerations to a system containing only a 'reference' spin I coupled to one spin S. We start once again from the expression for the quantum-mechanical spectral density $K_{1,1}^{DD}(-\omega_I)$, given by Eq.4.21, however with a very important adjustment [1-2]:

$$K_{11}^{DD}(-\omega_I) = 30\left(\frac{\mu_0 \hbar \gamma_I \gamma_S}{4\pi}\right)^2 \sum_{p,q=-1}^{1} \begin{pmatrix} 2 & 1 & 1 \\ 1-q & q & -1 \end{pmatrix} \begin{pmatrix} 2 & 1 & 1 \\ 1-p & p & -1 \end{pmatrix} \times \quad (7.1)$$

$$\int_0^\infty Tr_L \left\{ S_q^{1(L)+} \frac{D_{0,1-q}^{2*}(\Omega_{DDL}(\tau))}{r_{IS}^3(\tau)} \left[\exp(-i\hat{L}_L \tau) S_p^{1(L)} \frac{D_{0,1-p}^2(\Omega_{DDL}(0))}{r_{IS}^3(0)}\right] \rho_L^{eq} \right\} \exp(-i\omega_I \tau) d\tau$$

This formula allows for time fluctuations of the interspin distance r_{IS}. Jump diffusion of the ions, carrying the I and (or) the S spins, between available crystallographic positions changes not only the orientation of the $\vec{r}_{IS}$ vector encoded in the angle Ω_{DDL} (the index L refers to the laboratory frame) but also its length. To factorize the correlation function into the 'structural' part, $Tr_{L-S}\left\{\frac{D_{0,1-q}^{2*}(\Omega_{DDL}(\tau))}{r_{IS}^3(\tau)}\left[\exp(-i\hat{L}_{L-S}\tau)\frac{D_{0,1-p}^2(\Omega_{DDL}(0))}{r_{IS}^3(0)}\right]\rho_{L-S}^{eq}\right\}$, and the quadrupolar spin correlation function $Tr_S\left\{S_q^{1(L)+}\left[\exp(-i\hat{L}_S\tau)S_p^{1(L)}\right]\rho_S^{eq}\right\}$, one must be sure that the first one is determined by motional processes not related to any degrees of freedom of the spin S included in the Liouville operator $\hat{L}_S$. The lattice Liouvillian $\hat{L}_L$ includes the Zeeman coupling of the spin S, $\hat{L}_Z(S)$, the averaged (static) quadrupolar interaction $\hat{L}_Q^S(S)$ (they determine together the quadrupolar spin energy levels), the fluctuating part of the quadrupolar interaction, $\hat{L}_Q^T(S)$, the associated operator $\hat{L}_V$ reflecting the fluctuations of the electric field gradient tensor and two operators, denoted as $\hat{L}_M^S$ and $\hat{L}_M^I$, and describing the jump diffusion of the S and I ions, respectively:

$$\hat{L}_L = \hat{L}_Z(S) + \hat{L}_Q^S(S) + \hat{L}_Q^T(S) + \hat{L}_V + \hat{L}_M^S + \hat{L}_M^I = \hat{L}_S + \hat{L}_M^S + \hat{L}_M^I \quad (7.2)$$

In fact, the last two terms are responsible for the fluctuations of the $\vec{r}_{IS}$ vector. Fig. 7.1 shows how the individual parts of the lattice Liouvilian are associated with some degrees of freedom of the spin system. If the quadrupolar spins are fixed on their crystallographic positions, the 'structural' correlation functions are mediated only by the jump diffusion of the dipolar spins I.

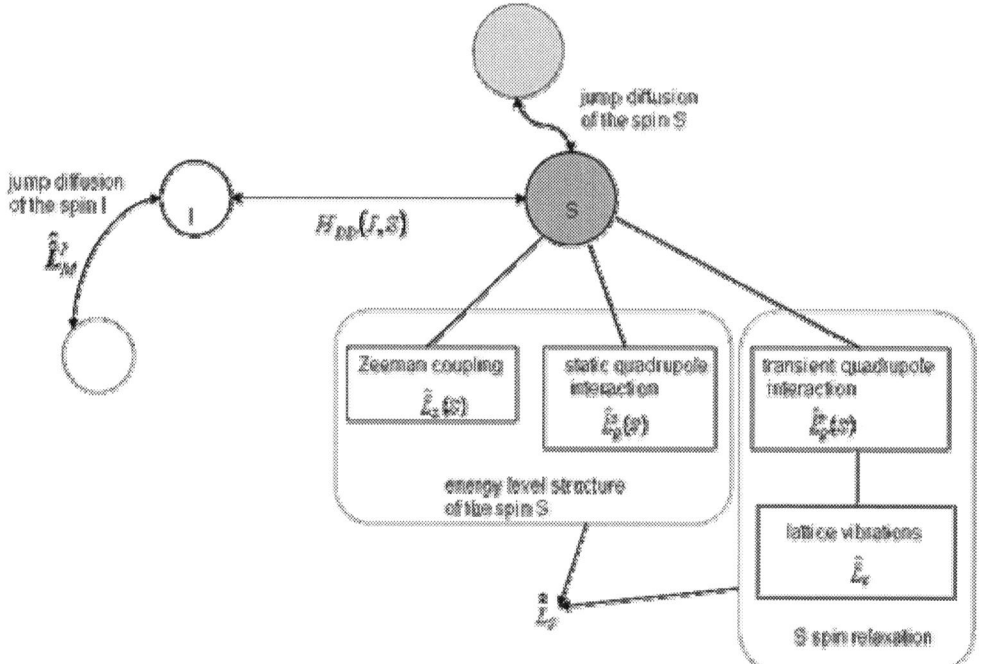

Fig.7.1

Lattice degrees of freedom for a dipolar spin – quadrupolar spin system. The dipolar spin I as well as the quadrupolar spin S jump between available lattice sites. It is assumed that all positions available for the spin S are equivalent and therefore the jump diffusion does not affect the quadrupolar interaction. In consequence, the energy level structure of the quadrupolar spin is determined by the Zeeman interaction and the static quadrupolar coupling (the jump diffusion does not change neither its value nor its orientation), while the transient quadrupolar interaction, modulated by lattice vibrations, causes the quadrupolar spin relaxation.

Since in this case the dynamics of the quadrupolar spins is independent of the presence of the dipolar ones, one can obviously separate the two correlation functions. Nevertheless, even if the quadrupolar spins change their sites, affecting in this way the structural correlation function, we are allowed to separate the two quantities as well. This is so due to the fact that the quadrupolar spins jump among equivalent sites in the crystal lattice and therefore they sense always the same electric field gradient tensor. In other words, their relaxation dynamic is independent of the position which they actually occupy, because all the positions are equivalent. For this reason one can write:

$$K_{1,1}^{DD}(-\omega_I) = 30\left(\frac{\mu_0 \hbar \gamma_I \gamma_S}{4\pi}\right)^2 \sum_{p,q=-1}^{1} \begin{pmatrix} 2 & 1 & 1 \\ 1-q & q & -1 \end{pmatrix} \begin{pmatrix} 2 & 1 & 1 \\ 1-p & p & -1 \end{pmatrix} \times \quad (7.3)$$

$$\int_0^\infty Tr_S\left\{S_q^{1(L)+}\left[\exp\left(-i\hat{L}_S\tau\right)\right]S_p^{(L)1}\right\}\rho_S^{eq}\left\langle\frac{D_{0,1-q}^{2*}(\Omega_{DDL}(\tau))D_{0,1-p}^{2}(\Omega_{DDL}(0))}{r_{IS}^3(\tau)\, r_{IS}^3(0)}\right\rangle \exp(-i\omega_I\tau)d\tau$$

Finding a suitable form of the correlation function between relative positions of the interacting spins, determined in solid state phase by the relative jump diffusion, is a challenging and demanding problem by itself. The simplest attempt is to assume that it is single exponential [3-5]:

$$\left\langle\frac{D_{0,1-q}^{2*}(\Omega_{DDL}(\tau))D_{0,1-p}^{2}(\Omega_{DDL}(0))}{r_{IS}^3(\tau)\, r_{IS}^3(0)}\right\rangle = \frac{D_{0,1-q}^{2*}(\Omega_{DDL}(0))D_{0,1-p}^{2}(\Omega_{DDL}(0))}{r_{IS}^3(0)\, r_{IS}^3(0)} \exp\left(-\frac{\tau}{\tau_{IS}}\right) \quad (7.4)$$

where the correlation time results from the exchange (jump) rates of the participating spins (denoted as τ_I^{-1} and τ_S^{-1} for the spins I and S, respectively): $\tau_{IS}^{-1} = \tau_I^{-1} + \tau_S^{-1}$. This assumption leads to a straightforward mathematical treatment of the spectral density $K_{1,1}^{DD}(-\omega_I)$, however one should be aware that it can turn out to be an oversimplification. Some experimental systems are described well by this approximation. In Fig.7.2 fluorine (^{19}F) relaxation profiles for BaF_2 are presented. They can be satisfactorily interpreted in terms of single exponential correlation functions describing the jump diffusion of the fluorine spins.

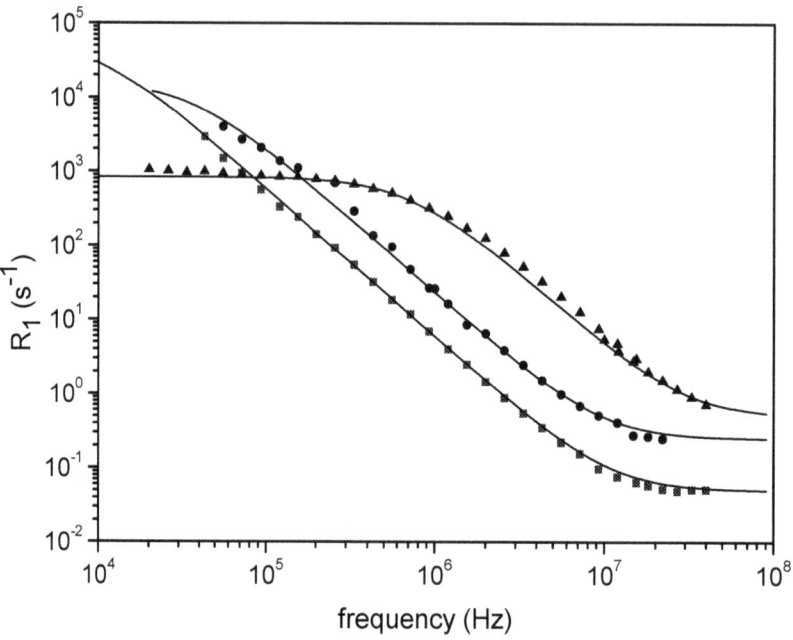

Fig.7.2
Fluorine spin (^{19}F) relaxation profiles collected for a single crystal of BaF_2. They can be interpreted in terms of single exponential correlation function describing the jump diffusion of fluorine ions. The obtained correlation times are as follows: 650K: squares, $\tau_I = 9.0*10^{-6} s$; 700K: circles, $\tau_I = 2.3*10^{-6} s$; 800K: triangles, $\tau_I = 1.2*10^{-7} s$.

Employing the single exponential correlation function, one can reformulate the expression for the spectral density [1]:

$$K_{1,1}^{DD}(-\omega_I) = 30\left(\frac{\mu_0 \hbar \gamma_I \gamma_S}{4\pi}\right)^2 \sum_{q,o=-1}^{1}\begin{pmatrix}2 & 1 & 1\\ 1-q & q & -1\end{pmatrix}\begin{pmatrix}2 & 1 & 1\\ 1-p & p & -1\end{pmatrix}\frac{D_{0,1-q}^{2*}(\Omega_{DDL}) D_{0,1-p}^{2}(\Omega_{DDL})}{r_{IS}^3 \; r_{IS}^3} s_Q^{p,q} \quad (7.5)$$

where $s_Q^{pq}(-\omega_I)$ represents the quadrupolar spin spectral density, including the modulation of the mutual I-S dipole-dipole coupling through the jump diffusion, at the frequency corresponding to the difference between the two Zeeman energy levels of the spin I [1]:

$$s_Q^{pq}(-\omega_I) = \int_0^\infty Tr_S \left\{ S_q^{1(L)+} \exp\left[-\left(\left(i\hat{\hat{L}}_S + \frac{1}{\tau_{IS}} + i\omega_I\right)\tau\right) S_p^{1(L)}\right] \rho_S^{eq} \right\} d\tau \qquad (7.6)$$

where the quadrupolar spin operator $\hat{\hat{L}}_S$ contains the main interactions, encoded in the Liouvilian $\hat{\hat{L}}_S^0 = +\hat{L}_Z(S) + \hat{\hat{L}}_Q^S$, and the relaxation superoperator $\hat{\hat{R}}_S$: $\hat{\hat{L}}_S = \hat{\hat{L}}_S^0 - i\hat{\hat{R}}_S$. The explicit evaluations of the spectral densities $s_Q^{pq}(-\omega_I)$ have been discussed in detail in the previous chapter and therefore they do not need any further comments. Usually the I spin relaxation result from dipole-dipole couplings to many neighboring spins S. To take into account the effect of many quadrupolar spins one has to replace the structural factors $\lambda_{qp} = \dfrac{D_{0,1-q}^{2*}(\Omega_{DDL}) D_{0,1-p}^2(\Omega_{DD}^L)}{r_{IS}^3 \; r_{IS}^3}$ associated with the spectral densities $s_Q^{pq}(-\omega_I)$ by corresponding quantities $\Lambda_{qp} = \sum_i \lambda_{qp} = \sum_i \dfrac{D_{0,1-q}^{2*}(\Omega_{DD_iL}) D_{0,1-p}^2(\Omega_{DD_iL})}{r_{IS_i}^3 \; r_{IS_i}^3}$ (the angle Ω_{DD_iL} describes the orientation of the $I-S_i$ dipole – dipole axis with respect to the laboratory frame). The summation must be performed over all quadrupolar spins S_i involved in the relaxation of the spin I. In the case of crystalline solid state systems it is appropriate to use the name 'lattice sums' for the factors Λ_{qp}; symmetry properties of crystal lattices are very useful for their evaluation.

If the exchange motion characterized by the correlation time τ_{IS} is on a rapid time scale relative to the quadrupolar spin relaxation, it has a dominant contribution to the time fluctuations of the dipole-dipole coupling. The spread of correlation times caused by multiexponential quadrupolar spin relaxation becomes negligible in this motional limit and the quadrupolar relaxation may be omitted altogether. The off-diagonal elements of the Liouville supermatrix vanish in this case, while the diagonal part takes the simple form: $i\hat{\hat{L}}_S^0 + i\omega_I \hat{\hat{1}} + \dfrac{1}{\tau_{IS}} \hat{\hat{1}}$, with the only one, field-independent time constant τ_{IS}. Thus, the resulting dipolar spin relaxation rate R_{1I} can be expressed as a sum of spectral densities of the form $\frac{\tau_{IS}}{1+\omega^2\tau_{IS}^2}$, where the static quadrupolar coupling affects the frequencies ω of the joined $I-S$ transitions.

As usual in this book I wish to put some attention to a clear formulation of the validity regimes of the discussed treatment. They are determined obviously by the requirements of the second order perturbation theory. First of all, the quadrupolar spin subsystem must fulfill the Redfield condition: $\omega_Q^T \tau_Q \ll 1$ and the secular approximation condition: $\omega_Q^S \gg \omega_Q^T \cdot H_Q^T \tau_Q$ which permits one to treat the averaged quadrupolar interaction as a main Hamiltonian at low field (ω_Q^S and ω_Q^T are the amplitudes of the static and transient quadrupolar interactions, expressed in angular frequency units, while the correlation time τ_Q characterizes the stochastic fluctuations of the quadrupolar interaction). If the conditions are not fulfilled, one cannot define explicitly time independent quadrupolar spin relaxation rates. However, if they are satisfied, some caution must be exercised regarding the validity of Redfield perturbation approach for the I spins. The Redfield condition for the spin I requires that $\omega_{DD}^{IS} \tau_c \ll 1$, where τ_c is determined by the exchange correlation time τ_{IS} and the quadrupolar spin relaxation time R_S^{-1} corresponding to the slowest process: $\tau_c^{-1} = \tau_{IS}^{-1} + R_S$. The second limitation is: $\omega_I \gg \omega_{DD}^{IS} \cdot \omega_{DD}^{IS} \tau_c$. One should be aware of the fact that in the low field limit the dipole-dipole interaction can be strong enough to break down this relation. An additional condition relevant for the validity and completeness of this theoretical description is related to the dipolar contribution to the quadrupolar spin relaxation. I shall comment more on this subject in the last section.

Fig.7.3 and Fig.7.4 illustrate the effect of the static (averaged) quadrupolar interaction on the dipolar spin relaxation depending on the applied magnetic field. The calculations have been performed for a fluorine spin I coupled to a quadrupolar spin of $S = \frac{7}{2}$. The quadrupolar spin relaxation has been neglected; then one can see how the energy level structure of the quadrupolar spin influences the dipolar spin relaxation. The curves show the ratio between and the result of the 'classical' Solomon expression of Eq.4.11 [6] (with $\tau_{c1} = \tau_{c2} = \tau_{IS}$), neglecting the static quadrupolar interaction and the predictions of the present treatment.

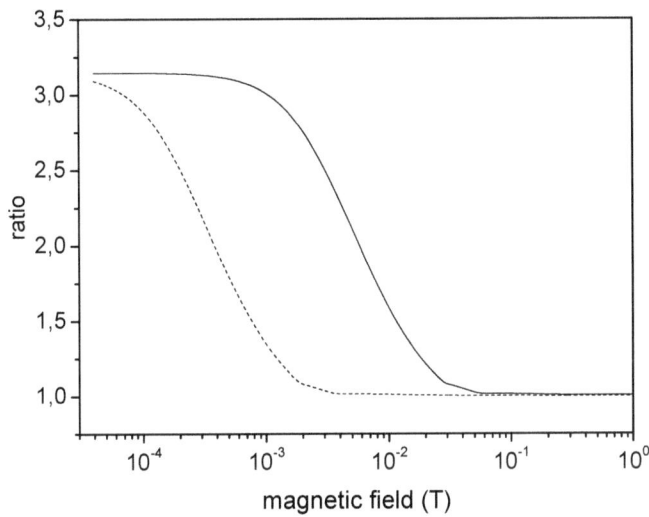

Fig.7.3
Ratios between fluorine (^{19}F) spin-lattice rates obtained from the Solomon expression and from the present treatment. The relaxation is caused by a dipole-dipole coupling of the fluorine spin to lanthanum quadrupolar spin, $S = \frac{7}{2}$. The quadrupolar spin relaxation has been neglected. Solid line corresponds to $a_Q^S = 15 MHz$, $\eta_S = 0.3$, while dashed line to $a_Q^S = 1 MHz$, $\eta_S = 0.3$. The correlation time τ_{IS} has been set to $\tau_{IS} = 10^{-6} s$. When the Zeeman interaction for the quadrupolar spins dominates over the quadrupolar coupling the present treatment converges to the Solomon-Boembergen-Morgan predictions. If the quadrupolar coupling is weaker, it takes place for relatively lower magnetic fields. However, a weaker static quadrupolar coupling does not imply that the discrepancies between the two descriptions are smaller at low magnetic field. It

has been assumed that the principal axis system of the quadrupolar interaction and the laboratory frame coincide.

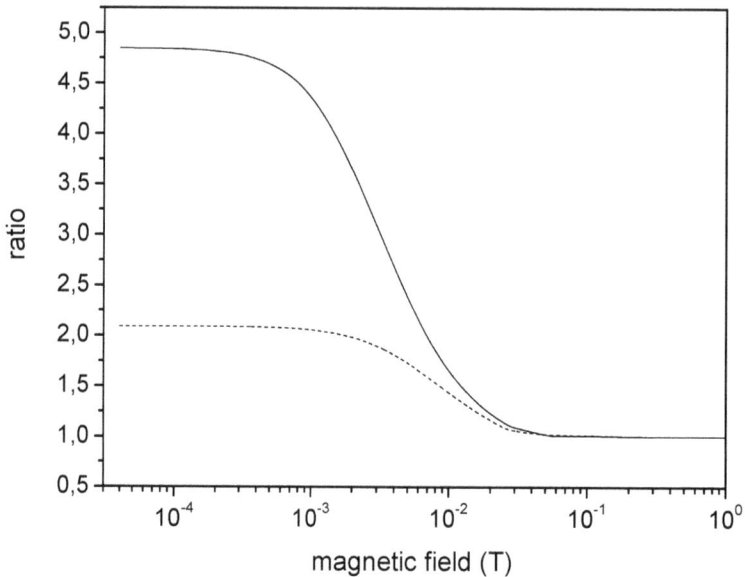

Fig.7.4
Ratios between fluorine (^{19}F) spin-lattice rates (due to a dipole-dipole coupling of the fluorine spin to lanthanum quadrupolar spin, $S = \frac{7}{2}$) predicted by the Solomon expression and by the present treatment. The quadrupolar spin relaxation has been neglected. Solid line corresponds to $a_Q^S = 15 MHz$, $\eta_S = 0$, while dashed line to $a_Q^S = 15 MHz$, $\eta_S = 0.8$. The correlation time τ_{IS} has been set to $\tau_{IS} = 10^{-6} s$. The asymmetry of the quadrupolar interaction influences significantly the dipolar spin relaxation. It has been assumed that the principal axis system of the quadrupolar interaction coincides with the laboratory frame.

The figure, Fig.7.5a,b, shows selected quadrupolar spin relaxation rates, illustrating the multiexponential relaxation [1,6-13] of the high spin $S = \frac{7}{2}$. The quadrupolar spin relaxation has been described in terms of the Lorentzian spectral densities determined by the amplitude of the fluctuating part of the quadrupolar interaction, a_Q^T, and the timescale of the fluctuations, τ_Q, according to Eq.4.59 (the anisotropy parameter η_T has been set to zero).

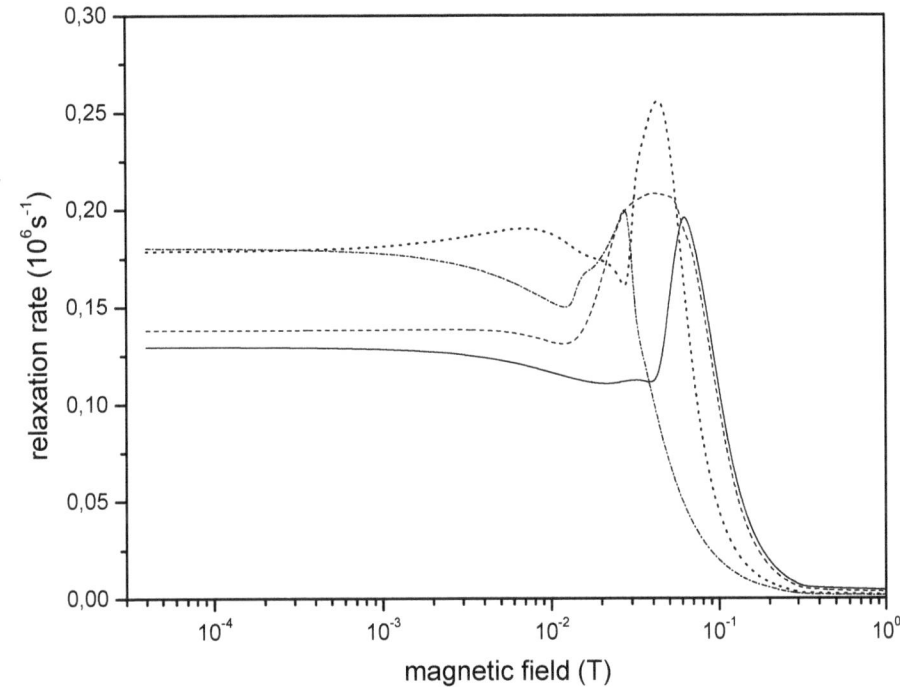

Fig.7.5a
Selected spin – lattice relaxation rates for lanthanum spin (^{139}La). The principal axis system of the static quadrupolar coupling and the laboratory frame coincide. The parameters are: $a_Q^S = 15 MHz, a_Q^T = 10 MHz, \tau_Q = 10^{-8} s$. Solid line: R_{1111}^S, dashed line: R_{2222}^S, dotted line: R_{3333}^S, dashed-dotted line: R_{4444}^S. The calculations have been performed using the following labeling of the Zeeman functions of the

quadrupolar spin, $|m_S\rangle$: $|1\rangle \equiv \left|\frac{7}{2}\right\rangle$, $|2\rangle \equiv \left|\frac{5}{2}\right\rangle$, $|3\rangle \equiv \left|\frac{3}{2}\right\rangle$, $|4\rangle \equiv \left|\frac{1}{2}\right\rangle$, $|5\rangle \equiv \left|-\frac{1}{2}\right\rangle$, $|6\rangle \equiv \left|-\frac{3}{2}\right\rangle$, $|7\rangle \equiv \left|-\frac{5}{2}\right\rangle$ *and* $|8\rangle \equiv \left|-\frac{7}{2}\right\rangle$.

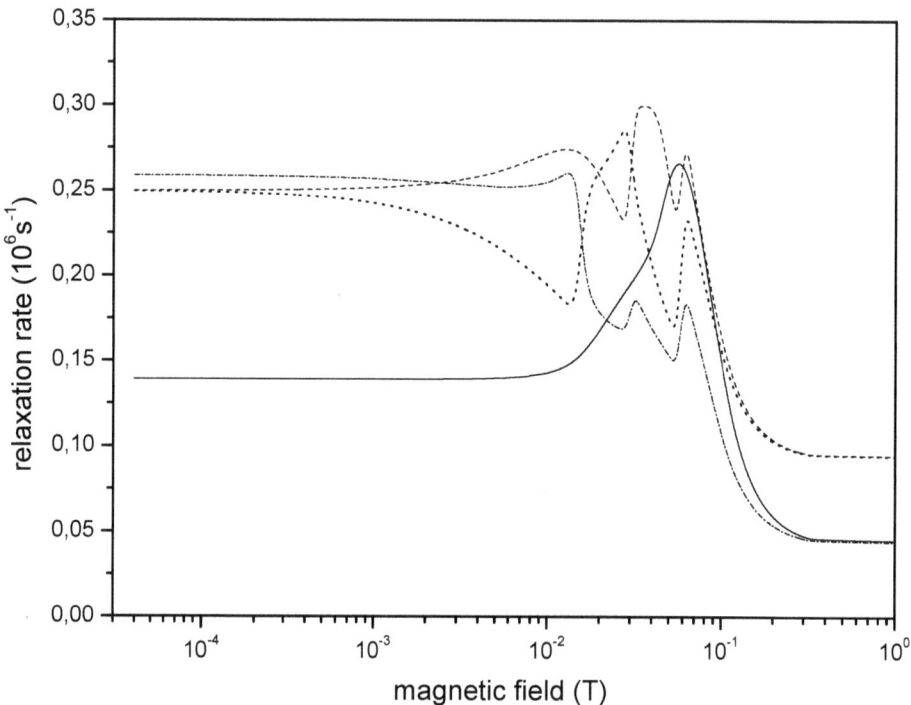

Fig.7.5b

Selected relaxation rates for lanthanum spin (^{139}La) calculated for the same set of parameters as used in Fig.5.7a. Solid line: R^S_{1212}, dashed line: R^S_{1313}, dotted line: R^S_{1414}, dashed-dotted line: R^S_{1515}.

I would like to remind the reader once again that, at this point, one has to be aware of the validity conditions of the perturbation theory applied to the quadrupolar spin.

VII.2. Examples of dipolar spin relaxation in the presence of neighboring quadrupolar spins

Let us consider an example of solid state systems for which the presented treatment is very suitable, namely LaF$_3$ [1,14]. The crystals have the tysonite structure ($P\bar{3}c1$), in which the fluorine ions occupy two distinct positions F_A and F_B [15]. The fact that we have to deal here with two types of fluorine spins complicates considerably the treatment. On the other hand this is a good opportunity to learn how to combine different elements of relaxation theories, presented so far, to describe a real spin system. Fluorine spins belonging to each of these sublattices are coupled by dipole-dipole interactions to fluorine spins from the same sublattice as well as from the second one, and they are also coupled to the lanthanum spins. These interactions are modulated by jumps of the fluorine ions between equivalent ($F_A - F_A, F_B - F_B$) and non-equivalent ($F_A - F_B$) lattice sites [1, 3, 4, 16-19]. The lanthanum spins remain on their positions. Lanthanum spins exhibit in the LaF$_3$ crystal structure a static quadrupolar coupling [20, 21]. The mutual dipole-dipole coupling between the non-equivalent fluorine spins from the distinct sublattices F_A and F_B implies that the time evolution of the quantities $\langle I_z^A \rangle$ and $\langle I_z^B \rangle$ follows the set of coupled equations (as described by Solomon [6] and discussed in Chapter IV, see Eq.4.8a,b):

$$\frac{d\langle I_z^A \rangle}{dt} = a_{AA}\left(\langle I_z^A \rangle - \langle I_z^A \rangle_{eq}\right) + a_{AB}\left(\langle I_z^B \rangle - \langle I_z^B \rangle_{eq}\right) \qquad (7.7a)$$

$$\frac{d\langle I_z^B \rangle}{dt} = a_{BA}\left(\langle I_z^A \rangle - \langle I_z^A \rangle_{eq}\right) + a_{BB}\left(\langle I_z^B \rangle - \langle I_z^B \rangle_{eq}\right) \qquad (7.7b)$$

The coefficients $a_{AA}, a_{BB}, a_{AB}, a_{BA}$ contain in addition to the specific dipolar relaxation rates, denoted respectively as $R_1^{AA}, R_1^{BB}, R_1^{AB}, R_1^{BA}$, the exchange rates $\tau_{AB}^{-1}, \tau_{BA}^{-1}$ between the non-equivalent lattice sites: $a_{AA} = -R_1^{AA} - \frac{1}{\tau_{AB}}$, $a_{AB} = -\frac{N_A}{N_B}R_1^{AB} + \frac{1}{\tau_{BA}}$ [1,3]. The remaining coefficient a_{BB}, a_{BA} can be obtained by exchanging the indices A and B, with the exchange lifetime τ_{AB} related to τ_{BA} through the ratio of the numbers of the fluorine spins in the subsystems F_A and F_B: $\tau_{AB} = \dfrac{N_A}{N_B}\tau_{BA}$. The relaxation rates R_1^{AA}, (R_1^{BB}) contain terms corresponding to the three relaxation pathways of fluorine spins, created respectively by dipole-dipole couplings within one sublattice $R_{1(A \to A)}^{AA}$, $\left(R_{1(B \to B)}^{BB}\right)$, couplings between spins from different fluorine sublattices $R_{1(A \to B)}^{AA}$, $\left(R_{1(B \to A)}^{BB}\right)$ and fluorine-lanthanum dipolar interactions $R_{1(A \to La)}^{AA}$, $\left(R_{1(B \to La)}^{BB}\right)$:

$R_1^{AA} = R_{1(A \to A)}^{AA} + R_{1(A \to B)}^{AA} + R_{1(A \to La)}^{AA}$ [1]. The relaxation channels are illustrated in Fig.7.6.

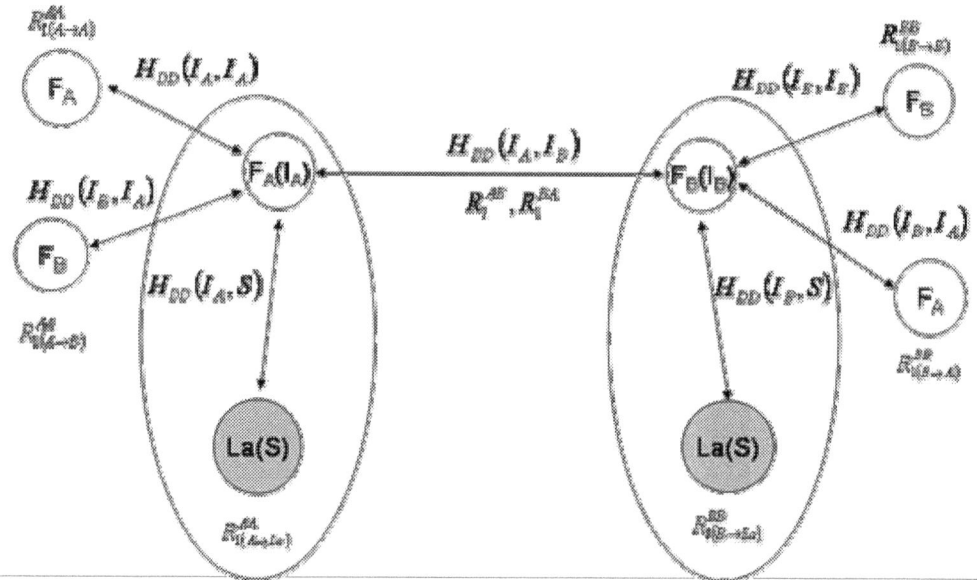

Fig.7.6

The dipole-dipole interactions and the relaxation channels in LaF$_3$. The fluorine-lanthanum relaxation channel is marked by ellipsoids. The $H_{DD}(I_A, I_B)$ dipole – dipole interaction causes cross-relaxation between the fluorine spins.

The first two contributions are given by the well known expressions for the dipolar spin-lattice relaxation rates for equivalent and nonequivalent spins $\frac{1}{2}$; Eq.4.4 and Eq.4.10, respectively. We write them down, adjusting for the present labeling:

$$R_{1(A \to A)}^{AA} = \left(\frac{\mu_0}{4\pi}\gamma_I^2 \hbar\right)^2 I(I+1)\left[J_1^{AA}(\omega_A) + 4J_2^{AA}(2\omega_A)\right] \tag{7.8a}$$

$$R_{1(A \to B)}^{AA} = \left(\frac{\mu_0}{4\pi}\gamma_I^2 \hbar\right)^2 I(I+1)\frac{1}{3}\left[J_0^{AB}(\omega_A - \omega_B) + 3J_1^{AB}(\omega_A) + 6J_2^{AB}(\omega_A + \omega_B)\right] \tag{7.8b}$$

The indices AA and AB indicate that the spectral densities J_m^{AA} and J_m^{AB} refer to the $I_A - I_A$ and $I_A - I_B$ dipole-dipole couplings, respectively. The frequencies ω_A and ω_B denote the transition frequencies for the F_A and F_B spins. However, to use the well known formulas we have to evaluate the spectral densities within a motional model. Modeling the ion diffusion as a two-site jumps process, we employ single exponential dipole-dipole correlation functions, as discussed above. It implies the Lorenzian forms of the spectral densities:

$$J_K^{\alpha\beta}(\omega) = \sum_i \left(\frac{D_{0,K}^2(\Omega_{I_\alpha I_L})}{r_{I_\alpha I_i}^3}\right)^2 \frac{2\tau_{\alpha\beta}}{1 + \omega^2 \tau_{\alpha\beta}^2} = \Lambda_{F_\alpha \to F_\beta}^{KK} \frac{2\tau_{\alpha\beta}}{1 + \omega^2 \tau_{\alpha\beta}^2} \tag{7.9}$$

where $\alpha, \beta = A, B$; $I_\alpha \neq I_i$. The summation is performed over spins I_i from the sublattice β involved in the relaxation processes of the spin I_α from the α sublattice. The fluoride lattice sums are given in Appendix A to [1], (the appendix contains also other lattice sums, which are needed for the further calculations). The correlation time $\tau_{\alpha\beta}$ describes the fluctuation of the dipole-dipole coupling due to the effective motion of both coupled spins and is obtained through the superposition of the two specific correlation rates τ_A^{-1}, τ_B^{-1}: $\tau_{\alpha\beta}^{-1} = \tau_\alpha^{-1} + \tau_\beta^{-1}$, containing the effects of ion jumps within the particular sublattices, (τ_{AA}, τ_{BB}) and between them: $\tau_A^{-1} = \tau_{AA}^{-1} + \tau_{AB}^{-1}$, $\tau_B^{-1} = \tau_{BB}^{-1} + \tau_{BA}^{-1}$ [1,3]. The relaxation rates R_1^{AB}, (R_1^{BA}) linking the evolution of the magnetizations $\langle I_z^A \rangle$ and $\langle I_z^B \rangle$ are given as (see Section IV.2):

$$R_1^{AB} = \left(\frac{\mu_0}{4\pi}\gamma_I^2\hbar\right)^2 I(I+1)\frac{1}{3}\left[-J_0^{AB}(\omega_A - \omega_B) + 6J_2^{AB}(\omega_A + \omega_B)\right] \tag{7.10}$$

Corresponding expressions for the relaxation rates R_1^{BB} and R_1^{BA} can be obtained by exchanging the indices A and B.

To evaluate the contributions to the fluorine spin relaxation, $R_{1(A\to La)}^{AA}$ and $R_{1(B\to La)}^{BB}$, resulting from the fluorine-lanthanum dipole-dipole couplings it is necessary to apply the above presented treatment. For example, the relaxation rate, $R_{1(A\to La)}^{AA}$, for the F_A fluorine spins should be evaluated from the formula [1]:

$$R_{1(A\to La)}^{AA} = 2\,\text{Re}\!\left[K_{1,1}^{DD}(-\omega_{F_A})\right] = \\ \frac{15}{2}\left(\frac{\mu_0 \hbar \gamma_F \gamma_{La}}{4\pi}\right)^2 (2S+1)\sum_{q,p}\begin{pmatrix}2 & 1 & 1\\ 1-q & q & -1\end{pmatrix}\!\begin{pmatrix}2 & 1 & 1\\ 1-p & p & -1\end{pmatrix}\Lambda_{F_A-La}^{1-q,1-p}\sigma_{La}^{qp}(-\omega_{F_A}) \tag{7.11}$$

where the spectral densities σ_S^{qp} are given as:

$$\sigma_S^{qp}(-\omega_{F_A}) = \frac{1}{2S+1}\int_0^\infty Tr_S\left\{S_q^{1(L)+}\exp\!\left[-\!\left(\left(i\hat{L}_S + \frac{1}{\tau_{ALa}} + i\omega_{F_A}\right)\tau\right)\right]S_p^{1(L)}\right\}\exp(-i\omega_{F_A}\tau)d\tau \tag{7.11a}$$

Since the lanthanum ions (La^{3+}) are fixed in their sublattice, the correlation time τ_{IS} introduced in Eq.7.4 is determined only by the motion of the fluorine ions and given by $\tau_{ALa}^{-1} = \tau_{AA}^{-1} + \tau_{AB}^{-1}$ and $\tau_{BLa}^{-1} = \tau_{BB}^{-1} + \tau_{BA}^{-1}$, for the fluorine spins belonging to the F_A and F_B sublattices, respectively. As predicted by Solomon, the evolution of the fluorine magnetization is biexponential. The two relaxation rates are given by Eq.4.8a,b. Now, one can rewrite them using the coefficients occurring in Eq.7.8a,b: $R_{1I}^\pm = -\frac{1}{2}\left[(a_{AA}+a_{BB})\pm\sqrt{(a_{AA}-a_{BB})^2+4a_{AB}a_{BA}}\right]$.

Several examples of experimental data illustrating the presented treatment, applied to LaF$_3$ are given in [1,14]. They correspond to the slower relaxation process. The quadrupolar spin relaxation (^{139}La) in lanthanum fluoride is slow compared to the timescale of the fluorine jump diffusion and can be neglected altogether. I shall come back to this statement later in more detail. It simplifies the analysis a lot and gives us the opportunity to observe the 'pure' effect of the energy level structure of the quadrupolar spins affecting relaxation of dipolar spins. Fig.7.7 shows two examples more.

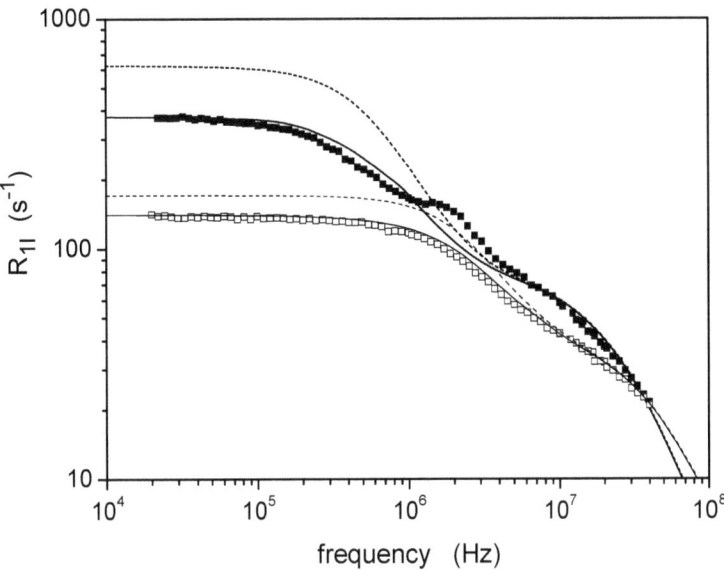

Fig.7.7

Experimental fluorine relaxation profiles (^{19}F) collected for single crystal of LaF$_3$ containing 0.01% Sr^{2+} admixtures; T=750K – solid squares, T=850K – open squares. Together with the experimental results the curves obtained as least-square fits of the theoretical model are presented. It has been obtained:
T=750K - $\tau_{AA} = 4.8*10^{-9}s, \tau_{AB} = 5.2*10^{-7}s, \tau_{BB} = 4.0*10^{-6}s$;
T=850K - $\tau_{AA} = 2.5*10^{-9}s, \tau_{AB} = 2.5*10^{-7}s, \tau_{BB} = 7.0*10^{-7}s$. The static quadrupolar coupling of the lanthanum spins (^{139}La) in the LaF$_3$ crystal lattice is $a_Q^S = 15 MHz$, while the anisotropy parameter is $\eta_S = 0.8$; the orientation of the electric field gradient tensor with respect to the c-axis (along which the magnetic field has been applied) is $(0,54°,0)$ [20,21].
To elucidate better the effect of the static quadrupolar coupling corresponding curves obtained by using the formula of Eq.7.8b, adapted to the lanthanum spins for the terms $R_{1(\alpha \to La)}^{\alpha\alpha}$ are presented. It is apparent from this figure that treatments based on this formula lead to misinterpretations of the data [18].

In the frequency range 10^6-$5*10^6$Hz, some of the relaxation profiles exhibit local maxima, caused by polarization transfer effects [14, 22-25] between fluorine (^{19}F) and lanthanum spins (^{139}La). Fluorine polarization can be taken over, under certain conditions, by lanthanum spins. In the experiment, this effect is detected as a faster decay of the fluorine magnetization and interpreted as a local increasing (maximum) of the fluorine relaxation rate. The fluorine polarization can be transferred to lanthanum when the transition energy of fluorine spins (determined by their Zeeman interaction) is equal to some transition energies of lanthanum spins (determined by the lanthanum Zeeman- and quadrupolar interactions) and if there is an efficient static I-S dipole-dipole coupling. The requirement that the mutual dipole-dipole interaction must be sensed by the participating spins as time independent means that the motional modulations of the relevant dipole-dipole coupling must be significantly slower than the fluorine spin relaxation. Chapter VIII is devoted to a detailed discussion of polarization transfer effects.

VII.3. Comments on dipolar contribution to the quadrupolar spin relaxation

It is of some interest to put more attention on the relaxation processes of the quadrupolar spins. Besides the relaxation resulting from the fluctuations of the electric field gradient tensor, the quadrupolar spin can also relax by the mutual $I-S$ dipole-dipole coupling or by dipolar couplings to other spins from its environment, denoted as I'. In fact, the dipolar contribution to the S spin relaxation can be relevant for various spin system containing quadrupolar and dipolar spins and, therefore, it requires more detailed explanations [14, 26, 27]. Lanthanum fluoride can be treated as a 'model' system for illustrating the quite complicated relaxation dynamics of the quadrupolar spins. Describing the relaxation of the F_A fluorine spins caused by their coupling to the lanthanum spins, $R_{A \to La}$, one should consider carefully the possible contributions to the lanthanum spin relaxation resulting from the $F_A - La$ and $F_B - La$ dipole-dipole couplings, mediated by the exchange motion of the fluorine spins (characterized according to our notation by the correlation times τ_{IS} and $\tau_{I'S}$, respectively). Fig.7.8 illustrates the relaxation pathways as well as some important points of the discussion below. The effective modulation of the $I-S$ dipole-dipole coupling (described by the correlation time τ_c) can result from the motion of the spins I and S, τ_{IS}, the quadrupolar relaxation mechanism of the spin S, and the dipolar contributions of the I and I' spins to the S spin relaxation: $\tau_c^{-1} = \tau_{IS}^{-1} + R_S^Q + R_S^{DD(I,S)} + R_S^{DD(I',S)}$.

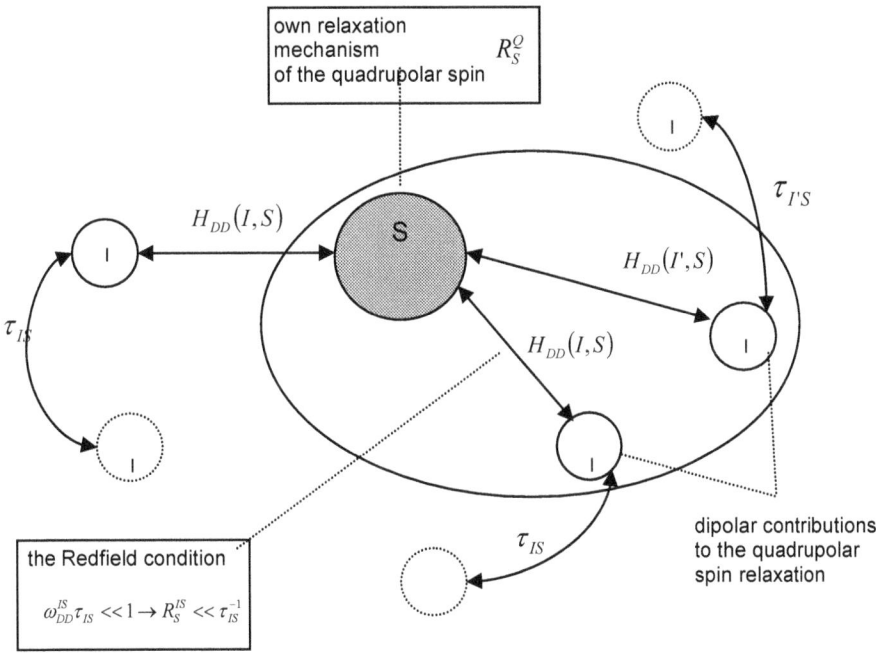

Fig.7.8

Schematic view of the effects of neighboring dipolar spins on the relaxation of the quadrupolar spin.

One of the basic assumptions of the present treatment is that the predominant relaxation mechanism of the quadrupolar spin is provided by the fluctuations of the electric field gradient. The formulas of Eq.5.5 and Eq.5.6 have been derived under the assumption that the dynamics of the quadrupolar spin, which is relevant for the I spin relaxation is independent of the presence of the dipolar spins, *i.e.* it is not affected by the couplings of the S spin to any other spins. If the dipolar contributions, $R_S^{DD(I,S)}$ and $R_S^{DD(I',S)}$, are non-negligible compared to the two other dynamic processes (τ_{IS}^{-1} and R_S^Q) the treatment presented in this chapter is not complete and cannot be applied to the interpretation of the I spin relaxation. The situation is particularly simple if the quadrupolar spin relaxation is much slower than the motion characterized by the correlation time τ_{IS}; in this case one can neglect it altogether as a source of the modulations of the $I-S$ dipole-dipole interaction. This is actually the case of lanthanum fluoride. To compare the timescale of the quadrupolar spin relaxation with the correlation time τ_{IS} and solve the problem in a straightforward manner, some independent information on the τ_{IS} dynamics must be available. For example, diffusion or conductivity measurements in ionic crystals (like LaF_3) can be considered as a source of information on the timescale of the ion jump diffusion [28]. It is important to realize that for the purpose of investigating independently the motional processes one should use experimental methods, the results of which are not affected by the quadrupolar spin relaxation. It can be useful and valuable to perform a complimentary NMR lineshape analysis. For lanthanum fluoride, the fluorine lineshape has been analyzed [29]. Being aware of the timescale of the quadrupolar spin relaxation, one can conclude whether the experimental lines indicate significantly faster fluctuations of the dipole-dipole couplings or not. Nevertheless, it can turn out that we do not have any supporting information and we have to critically review the applicability of the presented description only on the relaxation basis. In such a situation, it is advisable to consider carefully whether the fluctuations of the electric field gradient tensor can provide the dominating relaxation mechanism. However, we are aware of the fact that it can be difficult to give well motivated statements on this subject. Certainly, there is no straightforward way to separate the quadrupolar and dipolar contributions. Therefore, if there is no justification that the relaxation is caused by the fluctuations of the electric field gradient (due to, for example, lattice vibrations) it is reasonable to presume the less favorable case and attribute the total relaxation to the dipolar channel. Knowing the structure of the system under interest (it is necessary for calculating the strengths of the relevant dipole-dipole interactions), one can conclude whether it is possible to obtain the proper order of the quadrupolar relaxation assuming that the correlation times τ_{IS} and $\tau_{I'S}$ are much shorter. Nevertheless, the calculations are quite demanding from the theoretical point of view. It has been mentioned at the end of Chapter VI that despite the multiexponentiality of the quadrupolar spin relaxation, the combined effect of the averaged quadrupolar interaction and the Zeeman coupling on the energy level structure makes the description of the dipolar relaxation of the S spin even more complicated than for the spin I. I shall devote to this quite important aspect of relaxation a part of Chapter XI.

Let us neglect, for the time being, the spins I'. If it turns out that is possible to obtain the quadrupolar spin relaxation employing correlation time τ_{IS} within the Redfield limit, $\omega_{DD}^{IS} \tau_{IS} \ll 1$, we have arguments for using the presented approach (Eq.7.5 and Eq.7.6) excluding the quadrupolar spin relaxation can be applied, as it has been explained above, since $\tau_{IS}^{-1} \gg R_S^{DD(I,S)}$. Of course it does not guarantee that we will be able to interpret in this way the I spin relaxation profile. If the quadrupolar spin relaxation is too fast to be explained by assuming the correlation time within the limit $\omega_{DD}^{IS} \tau_{IS} \ll 1$, one has to take it into account. To preserve the validity of this description one has to attribute the relaxation to the fluctuations of the electric field gradient tensor, keeping in mind the restrictions for this relaxation channel: $\omega_Q^T \tau_Q \ll 1$ and $\omega_Q^S \gg \omega_Q^T \cdot \omega_Q^T \tau_Q$. It can also happen that the predominating relaxation mechanism for the quadrupolar spins is provided by the $S-I'$ dipole-dipole coupling. This statement can be well motivated, if we have at our disposal some independent experimental results, or based only on the calculations. Nevertheless, in such a case one cannot describe the I spin relaxation restricting the considerations only to the $I-S$ spin system. We shall deal with this problem in Chapter XI.

As long as the participating spins fulfill conditions of the Redfield relaxation theory, one is able to work out their relaxation dynamics undertaking the necessary computational effort. However, if the perturbation treatment breaks down for a part of a composite system, there is only one way to figure out its evolution: one must provide an appropriate solution of the Liouville equation. A general and elegant computational method, based on the superoperator formalism, suitable for any motional regime will be presented in Chapter X.

References:

1. D. Kruk, O. Lips, Field-dependent nuclear relaxation of spins ½ induced by dipole –dipole couplings to quadrupolar spins: LaF$_3$ crystals as an example, *J. Magn. Reson.* **179** (2006) 250 – 262
2. D. Kruk, T. Nilsson, J. Kowalewski, Outer-sphere nuclear spin relaxation in paramagnetic systems: a low field theory, *Mol. Phys.* **99** (2001) 1435-1445
3. G.A. Jaroszkiewicz, J.H. Strange, Motion on inequivalent lattice sites – NMR theory and application to LaF$_3$, *J. Phys. C: Solid State Phys.* **18** (1985) 2331-2349
4. A.F. Aalders, A.F.M. Arts, H.V. de Wijn, Vacancy distribution and ionic motion in LaF$_3$ studied by 19F NMR, *Phys. Rev. B* **32** (1985) 5412-5423
5. R. Kimmich, NMR - Tomography, Diffusometry, Relaxometry, *Springer – Verlag Berlin* (1997)
6. I. Solomon, Relaxation processes in a system of two spins, *Phys. Rev.* **99** (1955) 559-565
7. L.G. Werbelow, Relaxation theory for quadrupolar nuclei, in Encyclopedia of Nuclear Magnetic Resonance, D.M. Grant, R.K. Harris (Eds.) *Chichester; Willey* (1996) 4092-4101
8. P.P. Man, Quadrupolar interactions, in Encyclopedia of Nuclear Magnetic Resonance, D.M. Grant, R.K. Harris (Eds.) *Chichester; Willey* (1996) 3838-3847
9. B. Halle, H. Wennerstrom, Nearly exponential quadrupolar relaxation. A perturbation treatment, *J. Magn. Reson.* **44** (1981) 89-100
10. S. Wimperis, Relaxation of quadrupolar nuclei measured via multiple-quantum filtration, in Encyclopedia of Nuclear Magnetic Resoanace, D.M. Grant, R.K. Harris (Eds.) *Chichester; Willey* (1996) 4078-4084
11. J.R.C. Van der Maarel, Thermal relaxation and coherence dynamics of spin 3/2. I. Static and fluctuating quadrupolar interactions in the multipole basis, *Concepts Magn. Reson. Part A*, **19A** (2003) 97-116
12. G. Jaccard, S. Wimperis, G. Bodenhausen, Multiple-quantum NMR spectroscopy of S=3/2 spins in isotropic phase: a new probe for multiexponential relaxation, *J. Chem.Phys.* **85** (1986) 6282-6293

13. W.R. Carper, Direct determination of quadrupolar and dipolar NMR correlation times from spin-lattice and spin-spin relaxation rates, *Concepts Magn. Reson.* **11 (1)** (1999) 51-60

14. D. Kruk, O. Lips, Evolution of solid state systems containing mutually coupled dipolar and quadrupolar spins: a perturbation treatment, *Solid State Nucl. Magn. Reson.* **28 (2-4)** (2005) 180-192

15. B. Maximov, H. Schulz, Space group, crystal structure and twinning of lanthanum trifluoride, *Acta Cryst. B* **41** (1985) 88-91

16. A.F. Privalov, H.-M. Vieth, I.V. Murin, Nuclear magnetic resonance study of superionic conductors studied by ^{19}F NMR with homonuclear decoupling, *J. Phys. Chem. Solids* **50** (1989) 395-398

17. A.F. Privalov, H.-M. Vieth, I.V. Murin, Nuclear magnetic resonance study of superionic conductors with tysonite structure, *J. Phys.: Condens. Matter* **6** (1994) 8237-8243

18. A.F. Privalov, O. Lips, F. Fujara, Dynamic processes in the superionic conductor LaF_3 at high temperature as studied by spin-lattice relaxation dispersion, *J. Phys.: Condens. Matter* **14** (2002) 4515 – 4525

19. F. Wang, C.P. Grey, Probing the Mechanism of Fluoride – Ion conduction in LaF_3 and strontium-doped LaF_3 with high- resolution ^{19}F MAS NMR, *Chem. Mater* **9** (1997) 1068-1070

20. K. Lee, A. Sher, L.O. Andersson, W.G. Proctor, Temperature variation of La^{139} nuclear quadrupolar resonance in LaF_3, *Phys. Rev. 150 (1)* 1966 168-174

21. M. Matsushita, A. Mutoh, T. Kato, Coherent Raman spectroscopy of nuclear quadrupolar resonance of La around Pr^{3+} in LaF_3, *Phys. Rev. B* **58 (21)** (1998) 14372-14382

22. R. Kimmich, F. Winter, W. Nusser, K-H. Spohn, Interactions and fluctuations deduced from proton field-cycling relaxation spectroscopy of polypeptides, DNA, muscles, and algae, *J. Magn. Reson.* **68** (1986) 263-282

23. D. Kruk, J. Altmann, F. Fujara, A. Gadke, M. Nolte, A.F. Privalov, Analysis of ^{1}H-^{14}N polarization transfer experiments in molecular crystals, *J. Physc. Condensed Matter* **17 (3)** (2005) 519-533

24. F. Fujara, H-J. Stöckmann, H. Ackermann, W. Buttler, K. Dörr, H. Grupp, P. Heitjans, G. Kiese, A.J. Körblein, Cross-relaxation processes of polarized – β active nuclei in various crystalline solids, *Z. Phys. B* **37** (1980) 151-161

25. E. Jäger, B. Ittermann, G. Sultzer, K. Bürkermann, B. Fisher, H-P. Frank, H-J. Stöckmann, H. Ackermann, Cross relaxation of ^{12}B in single-crystal aluminium, *Z. Phys. B* **80** (1990) 87-94

26. L.G. Werbelow, J. Kowalewski, Nuclear spin relaxation of spin one-half nuclei in the presence of neighboring higher spin nuclei, *J. Chem.Phys.* **107** (1997) 2775-2786

27. L.G. Werbelow, G.A: Morris, P. Kumar, J. Kowalewski, Cross-correlated quadrupolar spin relaxation and carbon-13 lineshapes in the ^{13}CD$_2$ spin grouping, *J. Magn. Reson.* **140** (1999) 1-8

28. V.V. Sinitsyn, O. Lips, A.F. Privalov, F. Fujara, I.V. Murin, Transport properties of LaF$_3$ fast ionic conductor studied by field gradient NMR and impedance spectroscopy, *J. Phys. Chem. Solids* **64** (2003) 1201-1205

29. D. Kruk, O. Lips, P. Gumann, A. Privalov, F. Fujara, Dynamics of fluorine ions in LaF$_3$-type crystals investigated by NMR lineshape analysis, *J. Phys. Condens. Matter* **18** (2006) 1725-1741

CHAPTER VIII

Evolution of systems of mutually coupled spins under time independent interactions

Up to this point we have been concerned with several aspects of relaxation processes in rather complex systems containing electron, quadrupolar and dipolar spins. The considerations have been based on the Liouville - von Neumann equation (Eq.1.1a) with the Hamiltonian H including the 'pure' spin and lattice parts, H_I and H_L, respectively, and the coupling between them, represented by the interaction H_{IL}. The lattice has been regarded merely as a 'thermal reservoir' remaining at thermal equilibrium. The essence of the time evolution of the entire system can be captured by the statement that the energy from the spin subsystem is transferred immediately to the motional degrees of freedom of the lattice and does not return to the spins. In other words, the Liouville – von Neumann equation applied to this case describes an irreversible process. I have demonstrated how to solve the equation of motion through application of the second order perturbation theory, and discussed in detail validity regimes of this approach. In particular, it has been required for the lattice to provide a source of modulations for the H_{IL} coupling, fast enough to fulfil the condition $\omega_{IL}\tau_c \ll 1$. In this chapter we shall turn our attention to systems of mutually coupled dipolar and quadrupolar spins, which are not affected by any dynamic processes. So far, considering a two spin system, $I-S$, coupled by a dipole-dipole interaction, we have expected that the mutual dipole-dipole interaction is averaged out $\langle H_{DD}(I,S)\rangle = 0$ due to some motional processes occurring in the lattice. In this case the spins are mutually coupled by time independent interactions. Then we shall focus our attention on the essentially different mechanisms of the evolution of such systems and on the processes which can take place only under the condition that there is an efficient mutual coupling between the dipolar and quadrupolar spins.

VIII.1. Time evolution of spin observables

To describe the time evolution of systems of mutually coupled spins one has to set up an appropriate form of the Hamiltonian $H = H_{tot}$, which should cover all couplings relevant for the systems of interest. The entire density operator ρ_{tot} evolves in time following the Liouville – von Neumann equation based on this Hamiltonian (see Eq.2.1a):

$$\frac{d\rho_{tot}(t)}{dt} = -i[H_{tot}, \rho_{tot}(t)] \qquad (8.1)$$

I have introduced the index '*tot*' to point out that now one cannot divide the entire system of coupled spins into subsystems and establish a hierarchy of them, as has been done for some previous cases. One has to think about the dynamics of all the participating spins treating them as one system. This does not mean that we are not able to extract any information about the individual spins; I shall show how to follow the time evolution of observables associated with selected spins or groups of them. Nevertheless, the behaviour of the selected spins results from the state of the entire system, represented by the entire density operator $\rho_{tot}(t)$. From now on I shall omit the index '*tot*' to simplify the notation; however the forthcoming calculations concern the total spin system. Since all the interactions included in the Hamiltonian H are time independent, the solution of the Liouville – von Neumann equation can be obtained as [1-8]:

$$\rho(t) = \exp(-iHt)\rho(0)\exp(iHt) \qquad (8.2)$$

where the time point at the start of evolution has been denoted as $t=0$, and the density operator at that instant as $\rho(0)$ - the initial density operator. On this basis one can set up the explicit expression for the time evolution of the individual density matrix elements (coherences) ρ_{rs} between the eigenstates $|r\rangle$ and $|s\rangle$ of the Hamiltonian H:

$$\rho_{rs}(t) = \langle r|\rho(t)|s\rangle = \langle r|\rho(0)|s\rangle \exp\{-i\omega_{rs} t\} \quad (8.3)$$

The coherence ρ_{rs} oscillates at a frequency equal to minus the difference in the eigenvalues between the states, $\omega_{rs} = \omega_r - \omega_s$, whereas the populations ρ_{rr} remain stationary. The elements of the initial density operator $\rho_{rs}(0)$ represent an initial mixture of coherences. Then, the expectation value $\langle O(t)\rangle$ of an observable O can be obtained from the expression:

$$\langle O(t)\rangle = Tr\{\rho(t)O\} = \sum_{r,s=1}^{N}\langle r|\rho(t)|s\rangle\langle s|O|r\rangle = \sum_{r,s=1}^{N} a_{rs}\exp(-i\omega_{rs}t) \quad (8.4)$$

where N denotes the number of the eigenstates of the spin system. The amplitudes a_{rs} are given by a product of the corresponding matrix elements of the observable and of the initial density operator, both expressed in the eigenbasis of the Hamiltonian H:

$$a_{rs} = \langle r|\rho(0)|s\rangle\langle s|O|r\rangle \quad (8.5)$$

On the basis of Eq.8.4 and Eq.8.5 one can calculate the expectation value of an arbitrary observable O for any spin system characterised by a time independent Hamiltonian.

VIII.2. Polarization transfer processes within a two spin system

The general idea of this book is to illustrate the discussed problems starting from simplest examples and gradually increase the complexity of the considered spin systems, and in this way introducing some new relevant elements. In this chapter, I shall follow this method of presentation and begin with a quite detailed description of a two spin system containing a dipolar spin $I = \frac{1}{2}$, coupled by a dipole-dipole interaction to a quadrupolar spin $S \geq 1$. As pointed out in the previous section, it is essential for the whole development to set up a proper form of the interaction Hamiltonian H. Actually, our system evolves in time under the Hamiltonian H composed of the Zeeman interactions of the participating spins ($H_Z(I)$ and $H_Z(S)$ for the spins I and S, respectively) and internal spin interactions, *i.e.* the quadrupolar coupling of the spins S, $H_Q^0(S)$, and the mutual I-S dipole-dipole coupling, $H_{DD}^0(I,S)$:

$$H = H_Z(I) + H_Z(S) + H_Q^0(S) + H_{DD}^0(I,S) \tag{8.6}$$

The index '0' denotes explicitly that the interactions do not exhibit any time fluctuations (I do not like to use here the terminology 'static'; it could be confusing All interactions involved in the evolution of the spin system are now time-independent with respect to the molecular as well as the laboratory frames). The dipole - dipole interaction provides the coupling between the spins I and S (the Zeeman as well as the quadrupolar Hamiltonians represent interactions involving one spin). If this coupling is weak there is no communication between the two spins and they evolve in time independently of each other. Shortly I shall comment on the role of the dipole-dipole coupling in more detail. Dealing with several interactions, we must be always aware that all contributions to the total Hamiltonian have to be considered in the same reference frame. Representing the quadrupolar and dipole-dipole Hamiltonians in the laboratory frame (see Chapter III) we have to deal with two sets of Euler angles: the first one, Ω_{PL}, describes the orientation of the electric field gradient tensor at the position of the quadrupolar spin with respect to the laboratory frame, while the second one, Ω_{DDL}, gives us the orientation of the $I-S$ dipole-dipole axis. Considering the relaxation processes we have assumed for simplicity that $\Omega_{PL} = \Omega_{DDL}$ (*i.e.* that the (P) and (DD) frames coincide). This issue is of rather minor importance for the relaxation theories; if the frames do not coincide the calculations are just slightly more cumbersome [9, 10]. However, to describe properly the polarization transfer effects one cannot employ simplifying assumptions of this type. From now on, we will treat the principal axis system of the electric field gradient as the molecular frame and denote it as (M), that is: $\Omega_{PL} \equiv \Omega_{ML}$. In fact, the relationship between the two sets of angles is determined by the geometry of the molecule carrying the participating spins or by the crystal structure. Therefore, the laboratory representation of the dipole-dipole coupling can be obtained through the two steps transformations [9-14]:

$$F_{-m}^{2(L)} = \sum_{n=-2}^{2} \sum_{k=-2}^{2} F_{k}^{2(DD)} D_{k,n}^{2}(\Omega_{DDM}) D_{n,-m}^{2}(\Omega_{ML}) \tag{8.7}$$

This way of dealing with several interactions with different principal axis systems is very profitable and useful when applied to systems containing more than few spins. First, all internal interactions (quadrupolar- and dipole-dipole couplings) which are relevant for the considered spin system are 'grouped' in the molecular frame by appropriate transformations. The molecular frame representations of the individual internal interactions are independent of the molecular orientation; they are determined by the internal molecular geometry. Next, the bunch of the internal spin interactions undergoes a common transformation from the molecular axes system to the laboratory frame via the angle Ω_{ML} (which describes the molecular orientation). I shall provide further examples of this approach in the next section.

The time evolution of the system depends on its initial state described by the operator $\rho(0)$ (Eq.8.3). The initial state is established by an initial Hamiltonian H^{in} determined by the conditions at the time $t=0$. It is very important to realise that the initial conditions can be created by the experimentalist and the Hamiltonian H^{in} does not have to be equal to the Hamiltonian H, which is responsible for the further time evolution of the system [7,8]. To illustrate this statement let us consider the case of a high magnetic field applied for time long enough to polarize the dipolar as well as quadrupolar spins. The initial Hamiltonian contains in principle also four terms: the Zeeman couplings for the spins I and S, the quadrupolar coupling and the dipole-dipole interaction:

$$H^{in} = H_Z^{in}(I) + H_Z^{in}(S) + H_Q^0(S) + H_{DD}^0(I,S) \tag{8.8}$$

However, if the applied magnetic field is high enough, the Zeeman part of the Hamiltonian determines the initial state of the $I-S$ spin system and, in consequence $H^{in} = H_Z^{in}(I) + H_Z^{in}(S)$. In the limit of high temperature approximation, which is usually fulfilled at room temperature, the initial density operator is proportional to the linear term in the Taylor series expansion of the Boltzmann factor: $\exp\left(-\frac{H_{in}}{k_B T}\right) \cong 1 - \frac{H_{in}}{k_B T}$ [1]. The initial density operator has in this case a particularly simple form, it is just proportional to the combinations of the I_z and S_z operators weighted by the corresponding gyromagnetic factors γ_I and γ_S:

$$\rho(0) = \frac{1}{(2I+1)(2S+1)} \exp\left(-\frac{H_z^{in}(I) + H_z^{in}(S)}{k_B T}\right) \propto I_z + \frac{\gamma_S}{\gamma_I} S_z \qquad (8.9)$$

If the initial density operator is affected by other interaction, for example the quadrupolar coupling, it can be easily included to the formulation of Eq.8.8, however one must remember that all interactions relevant for the initial state of the system have to be expressed in the same reference frame. When the system reaches the required initial state (the full polarization in this case) the conditions can be changed, for example the magnetic field can be reduced. This means that the system evolves in time under the Hamiltonian H describing the new conditions (which can be completely different from the initial ones), and we ask about the expected values of observables after time t, knowing and taking into account that the evolution has started from the initial state generated by the Hamiltonian H^{in}.

The next question is how to describe the experimental results in terms of observables O. Generally this problem is difficult, but for some cases it is easy to answer this question. If, for example, the longitudinal magnetization of the dipolar spins is detected in an experiment, the observable O corresponds to the I_z operator.

To calculate explicitly the amplitudes a_{rs}, which tell us how the evolution of the density matrix element $\rho_{rs}(t)$ influences the observable, one needs to obtain the eigenstates and the corresponding eigenvalues (energy levels) for the entire system. It can be achieved, as usual, by setting up and diagonalizing a matrix representation of the Hamiltonian H in the Zeeman basis, $\{|n\rangle = |m_S, m_I\rangle\}$ (where m_S and m_I are the magnetic quantum numbers for the spins S and I, respectively). Using the representation of the eigenfunctions $|\psi_r\rangle$ of the

Hamiltonian H for a given molecular orientation Ω_{ML} in terms of the Zeeman functions $|n\rangle$, $|\psi_r\rangle = \sum_{n=1}^{N} c_{rn}(\Omega_{ML})|n\rangle$, one can write down the expression for the coefficient $a_{rs}(\Omega_{ML})$:

$$a_{rs}(\Omega_{ML}) \propto \sum_{n=1}^{N} \left\{ \langle n|I_z|n\rangle + \frac{\gamma_S}{\gamma_I}\langle n|S_z|n\rangle \right\} (c_{rn}^* c_{sn})(\Omega_{ML}) \times \sum_{n=1}^{N} \langle n|I_z|n\rangle (c_{rn}^* c_{sn})(\Omega_{ML}) \quad (8.10)$$

Thus, the magnetization generated by the dipolar spins I belonging to a molecule or a crystal oriented with respect to the laboratory frame by the angle Ω_{ML} is given by the expression [8, 9]:

$$\langle I_z(t)\rangle(\Omega_{ML}) \propto \sum_{r,s=1}^{N} \left(\sum_{n=1}^{N} \langle n|I_z|n\rangle (c_{rn}^* c_{sn}) \right)^2 \exp(-i\omega_{rs}t) + \\ \frac{\gamma_S}{\gamma_I} \sum_{r,s=1}^{N} \left\{ \left(\sum_{n=1}^{N} \langle n|I_z|n\rangle (c_{rn}^* c_{sn}) \right) \times \left(\sum_{n=1}^{N} \langle n|S_z|n\rangle (c_{rn}^* c_{sn}) \right) \right\} \exp(-i\omega_{rs}t) \quad (8.11)$$

It has been taken into account here that the operators I_z and S_z are diagonal in the Zeeman representation, and therefore one can set $n=m$. When the final magnetization of the dipolar spins results from molecules with different orientations in the sample, in the last step one needs to perform an averaging over all molecular orientations.

Since the Zeeman states with the magnetic quantum number m_S and $-m_S$ contribute to the eigenstates of the I-S system with the same 'weight' factors, the second term in Eq.8.11 (representing the explicit influence of the initial state of the spin S on the time evolution of the spin I) vanishes. I have assumed that the initial density operator does not contain any two-spin I-S terms. Of course, this is a natural consequence of the high magnetic field applied at the initial stage, it makes all other couplings negligible compared to the Zeeman ones. However, this does not mean that the spins S do not influence the evolution of the spins I. As indicated above, the coupling between the two spin subsystems is provided by the I-S dipole-dipole interaction, which influences the eigenvectors of the entire system. The role of the dipole-dipole coupling becomes apparent if one looks at the matrix representation of the Hamiltonian H for a dipolar spin – quadrupolar spin system of $S=1$:

$$[H] = \begin{bmatrix} & |1\rangle & |2\rangle & |3\rangle & |4\rangle & |5\rangle & |6\rangle \\ & \frac{1}{\sqrt{6}}\tilde{V}_0^2 + \frac{1}{\sqrt{6}}\tilde{F}_0^2 + \frac{1}{2}\omega_I + \omega_S & \frac{1}{\sqrt{2}}\tilde{V}_{-1}^2 + \frac{1}{2\sqrt{2}}\tilde{F}_{-1}^2 & \tilde{V}_{-2}^2 & \frac{1}{2}\tilde{F}_{-1}^2 & \frac{1}{\sqrt{2}}\tilde{F}_{-2}^2 & 0 \\ & & -\frac{2}{\sqrt{6}}\tilde{V}_0^2 + \frac{1}{2}\omega_I & -\frac{1}{\sqrt{2}}\tilde{V}_{-1}^2 + \frac{1}{2\sqrt{2}}\tilde{F}_{-1}^2 & -\frac{1}{2\sqrt{3}}\tilde{F}_0^2 & 0 & \frac{1}{\sqrt{2}}\tilde{F}_{-2}^2 \\ & & & \frac{1}{\sqrt{6}}\tilde{V}_0^2 - \frac{1}{\sqrt{6}}\tilde{F}_0^2 + \frac{1}{2}\omega_I - \omega_S & 0 & -\frac{1}{2\sqrt{3}}\tilde{F}_0^2 & -\frac{1}{2}\tilde{F}_{-1}^2 \\ & & & & \frac{1}{\sqrt{6}}\tilde{V}_0^2 - \frac{1}{\sqrt{6}}\tilde{F}_0^2 - \frac{1}{2}\omega_I + \omega_S & \frac{1}{\sqrt{2}}\tilde{V}_{-1}^2 - \frac{1}{2\sqrt{2}}\tilde{F}_{-1}^2 & \tilde{V}_{-2}^2 \\ & & & & & -\frac{2}{\sqrt{6}}\tilde{V}_0^2 - \frac{1}{2}\omega_I & -\frac{1}{\sqrt{2}}\tilde{V}_{-1}^2 + \frac{1}{2\sqrt{2}}\tilde{F}_{-1}^2 \\ & & & & & & \frac{1}{\sqrt{6}}\tilde{V}_0^2 + \frac{1}{\sqrt{6}}\tilde{F}_0^2 - \frac{1}{2}\omega_I - \omega_S \end{bmatrix}$$

(8.12)

The Zeeman functions $|m_I, m_S\rangle$ are here labelled as follows: $|1\rangle = \left|\frac{1}{2}, 1\right\rangle$, $|2\rangle = \left|\frac{1}{2}, 0\right\rangle$, $|3\rangle = \left|\frac{1}{2}, -1\right\rangle$, $|4\rangle = \left|-\frac{1}{2}, 1\right\rangle$, $|5\rangle = \left|-\frac{1}{2}, 0\right\rangle$ and $|6\rangle = \left|-\frac{1}{2}, -1\right\rangle$, while the quadrupolar and the dipole–dipole Hamiltonian are expressed, according to Chapter III, as:

$$H_Q^{0(L)} = \sum_{m=-2}^{2} (-1)^m \tilde{V}_{-m}^{2(L)} T_m^2(S) \qquad \text{and} \qquad H_{DD}^{0(L)}(I,S) = \sum_{m=-2}^{2} (-1)^m \tilde{F}_{-m}^{2(L)} T_m^2(I,S),$$

respectively. In Eq.8.12 I have simplified the notation omitting the index (L) for the spatial functions. The dipole-dipole part of the total Hamiltonian provides a coupling between the two sub-sets of the basis functions corresponding to the different values of the m_I quantum number, so it connects the states with $|\Delta m_I| = 1$. This coupling becomes especially efficient under certain conditions. If the magnetic field is set to a value which leads to the Zeeman splitting of the dipolar spin matching the energy splitting of the quadrupolar spin (determined by the quadrupolar and Zeeman interactions), the dipole-dipole coupling causes polarization transfer processes. The mutual dipole-dipole coupling links transitions of the dipolar spin to some transitions of the quadrupolar spin, so they cannot occur independently. However, the entire system conserves the total energy. The transitions of the dipolar spin between the states $m_I = \frac{1}{2}$ and $m_I = -\frac{1}{2}$ can occur only if the quadrupolar part of the system is ready to use the released energy for its transitions. The splitting of the dipolar spin may be equal only to the splitting of one particular pair of energy levels for the quadrupolar spin, affected in addition by the molecular orientation (since the energy level structure for the quadrupolar spin results from the superposition of the quadrupolar and Zeeman couplings). The condition of the equal energy splitting for both the subsystems can be formulated in terms of the energy level structure for the total system. It is equivalent to the statement that there is a crossing of two energy levels (eigenstates) E_r and E_s associated with the eigenvectors $|\psi_r\rangle$ and $|\psi_s\rangle$. Strictly speaking, the dipole-dipole coupling introduces, for symmetry reasons, an anti-crossing of the energy levels which are separated by a small energy difference determined by its amplitude. The polarization of the dipolar spins is transferred to the quadrupolar subsystem with an efficiency directly related to the probability of the joint transitions. The probability is determined by a square of the matrix element of the dipole-dipole coupling taken between the involved eigenstates $|\langle r|H_{DD}(I,S)|s\rangle|^2$.

To illustrate the considerations I choose a system containing a proton spin ($I = 1/2$) coupled to a nitrogen spin ^{14}N ($S = 1$). Fig.8.1a,b, Fig.8.2 and Fig.8.3 show several curves corresponding to the observable $\langle I_z \rangle$ (the proton magnetization) calculated versus the magnetic field determining the Zeeman coupling in the Hamiltonian H, under the assumption that initially both the participating spins have been fully polarized by a strong magnetic field. The magnetization dips reflect the effects of the dipolar spin polarization taken over by the quadrupolar spin under the above discussed conditions. After the polarization the magnetic field has been switched down for a certain time and varied for searching for the ^1H-^{14}N polarization transfer effects. The remaining proton magnetization has been recorded at the proton resonance frequency after a 90^0 pulse [7].

a)

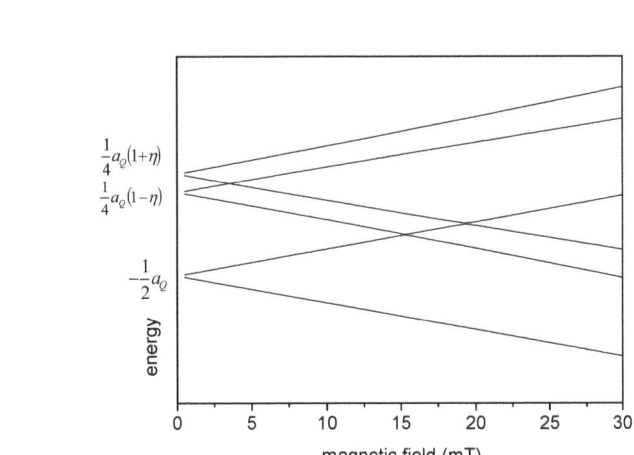

b)

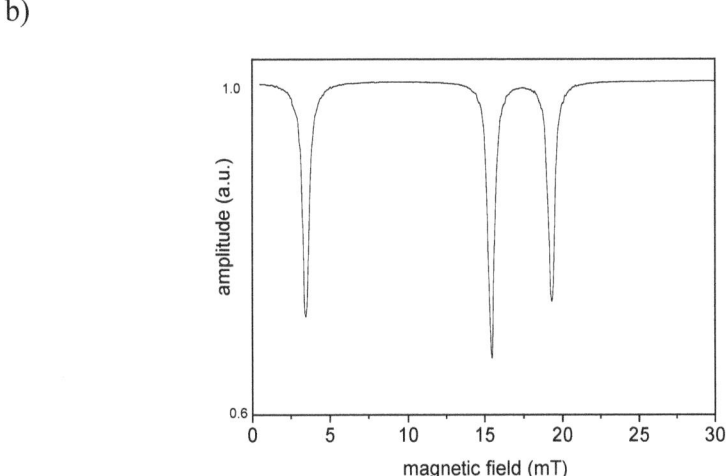

Fig.8.1
a) Energy levels for ^1H-^{14}N spin system versus the magnetic field. The quadrupolar coupling parameters for nitrogen spin are $a_Q = 1MHz \, (\cong 3{,}3*10^{-5}\,cm^{-1})$, $\eta = 0.3$. It has been assumed that the principal axis system of the electric field coincides with the direction of the applied magnetic field.

b) *Proton magnetization (the $\langle I_z(t)\rangle$ quantity) versus the magnetic field. The interspin distance is 250 pm. The curve is obtained as a result of powder averaging. The positions of the polarization transfer dips correspond to the energy level crossings.*

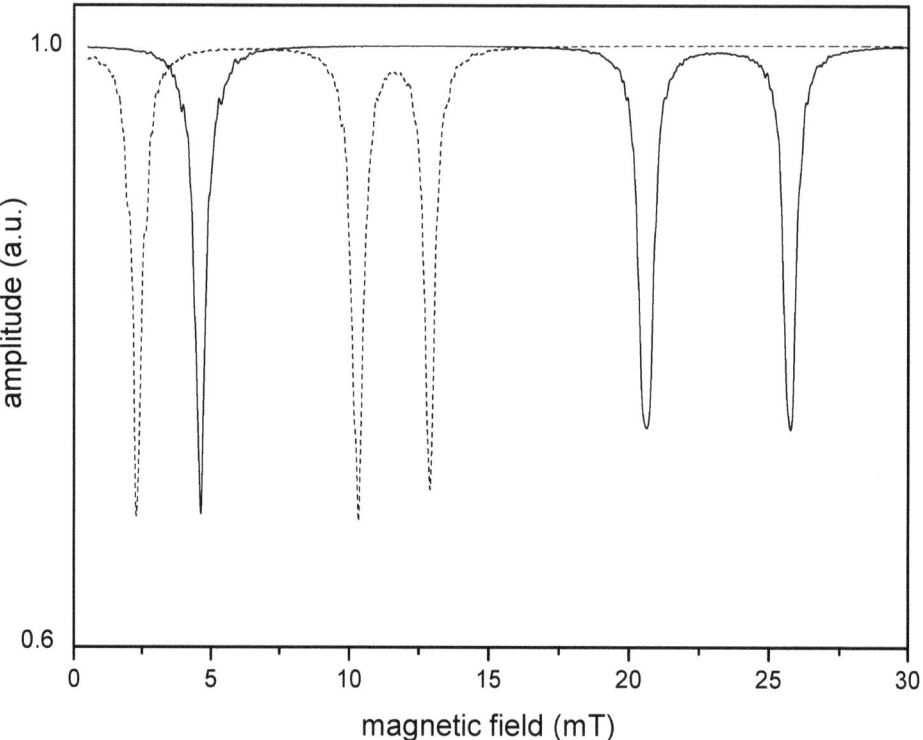

Fig.8.2
Polarization transfer effects for ^1H-^{14}N spin system for different quadrupolar coupling constants. Solid line corresponds to $a_Q = 1.3 MHz$ ($\cong 4.4*10^{-5} cm^{-1}$), dashed line to $a_Q = 0.65 MHz$ ($\cong 2.2*10^{-5} cm^{-1}$). The asymmetry parameter has been set to $\eta = 0.3$, the interspin distance is 250pm, the principal axis system of the quadrupolar interaction coincides with the dipole-dipole axis. The curves result from averaging over molecular orientations.

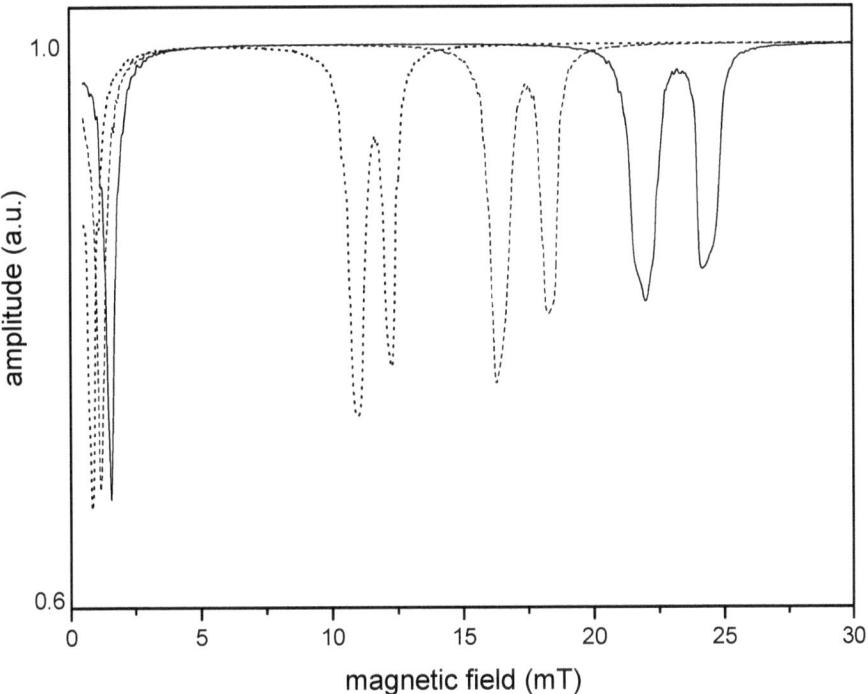

Fig.8.3
Polarization transfer effects for ^1H-^{14}N spin system for different values of the quadrupolar coupling constant. The asymmetry parameter is $\eta = 0.1$, the interspin distance is 250pm. It is interesting to compare this figure with Fig.8.1b and Fig.8.2. Solid line: $a_Q = 1.3 MHz$, dotted line $a_Q = 1.0 MHz$, dashed line $a_Q = 0.65 MHz$. The curves have been obtained by averaging over molecular orientations.

If the initial Zeeman couplings do not dominate over the quadrupolar interaction, the initial density operator takes a more complicated form:

$$\rho(0) = \frac{1}{Z}\exp\left(-\frac{H_Z^{in}(I) + H_Z^{in}(S) + H_Q^0(S)}{k_B T}\right) \propto I_z + \frac{\gamma_S}{\gamma_I}S_z + \frac{a_Q^0}{\gamma_I}\left(\frac{1}{a_Q^0}H_Q^0(S)\right) \quad (8.13)$$

The contribution of the quadrupolar interaction to the initial state of the $I-S$ system can influence considerably the magnetization of the dipolar spin.

If more dipolar spins contribute to the polarization transfer, their magnetization can be taken over not only through single quantum transitions $|\Delta m_I|=1$, but also through the multi-quantum channels. The next section is devoted to this issue.

VIII.3. Multi-quantum polarization transfer effects in systems containing several dipolar and quadrupolar spins

In this section, I shall extend the considerations to systems containing N_I dipolar spins $I=\frac{1}{2}$, coupled through dipole-dipole interactions to N_S quadrupolar spins [8,9]. The approach presented in this chapter is very flexible. On this theoretical basis one can describe time evolution of any observable relevant for the system of interest, assuming arbitrary initial conditions. Presenting appropriate examples, I aim to show how this approach can be adapted in a straightforward manner to more complicated and general problems, like multi-quantum processes occurring in systems containing several mutually coupled spins. In this case, the Hamiltonian H contains, beside the Zeeman and qudrupole couplings, three types of the dipole-dipole interactions: H_{DD}^{II}, H_{DD}^{SS} and H_{DD}^{IS} between the $I_{i_1}-I_{i_2}$, $S_{j_1}-S_{j_2}$ and I_i-S_j pairs of the spins, respectively:

$$H = \sum_{i=1}^{N_I} H_z(I_i) + \sum_{j=1}^{N_S} H_z(S_j) + \sum_{j=1}^{N_S} H_Q^0(S_j) + \\ \sum_{i_1,i_2=1, i_2<i_1}^{N_I} H_{DD}^0(I_{i_1}, I_{i_2}) + \sum_{i=1}^{N_I}\sum_{j=1}^{N_S} H_{DD}^0(I_i, S_j) + \sum_{j_1,j_2=1, j_2<j_1}^{N_S} H_{DD}^0(S_{j_1}, S_{j_2})$$

(8.13)

At this point some caution must be exercised regarding the proper representation of the individual interactions. In the previous section we have chosen as the molecular frame the principal axis system of the electric field gradient tensor (P) at the position of the spin S. If one wishes to follow this approach dealing with systems containing more spins, one has to select one quadrupolar spin (let us denote it as S_1) and associate the molecular frame (M) with the principal axes of the quadrupolar coupling $H_Q^S(S_1)$: $(P_1) \equiv (M)$. The laboratory representations of the other quadrupolar interactions can be also obtained by the two-step transformations, in analogy to the procedure applied to the dipole-dipole coupling in Eq.8.7:

$$A_{-m}^{2(L)}(S_i) = \sum_{n=-2}^{2}\sum_{k=-2}^{2} A_k^{(P_i)} D_{k,n}^2(\Omega_{P_iM}) D_{n,-m}^2(\Omega_{ML}) \qquad (8.14)$$

The first set of the Wigner rotation matrices, $D_{k,n}^2(\Omega_{P_iM})$, gives the representation of the quadrupolar interaction for a spin S_i in the principal axes system of the selected spin S_1. The quadrupolar Hamiltonian is transformed through the second set of the Wigner rotation matrices $D_{n,m}^2(\Omega_{ML})$ from the molecular representation to the laboratory frame. The laboratory representations of the individual dipole-dipole couplings $H_{DD}^{IS}(I_i, S_j)$ can be obtained directly from Eq.8.7:

$$F_{-m}^{2(L)}(I_i, S_j) = \sum_{n=-2}^{2}\sum_{k=-2}^{2} F_k^{2(DD)} D_{k,n}^2(\Omega_{DD_{I_iS_j}M}) D_{n,-m}^2(\Omega_{ML}) \qquad (8.15)$$

with the specific angles $\Omega_{DD_{I_iS_j}M}$ describing the orientation of the $I_i - S_j$ dipole-dipole coupling relative to the molecular frame. Of course, the transformation of Eq.8.15 written explicitly for a pair of spins $I_i - S_j$ is fully applicable for the $I_i - I_j$ and $S_i - S_j$ dipole-dipole interactions. The transformations are presented graphically in Fig.8.4.

The density operator $\rho(0)$ describing the initial state of the entire spin systems is given by a superposition of the I_{iz} and S_{jz} operators:

$$\rho(0) = \frac{1}{Z}\exp\left(-\left(\sum_{i=1}^{N_{Ii}} H_Z(I_i) + \sum_{j=1}^{N_S} H_Z(S_j)\right)\Big/k_BT\right) \propto \sum_{i=1}^{N_I} I_{iz} + \frac{\gamma_S}{\gamma_I}\sum_{j=1}^{N_S} S_{jz} \qquad (8.16)$$

where the ensemble partition function is $Z = (2I+1)^{N_I}(2S+1)^{N_S}$. The expression of Eq.8.16 corresponds to the case when a high magnetic field is applied in order to polarize fully all spins of the considered system. I have assumed, in addition, that the dipolar spins, I_i, as well as the quadrupolar ones, S_j, are equivalent. In the present case, the longitudinal magnetization of the dipolar spins is represented by a superposition of the operators I_{iz}: $O = \sum_{i=1}^{N_I} I_{iz}$.

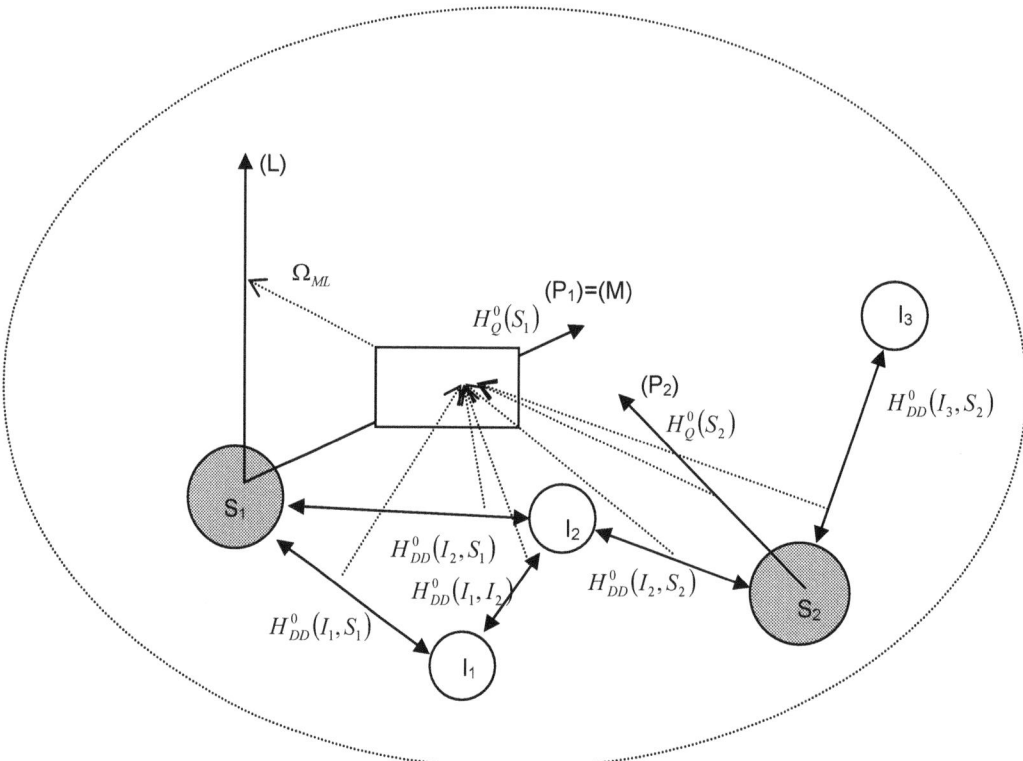

Fig.8.4

Schematic view of the two-step transformation in case of several spin interactions. First, all relevant interactions are represented in the principal axis system of the electric field gradient tensor at the position of the selected quadrupolar spin S_1, denoted as the molecular frame (M). These transformations are independent of the molecular orientation; they are determined by the internal geometry of the molecule. Next, all the interactions are transformed from the molecular frame to the laboratory axis system via the angle Ω_{ML} which describes the molecular orientation.

Further calculations require diagonalization of the matrix representation of the Hamiltonian H. The appropriate basis is now formed by the Zeeman functions $|n\rangle = |m_{S_1},...,m_{S_{N_S}}, m_{I_1},...,m_{I_{N_I}}\rangle$ where m_{S_j} and m_{I_i} are the magnetic quantum numbers for the quadrupolar spin S_j and the dipolar spin I_i, respectively. The number of the eigenstates for the entire system is $N = 2^{N_I}(2S+1)^{N_S}$. By replacing the operators I_z and S_z in Eq.8.10 and Eq.8.11 by the corresponding sums:

$$I_z \to \sum_{i=1}^{N_I} I_{iz}, \quad S_z \to \sum_{j=1}^{N_S} S_{jz},$$

one obtains the time evolution of the magnetization of the dipolar spins. More dipolar spins contributing to the polarization transfer implies that the quantum number $\Delta m_I = \sum_i \Delta m_{I_i}$ can change by more than one, due to simultaneous transitions of many spins. The probability of the simultaneous, multi-quantum transitions is smaller, so they provide relatively less efficient pathways for the polarization transfer. The dipole-dipole couplings between the I spins involved in this process play an important role in the multi-quantum transitions. A strong coupling between them increases the probability of the joint transitions. The order of the polarization transfer transitions is restricted by the maximum possible change of the quadrupolar spin quantum number, $1 \leq |\Delta m_S| \leq 2S+1$. Fig.8.5a,b shows examples of the I spin magnetization versus the magnetic field when the single as well as the double quantum pathways for the polarization transfer are involved. The calculations have been performed for a $I-S-I$ spin system containing two equivalent protons and one nitrogen spin. The Hamiltonian H on which the calculations have been based contains six terms, defined above:

$$H(I_1, I_2, S) = \qquad\qquad\qquad\qquad\qquad\qquad\qquad\qquad\qquad\qquad (8.17)$$
$$H_z(I_1) + H_z(I_2) + H_z(S) + H_Q^0(S) + H_{DD}^0(I_1, I_2) + H_{DD}^0(I_1, S) + H_{DD}^0(I_2, S)$$

If the mutual $I_1 - I_2$ dipole–dipole coupling is weak, the polarization of the dipolar spins is transferred to the quadrupolar one independently as a single-quantum process. In other words, in this case we rather deal with two independent $I-S$ spin systems, instead with the $I_1 - S - I_2$ one.

a)

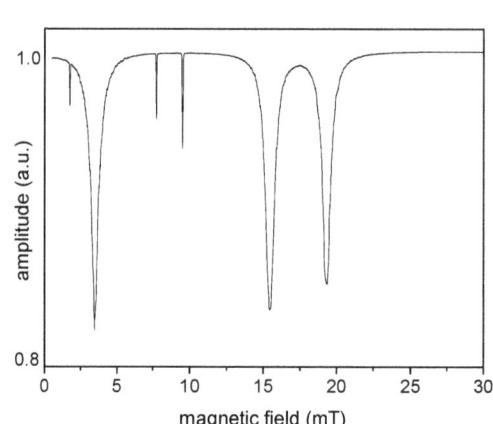

b)

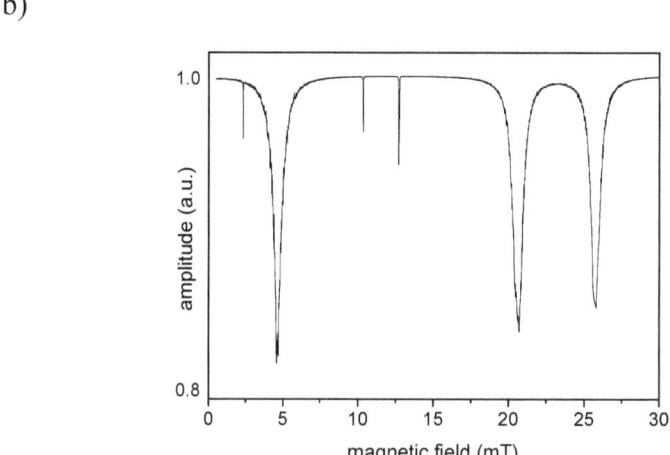

Fig.8.5

Polarization transfer effects for ^1H-^{14}N-^1H spin system for different values of the quadrupolar coupling constant:

a) $a_Q = 1.0 MHz$, $\eta = 0.3$ (it corresponds to Fig.8.1b)

b) $a_Q = 1.3 MHz$, $\eta = 0.3$ (it corresponds to Fig.8.2)

One can see single as well as double – quantum polarization transfer dips, predicted by the presented theory. The dipole-dipole interactions coincide with the principal axis system of the quadrupolar coupling. The curves result from averaging over molecular orientations. The proton spin – nitrogen spin distances are 250 pm.

A higher ratio $\frac{N_S}{N_I}$ between the quadrupolar and dipolar spins increases the effectiveness of the magnetization transfer because there is more quadrupolar spin with the capability to take over the dipolar spin polarization per one dipolar spin.

It is of some interest to analyse the polarization transfer processes in systems containing non-equivalent dipolar spins. The simplest system of this type is formed by two non-equivalent spins I_1 and I_2 coupled to a quadrupolar spin S. The already presented description can be straightforwardly adapted to this case. In particular, the initial density operator must be modified by taking into account different gyromagnetic factors of the dipolar spins:

$$\rho(0) \propto \gamma_{I_1} I_{1z} + \gamma_{I_2} I_{2z} + \gamma_S S_z \tag{8.18}$$

It should be understood from the form of the $\rho(0)$ operator that we have assumed that the participating spins have been completely polarized at the beginning. Eq.8.11 gives the expectation value of the I_z operator. Depending on the spin we are interested in, one can set: $I_z \equiv I_{1z}$, $I_z \equiv I_{2z}$ or $I_z \equiv \gamma_{I_1} I_{1z} + \gamma_{I_2} I_{2z}$.

Several experiments showing polarization transfer processes for various spin systems have been presented in [8,9, 15-23]. A very interesting example of the polarization transfer via multiquantum pathways in LaF$_3$ crystals can be found in [9]. The polarization is transferred between the high quadrupolar spin of lanthanum, $S = \frac{7}{2}$, and several mutually coupled neighbouring fluorine spins. Therefore the detected magnetization curve exhibits many polarization transfer dips, see Fig.8.6.

VIII.4. Motional conditions of polarization transfer effects and relaxation enhancement

Even though I have pointed out that the fundamental condition necessary for polarization transfer between two spin subsystems is that there is a static (time independent) coupling between them, I would like to turn attention once again to this issue. Polarization transfer processes between spins I and S can arise if the $I-S$ dipole – dipole coupling fulfils the condition $\omega_{DD}^{IS} \tau_c \gg 1$. In the

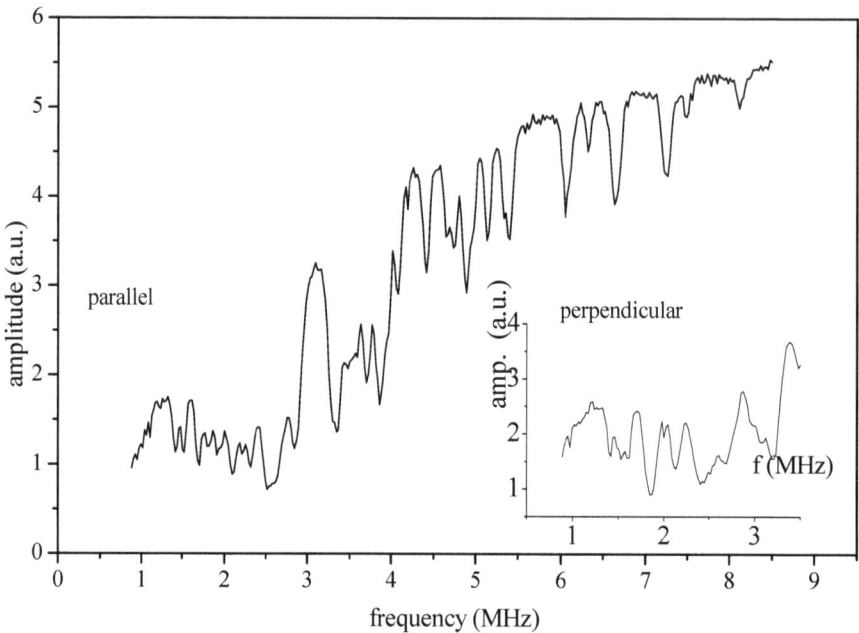

Fig.8.6
Experimentally detected polarization transfer effects for LaF$_3$ for two orientations of the c-axis of the single crystal with respect to the applied magnetic field. The fluorine longitudinal magnetization has been recorded.

opposite motional regime when stochastic fluctuations of the dipole-dipole interaction are fast enough to satisfy the Redfield condition $\omega_{DD}^{IS}\tau_c \ll 1$, this interaction leads to relaxation processes. Interactions which are averaged out because of fast fluctuations cannot cause any polarization transfer. The different motional regimes are shown in Fig.8.7. However, from the perspective of the dipolar spin, both the processes (relaxation and polarization transfer) lead to the same effect: a faster decay of the magnetization. Therefore it is possible to reproduce the polarization transfer features of fluorine (^{19}F) relaxation profiles attributing them to the enhancement of the dipolar spin relaxation caused by fast quadrupolar spin relaxation, employing the theory discussed in Chapter VI, in full analogy to the local enhancement of nuclear spin relaxation in the presence of fast relaxing electron spins, characterised by a non-zero static ZFS. Such an interpretation is, however, incorrect. This has been explained in detail in [9] (see Fig.5 of this reference).

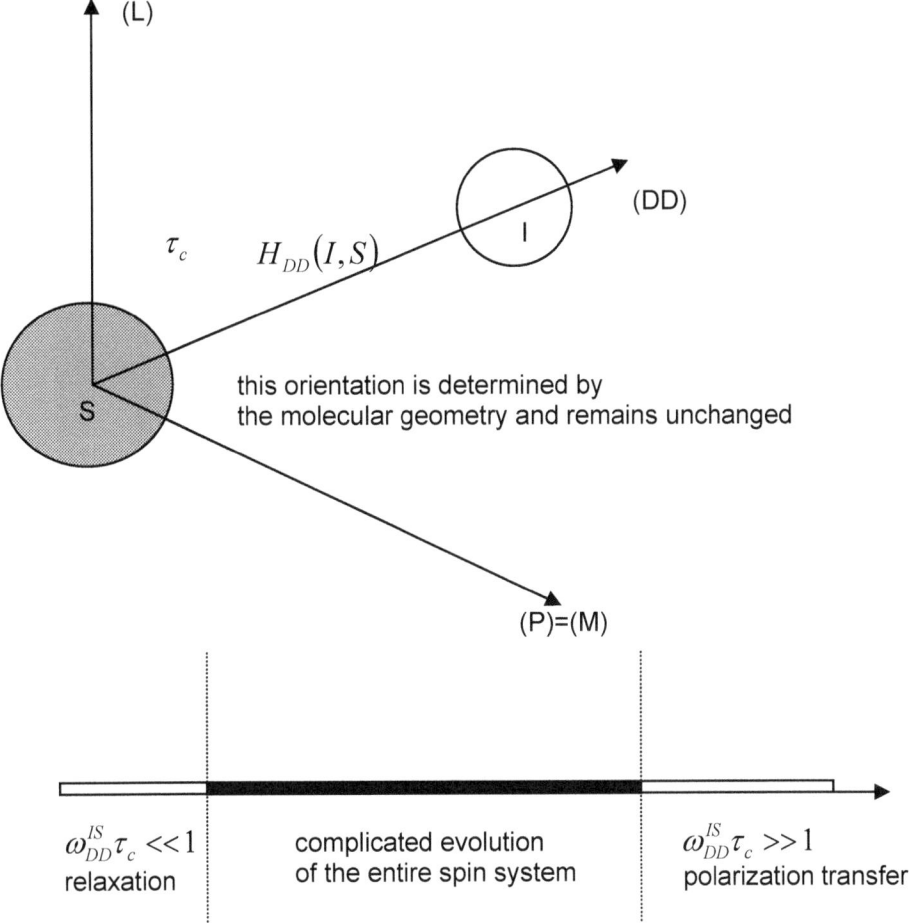

Fig.8.7
Motional conditions required for relaxation and polarization transfer processes. The orientation of the principal axis system of the quadrupolar interaction is determined by the molecular geometry and it is fixed (we do not consider here any changes in the internal geometry of the molecular system). Depending on the strength of the I – S dipole – dipole coupling and the timescale of its modulation different processes occur.

This example should be kept in mind since it can turn out to be very useful to avoid improper conclusions concerning dynamic processes in complex systems. Aiming for a proper description of spin systems containing quadrupolar and dipolar spins, it is of primary importance to investigate carefully the relaxation of the quadrupolar ones.

References:

1. C.P. Slichter, Principles of magnetic resonance, *Springer - Verlag, Berlin* (1990)
2. R.R. Ernst, G. Bodenhausen, A, Wokaun, Principles of nuclear magnetic resonance in one and two dimensions, *Clarendon Press, Oxford* (1994)
3. R. Kimmich, NMR - Tomography, Diffusometry, Relaxometry, *Springer – Verlag Berlin* (1997)
4. J. Kowalewski, L. Mäler, Nuclear spin relaxation in liquids: theory, experiments and applications, Series in Chemical Physics, *Taylor & Francis Group* (2006)
5. A. Kumar, R. C. R. Grace, P. K. Madhu, Cross-correlations in NMR, *Prog. Nucl. Magn. Reson. Spectr.* **37** (2000) 191-319
6. M. Goldman, Formal theory of spin-lattice relaxation, *Adv. Magn. Reson.* **149** (2001) 160-187
7. D. Kruk, J. Altmann , F. Fujara , A. Gadke, M. Nolte, A.F. Privalov, Analysis of ^{1}H-^{14}N polarization transfer experiments in molecular crystals, *J. Physc. Condensed Matter* **17 (3)** (2005) 519-533
8. D. Kruk, O. Lips, Evolution of solid state systems containing mutually coupled dipolar and quadrupolar spins: a perturbation treatment, *Solid State Nucl. Magn. Reson.* **28 (2-4)** (2005) 180-192
9. T. Nilsson, J. Kowalewski, *Mol. Phys.* **98** (2000) 1617-1638, Erratum: *Mol. Phys.* **99** (2001) 369
10. D. Kruk, T. Nilsson, J. Kowalewski, 'Nuclear spin relaxation in paramagnetic systems with zero-field splitting and arbitrary electron spin', *Phys. Chem. Chem. Phys.* **3** (2001) 4907-4917
11. D.A. Varshalovich, A.N. Moskalev, V.K. Khersonkii, Quantum theory of angular momentum, *Word Scientific Publishing, Singapore* (1988)
12. D. M. Brink, G.R. Satchler, Angular momentum, *Clarendon Press, Oxford* (1979)
13. A. R. Edmunds, Angular momentum in quantum mechanics, *Princeton University Press, Princeton* (1974)
14. M. E. Rose, Elementary theory of angular momentum, *Wiley, New York* (1957)

15. F. Fujara, H-J. Stöckmann, H. Ackermann, W. Buttler, K. Dörr, H. Grupp, P. Heitjans, G. Kiese, A.J. Körblein, Cross-relaxation processes of polarized – β active nuclei in various crystalline solids, *Z. Phys. B* **37** (1980) 151-161

16. F. Winter, R. Kimmich, NMR field – cycling relaxation spectroscopy of bovine serum-albumin, muscle – tissue, micrococcus – luteus and yeast – ^{14}N - H^1- quadrupolar dips, *Biochemica et Biophysica Acta* **719 (2)** (1982) 292-298

17. R. Kimmich, F. Winter, W. Nusser, K-H. Spohn, Interactions and fluctuations deduced from proton field-cycling relaxation spectroscopy of polypeptides, DNA, muscles, and algae, *J. Magn. Reson.* **68** (1986) 263-282

18. E. Jäger, B. Ittermann, G. Sultzer, K. Bürkermann, B. Fisher, H-P. Frank, H-J. Stöckmann, H. Ackermann, Cross relaxation of ^{12}B in single-crystal aluminium, *Z. Phys. B* **80** (1990) 87-94

19. E. Anoardo, D. J. Pusiol, N-14 nuclear quadrupolar dips in the proton spin – lattice relaxation dispersion in the smectic-C phase of HpAB, *Phys. Rev. Lett.* **76 (21)** (1996) 3983-3986

20. E. Anoardo, D. J. Pusiol, Nuclear quadrupolar magnetic cross relaxation spectra analysis in a smectic A mesophase, *J. Mol. Structure* **516 (2-3)** (2000) 273-282

21. M. Nolte, A. Privalov, J. Altmann, V. Anferov, F. Fujara, 1H-14N cross-relaxation in trinitrotoluene – a step toward improved landmine detection, *J. Phys. D: Appl. Phys.* **35** (2002) 939-942

22. J. Seliger, V. Zagar, N-14 nuclear quadrupolar resonance in solid 4,4'-azoxyanisole and 4,4'-bis(heptyloxy)azoxybenzene, *Chem. Phys.* **306 (1-3)** (2004) 309-314

23. R. Blinc, T. Apih, P. Jeglic, I. Emri, T. Prodan, Proton NMR study of molecular motion in bulk and in highly drawn fiber polyamide -6, *Appl. Magn. Reson.* **29 (4)** (2005) 577-578

CHAPTER IX

Motional conditions and relaxation processes in multi-spin systems: examples and warnings

In this chapter I shall to discuss how motional conditions combined with relative strengths of the quadrupolar (or ZFS) interactions and of the S spin Zeeman coupling influence the relaxation dynamics of quadrupolar nuclear spin (or, alternatively, the electron spin) as well as the dipolar spins. I shall consider, in particular, various timescales of the modulations of the molecular axis system with respect to the laboratory frame and consequences of the transient parts of the quadrupolar (ZFS) coupling dominating over the static one. In the last section I shall give, in outline, a discussion on the validity regimes of the perturbation theory applied the calculation of the thermal relaxation times of the quadrupolar (electron) spin, and in consequence to the dipolar spin. The chapter will end up with the rather frustrating statement, that the applicability of the treatments of the relaxation processes, presented so far, are strongly limited by the requirements of the second order perturbation theory. This chapter is meant to be a warning that in the case of several interactions affected by different motional processes the validity conditions must be examined with special caution at every stage of the calculations; otherwise the whole description can very easily break down. The discussion and the presented examples underline the necessity of a much more general approach, which is based on a full solution of the Liouville – von Neumann equation. I shall present the general treatment in Chapter X.

IX.1. Effects of fast modulations of ZFS tensor on electron as well as nuclear spin dynamics

If the Zeeman coupling dominates over the second order interactions (the quadrupolar or the ZFS couplings), it is considered as the main interaction describing the energy level structure of the spin S, independently of the motional conditions. The role of the transient part of the ZFS or of the quadrupolar coupling is also well defined. The transient interactions provide the relaxation mechanism for the electron or the quadrupolar spin, respectively. However, the timescale of the fluctuations of the permanent ZFS tensor (and analogously, of the permanent quadrupolar interaction) is crucial for the effects of these interactions on the S spin dynamics. As has been already described, if the molecular frame (which, in particular, can coincide with the principal axis system of the static components of the second order interactions) does not change its orientation with respect to the laboratory frame [1-6], the interactions H_{ZFS}^S and H_Q^S both contribute to the energy level structure of the spin S. Basically it is so for an arbitrary magnetic field, however in the high field limit the non-Zeeman contribution becomes negligible. In the opposite motional limit when the modulations of the molecular frame with respect to the laboratory axis (due to, for example, the reorientational motion) is fast enough to fulfill the Redfield conditions: $\omega_{ZFS}^S \tau_c \ll 1$ or $\omega_Q^S \tau_c \ll 1$, the static interactions provide a relaxation channel for the S spin [7-10]. It is important to remind that even though usually the relaxation mechanism provided by the static interactions is associated with their rotational modulations, the interactions can be also affected by another motional process. For example the jump diffusion (exchange motion) of the spins S between non-equivalent crystal sites can lead to a different local electric field gradient sensed momentary by the quadrupolar spin, as explained in Chapter VII. Independently of the nature of the motion modulating the static parts of the second order interactions, if it is fast enough to fulfill the Redfield condition, there are two relaxation pathways for the spin S at high magnetic field. The first one is due to the transient interactions connected with distortions (vibrations), while the second one is provided by the permanent counterparts (the static ZFS and static quadrupole coupling).

Let us turn attention to fast rotating molecules carrying an electron spin. If both the perturbing interactions $H_{ZFS}^{S}(t)$ and $H_{ZFS}^{T}(t)$ fulfill the Redfield condition, ($\omega_{ZFS}^{S}\tau_R \ll 1$ and $\omega_{ZFS}^{T}\tau_D \ll 1$), the electron spin relaxation can be described in terms of the combined spectral densities $J_m^S(\omega) = J_m^{S(S)}(\omega) + J_m^{S(T)}(\omega)$, defined as: $J_m^{S(S)}(\omega) = \int_0^\infty \langle \widetilde{V}_m^{2S(L)*}(\tau)\widetilde{V}_m^{2S(L)}(0)\rangle \exp(-i\omega\tau)d\tau$ and

$J_m^{S(T)}(\omega) = \int_0^\infty \langle \widetilde{V}_m^{2T(L)*}(\tau)\widetilde{V}_m^{2T(L)}(0)\rangle \exp(-i\omega\tau)d\tau$ (see Eq.3.16 and Eq.3.18 for the corresponding correlation functions). The spectral densities $J_m^S(\omega)$, defined in Section IV.3, are now denoted as $J_m^{S(T)}(\omega)$ to point out that they describe the transient ZFS mechanism of the S spin relaxation. In fact, one can rewrite all expressions derived for the electron spin in the high field limit with the transient ZFS as the relaxation mechanism (Sections IV.3 and IV.4) replacing the spectral densities occurring in these formulas by their combined counterparts including the transient as well as the static ZFS. According to the motional models reflected by the correlation functions of Eq.3.18 and Eq.3.16, the spectral densities take the form: $J_m^{S(S)}(\omega) = J^{S(S)}(\omega) = \frac{1}{5}\Delta_S^2 \frac{\tau_R}{1+\omega^2\tau_R^2}$ and

$J_m^{S(T)}(\omega) = J^{S(T)}(\omega) = \frac{1}{5}\Delta_T^2 \frac{\tau_{eff}}{1+\omega^2\tau_{eff}^2}$ (as we know already from Section IV.3). Taking into account, that the pseudorotational model is obviously a simplification, and its task is to reflect the time scale of the transient ZFS fluctuations, one does not need to consider the rotational and distortional contributions to the effective correlation time $\tau_{eff}^{-1} = \tau_D^{-1} + \tau_R^{-1}$ in detail. Thus, defining the spectral density $J^{S(T)}(\omega)$ one could even use the correlation time τ_D, knowing that it reflects the combined effect of rotation and distortion. However, the expression for this spectral density is based on a very important assumption that there is no correlation between time fluctuations of the angles $\Omega_{P_T P_S}(t)$ and $\Omega_{P_S L}(t)$, as it has been pointed out in Section III.2. I have referred to this issue in Section IV.3. To avoid a relatively complicated discussion, I have just stated that the distortional (vibrational) motion is usually on a very rapid time scale relative to the molecular tumbling. This implies that the reorientational and vibrational motions can be considered as statistically uncorrelated because of different timescales. Nevertheless, the two types of motion have completely different physical origins and therefore there is no correlation between them, anyway. This assumption simplifies the description of the electron spin relaxation; however at this stage it is not of primary importance. Describing the S spin dynamics one could take into account eventual correlations between these motions by considering interference terms between the static and the transient ZFS Hamiltonians. Nevertheless it leads to very important consequences for the I spin relaxation. This has been clearly pointed out in Section IV.4. The calculations presented in this section require the assumption that one can separate the rotational motion and the dynamics of the S spin. This is relatively easy if there is no contribution to the electron spin relaxation from the static ZFS, as we have already seen. If we cannot avoid including the static ZFS to the description of the electron spin relaxation, this problem becomes more complicated and we need to discuss it in detail.

Generally, the dipolar spin I relaxes due to modulations of the $I-S$ dipole-dipole interaction by molecular rotation, the S spin relaxation and, possibly, chemical exchange or relative translation diffusion of the participating spins. Rotational motion influences the I spin relaxation in a 'direct' manner (as the origin of the stochastic modulations of the $I-S$ dipole-dipole interaction) and through the S spin dynamics affected by the static zero-field splitting or the static quadrupolar coupling, modulated by the molecular tumbling as well. Analytical descriptions of the dipolar spin relaxation require an explicit treatment of the S spin relaxation with well defined relaxation rates. In addition, the dynamics of the spin S must be independent of the presence of the spin I. Only under the last condition the spin S can be treated as a part of the lattice for the dipolar spin. In consequence, only under this condition the I spin relaxation can be described by the quantum – mechanical dipole-dipole spectral density $K_{1,1}^{DD}(-\omega_I)$ given by Eq.4.21 and generalized by Eq.7.1. Therefore, the S spin dynamics may not be correlated with any kinds of motion modulating directly the $I-S$ dipole-dipole interaction. In principle, the requirement that the S spin subsystem can be treated separately leads directly to the statement that the S spin dynamics is not affected by any degrees of freedom, relevant for the spin I. In fact I solved the problem in this 'crude' manner, by stating that the rotational contribution to the modulations of the transient ZFS with respect to the laboratory frame is negligible. Nevertheless, to decompose the S spin dynamics from other degrees of freedom modulating the $I-S$ dipole-dipole coupling it is enough if the processes are uncorrelated. Let us take once again as an example a fast rotating molecule with an electron spin S. In this case the well described electron spin dynamics has to be incorporated, as a part of the lattice, to the general form of the spectral density $K_{1,1}^{DD}(-\omega_I)$ of Eq.4.21. It can be rewritten as [7]:

$$K_{1,1}^{DD}(-\omega_I) = 30\left(\frac{\mu_0 \hbar \gamma_I \gamma_S}{4\pi r_{IS}^3}\right)^2 \sum_{p,q=-1}^{1}\begin{pmatrix} 2 & 1 & 1 \\ 1-q & q & -1 \end{pmatrix}\begin{pmatrix} 2 & 1 & 1 \\ 1-p & p & -1 \end{pmatrix} \times$$
$$\int_0^\infty Tr_R\left\{D_{0,1-q}^{2*}(\Omega_{DDL}(\tau))Tr_S\left\{S_q^{1(L)+}\left[\exp(-i\hat{\hat{L}}_S\tau)S_p^{1(L)}\right]\rho_S^{eq}\right\}D_{0,1-p}^{2}(\Omega_{DDL}(0))\right\}\exp(-i\omega_I\tau)d\tau \quad (9.1)$$

This expression results from the obvious fact that the orientation of the $I-S$ dipole-dipole axis relative to the laboratory frame depends only on the rotational motion and is not affected by the electron spin dynamics, encoded in the Liouville operator $\hat{\hat{L}}_S$. This operator contains the Zeeman interaction, the static ZFS modulated by the rotational motion and the transient ZFS modulated mainly by the distortions: $\hat{\hat{L}}_S = \hat{\hat{L}}_Z + \hat{\hat{L}}_{ZFS}^S(t) + \hat{\hat{L}}_R + \hat{\hat{L}}_{ZFS}^T(t) + \hat{\hat{L}}_D$. From this expression one can see clearly the double role of the rotational motion; one can see it also from Fig.9.1a. An analogous picture one can draw for solid state systems containing dipolar and quadrupolar spins. If the quadrupolar spin jumps between non-equivalent lattice sites it senses different averaged (static) electric field gradient tensor. Thus, the exchange motion (jump diffusion) which contributes directly to the modulations of the $I-S$ dipole-dipole coupling is at the same time responsible for the fluctuations of the static quadrupolar coupling (which contributes to the quadrupolar spin relaxation); this problem is illustrated in Fig.9.1b.

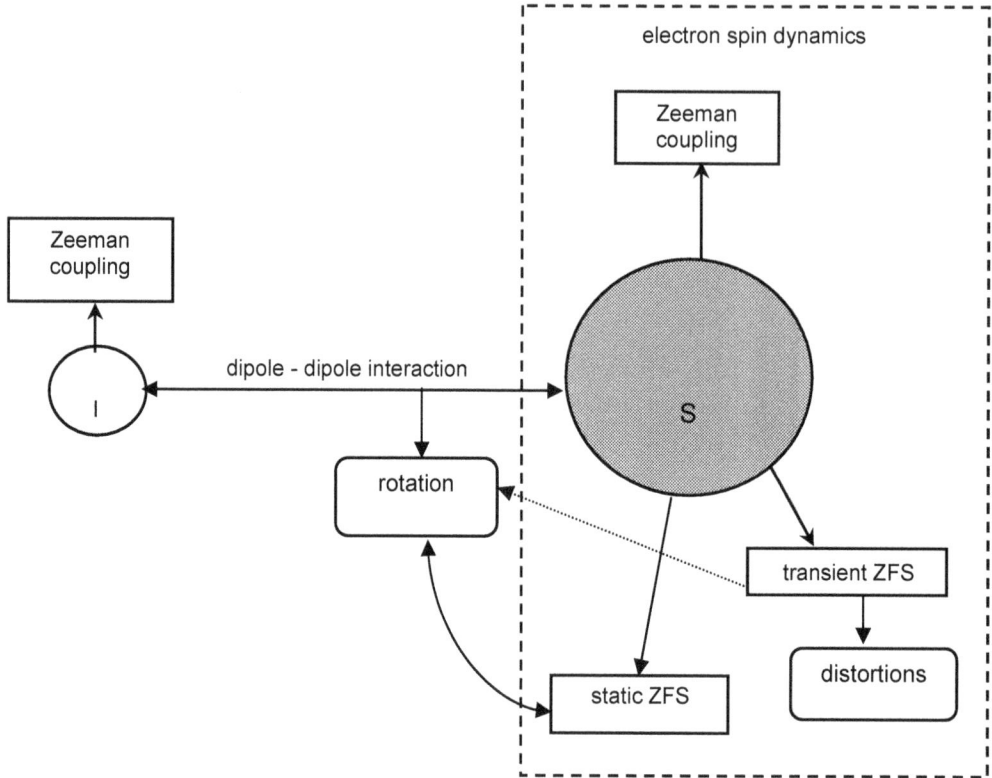

Fig.9.1a

The rotational motion modulates directly the I – S dipole – dipole coupling, which causes the nuclear spin relaxation, modulating at the same time the static ZFS interaction which contributes to the electron spin relaxation (which, in turn, influences the nuclear spin relaxation). As discussed in the text one can, in principle, neglect the effect of the rotational motion on the transient ZFS and therefore it is presented here as 'weak' dotted line.

Chapter IX - Motional conditions and relaxation processes in multi-spin systems

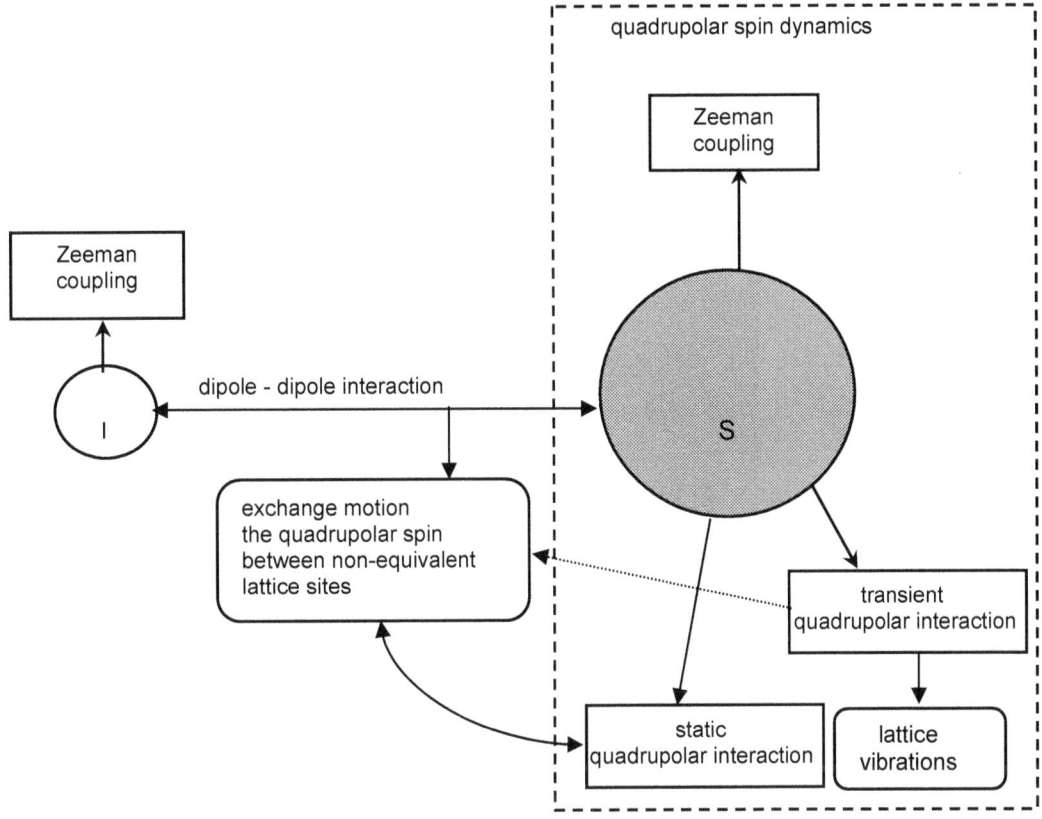

Fig.9.1b
The exchange motion of the quadrupolar spin modulates directly the I – S dipole – dipole coupling and modulates at the same time (if it occurs between non-equivalent lattice postions) the averaged electric field gradient tensor which is sensed momentary by the quadrupolar spin. Since one can consider the fluctuations of the electric field gradient tensor around its averaged value at every position as modulated mainly by lattice vibrations, the effect of the jump diffusion on the transient quadrupolar interaction is rather negligible.

To understand the resulting difficulties let us consider the electron spin correlation function: $G_{q,p}(\tau) = \frac{1}{2S+1} Tr_S \left\{ S_q^{1(L)+} \left[\exp(-i\hat{L}_L \tau) \right] S_p^{1(L)} \right\} \equiv \left\langle S_q^{1(L)+}(\tau) S_p^{1(L)}(0) \right\rangle$ (see Eq.4.25, and Eq.4.30a,b). In principle, we can neglect the influence of the

239

rotational motion on the transient component of the ZFS tensor, as has been done before. However, because of the static ZFS being modulated by the rotational motion, the electron spin correlation function $G_{q,p}(\tau)$ depends, in general, on the angle $\Omega_{DDL}(\tau)$ at the time τ. From the perspective of the spin I, this quantity (which is dependent on the momentary orientation of the molecular frame) has to be split with the explicit rotational modulation of the dipole-dipole axis encoded in the Wigner rotation matrices $D^2_{0,1-q}(\Omega_{DDL}(\tau))$ and averaged over the rotational degrees of freedom (Tr_R), according to Eq.9.1. Despite computational difficulties, it is quite obvious that such a treatment cannot lead to any closed form expressions for the nuclear spin relaxation. Fortunately, the condition that to describe the I spin relaxation by the spectral density $K^{DD}_{1,1}(-\omega_I)$ the dynamics of the spin S subsystem has to be independent of the presence of the I spin does not imply that there are no common degrees of freedom relevant for the dynamics of both the spins. The rotational motion influences the electron as well as the nuclear spin relaxation, but the principal axis system of the static ZFS tensor changes its orientation independently of the presence of the nuclear spin. Therefore, the nuclear spin relaxation is properly described by the general form of the quantum-mechanical spectral density (Eq.4.21). The point is that we wish to formulate an analytical description of the I spin relaxation, by evaluating explicitly this spectral density. For this purpose we need to decompose the correlation function for the orientation of the $I-S$ dipole-dipole axis, $\langle D^{2*}_{0,1-q}(\Omega_{DDL}(\tau)) D^2_{0,1-p}(\Omega_{DDL}(0)) \rangle$, from other elements contributing the entire correlation function.

We should appreciate, at this stage, the assumption that the electron spin fulfills the conditions of the Redfield relaxation theory [11-14]. As it has been discussed in detail in Chapter II, the relaxation rates are time independent because the condition $\omega_1 \tau_c \ll 1$ (in this case $\omega^S_{ZFS} \tau_R \ll 1$) permits one to extend the upper limit of integration in Eq.2.9a,b to infinity. In consequence, the electron spin correlation functions $G_{q,p}(\tau)$ decay multiexponentialy (with the individual relaxation rates as the decay coefficients) and are independent of the actual orientation of the molecular frame with respect to the laboratory axis, *i.e.* of the angle $\Omega_{DDL}(\tau)$. This 'mathematical' argumentation can be also expressed by words. Since we have assumed that the electron spin fulfills the Redfield condition, it means that the molecular tumbling is much faster than the electron spin relaxation caused by the rotationally modulated static ZFS. The time scale separation permits us to separate the correlation function describing the explicit effect of the rotational motion on the dipole-dipole axis, from the implicit rotational effect encoded in the electron spin relaxation. Thus, the nuclear spin senses the modulations of the $I-S$ dipole-dipole coupling as a result of two uncorrelated processes: the rotational motion and the electron spin relaxation, so that $\tau_{c,i}^{-1} = \tau_R^{-1} + R_i^S$, where $\tau_{c,i}^{-1}$ is the effective correlation time associated with the electron spin relaxation rate R_i^S. Therefore, the entire correlation function can be decomposed into the two parts [7]:

$$Tr_R\left\{D_{0,1-q}^{2*}(\Omega_{DDL}(\tau))Tr_S\left\{S_q^{1(L)+}\left[\exp(-i\hat{L}_S\tau)S_p^{1(L)}\right]\rho_S^{eq}\right\}D_{0,1-p}^2(\Omega_{DDL}(0))\right\} = $$

$$\frac{1}{2S+1}\langle D_{0,1-q}^{2*}(\Omega_{DDL}(\tau))D_{0,1-p}^2(\Omega_{DDL}(0))\rangle Tr_S\left\{\left[S_q^{1(L)+}\exp\left(-\left(i\hat{L}_Z+\hat{\hat{R}}^S\right)\tau\right)S_p^{1(L)}\right]\right\}$$

(9.2)

where the relaxation superoperator $\hat{\hat{R}}^S = \hat{\hat{R}}^{S(S)} + \hat{\hat{R}}^{S(T)}$ includes the two relaxation pathways for the electron spin. In consequence, for example for the electron spin quantum number , $S=1$ the nuclear spin-lattice relaxation rate is given by the expression of Eq.4.39 with the electron spin spectral densities $s_{m,m}$ including the modified (extended) relaxation rates: $R_{\alpha\alpha'\beta\beta'}^S = R_{\alpha\alpha'\beta\beta'}^{S(S)} + R_{\alpha\alpha'\beta\beta'}^{S(T)}$.

It is worthwhile to notice that even though this description is based on the assumption that the rotational motion is much faster than the relaxation caused by the static ZFS, the transient ZFS can act as a very efficient predominant relaxation mechanism and the effective electron spin relaxation can be in some cases significantly faster than the rotational motion.

The factorization problem does not exist for high symmetry molecular systems with zero static ZFS. In this case the electron spin dynamics is just independent of the rotational motion.

All the arguments invoked in this section concern also other 'equivalent' spin systems. One can consider, for example, a crystal lattice containing dipolar and quadrupolar spins. The latter can jump between non-equivalent crystallographic sites, characterized by different static electric field gradients. In this case the jump motion plays the role of the molecular tumbling; it leads to time fluctuations of the orientation of the electric field gradient tensor at the S spin site relative to the laboratory axis. In fact, the situation is much more complicated because distinct crystal sites are usually characterized also by different values of the electric field gradient tensor, not only the orientation of its main frame. However, this is a matter of an appropriate model of the quadrupolar interaction and in consequence of the spectral densities describing the quadrupolar spin relaxation. The treatment of the dipolar spin relaxation is essentially the same; the spectral density $K_{1,1}^{DD}(-\omega_I)$ can be written as:

$$K_{1,1}^{DD}(-\omega_I) = 30\left(\frac{\mu_0}{4\pi}\gamma_I\gamma_S\hbar\right)^2 \sum_{p,q=-1}^{1}\begin{pmatrix}2 & 1 & 1\\ 1-q & q & -1\end{pmatrix}\begin{pmatrix}2 & 1 & 1\\ 1-p & p & -1\end{pmatrix} \times \tag{9.3}$$

$$\int_0^\infty Tr_M\left\{\frac{D_{0,1-q}^{2*}(\Omega_{DDL}(\tau))}{r_{IS}^3(\tau)}Tr_S\left\{S_q^{1(L)+}\left[\exp\left(-i\hat{L}_S\tau\right)S_p^{1(L)}\right]\rho_S^{eq}\right\}\frac{D_{0,1-p}^2(\Omega_{ML}(0))}{r_{IS}^3(0)}\right\}\exp(-i\omega_I\tau)d\tau$$

We have denoted the averaging over the degrees of freedom related to the jump diffusion as Tr_M. The Liouville operator $\hat{\hat{L}}_S$ contains now the exchange Liouvilian instead of the rotational one: $\hat{\hat{L}}_S = \hat{\hat{L}}_Z + \hat{\hat{L}}_{ZFS}^S(t) + \hat{\hat{L}}_M + \hat{\hat{L}}_{ZFS}^T(t) + \hat{\hat{L}}_D$. Invoking the argument that the exchange motion is much faster than the quadrupolar spin relaxation caused by fluctuations of the static part of the ZFS tensor, which senses the quadrupolar spin changing its positions, one can write in full analogy to Eq.9.2 that:

$$Tr_M\left\{\frac{D_{0,1-q}^{2*}(\Omega_{DDL}(\tau))}{r_{IS}^3(\tau)}Tr_S\left\{S_q^{1(L)+}\left[\exp\left(-i\hat{L}_S\tau\right)S_p^{1(L)}\right]\rho_S^{eq}\right\}\frac{D_{0,1-p}^2(\Omega_{DDL}(0))}{r_{IS}^3(0)}\right\} = \tag{9.4}$$

$$\frac{1}{2S+1}\left\langle\frac{D_{0,1-q}^{2*}[\Omega_{DDL}(\tau)]}{r_{IS}^3(\tau)}\frac{D_{0,1-p}^2[\Omega_{DDL}(0)]}{r_{IS}^3(0)}\right\rangle Tr_S\left[S_q^{1(L)+}\exp\left(-\left(i\hat{L}_Z+\hat{\hat{R}}^S\right)\tau\right)S_p^{1(L)}\right]$$

Now, we turn attention to the low field limit. In this regime the energy levels for the spin S are determined by the static second order couplings: H_{ZFS}^S

or H_Q^S, while the transient components H_{ZFS}^T or H_Q^T, respectively, cause the relaxation independently of the timescale of the motion modulating the orientation of the molecular frame. This statement is correct if the static interactions dominate over their transient counterpart. We shall discuss the opposite case in some detail in the next section. From the perspective of the molecular frame the Zeeman coupling fluctuates in time due to the same degrees of freedom, which are responsible for the fluctuations of the static ZFS (or of the averaged quadrupolar coupling) with respect to the laboratory frame in the high field limit. Therefore, with the static second order interactions as the main Hamiltonian, one might consider the Zeeman coupling as a perturbation contributing also to the S spin relaxation. Nevertheless, as it has been pointed out in Chapter II that perturbation theory requires that the perturbing Hamiltonian to be characterized by a zero average. The Zeeman Hamiltonian does not fulfill this condition and therefore one can describe the relaxation of the spin S within the framework of the perturbation treatment only at very low magnetic field, where the effect of the Zeeman interaction can be neglected altogether. We have worked out in detail in Chapter IV the low field expressions for the electron spin of $S = 1$ as well as the nuclear spin relaxation, telling nothing about the timescale of the rotational motion, because it does not influence the role of the static and the transient ZFS interactions and, in consequence, the physical picture of the relaxation processes. Closed form expressions for higher electron spin quantum numbers one can find in [15]. This paper has been dedicated in principle to slowly rotating systems, however the derived expressions for the nuclear spin relaxation remain unchanged if the rotation becomes faster [7]. Fig.9.2 shows the effect of the rotational motion on the nuclear spin relaxation at low field for the electron spin quantum numbers $S = 1$ and $S = \frac{7}{2}$, respectively.

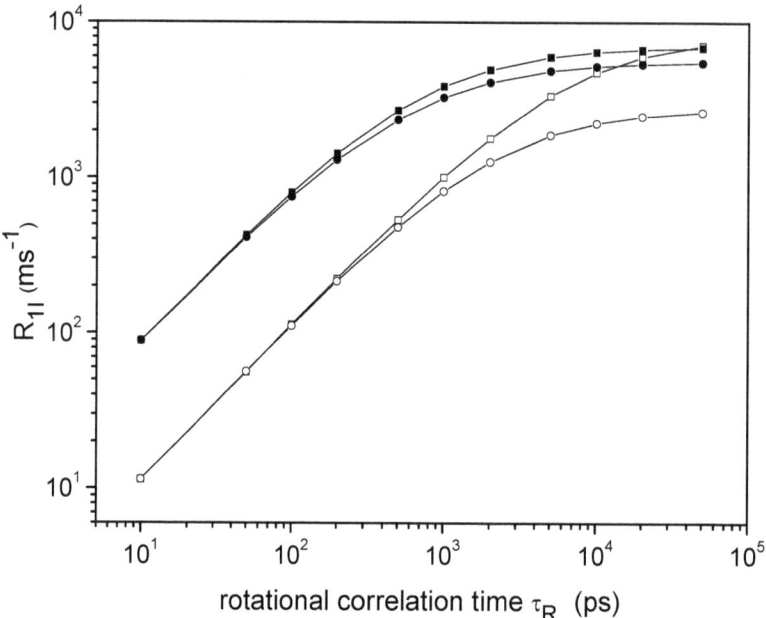

Fig.9.2
Effects of the rotational motion on the proton spin – lattice relaxation at low field. Lines with open squares and open circles correspond to the spin quantum number $S = 1$ and the sets of parameters: $D_S = 0.05 cm^{-1}$, $D_T = 0.02 cm^{-1}$, $\tau_D = 10 ps$ (open squares) and $D_S = 0.05 cm^{-1}$, $D_T = 0.02 cm^{-1}$ $\tau_D = 30 ps$ (open circles). Lines with solid squares and solid circles have been calculated for the same sets of the parameters but for $S = \frac{7}{2}$. The interspin distance has been set to 300pm. For fast molecular tumbling the electron spin relaxation does not matter (lines with square and circles coincide). For slower rotational motion one can see clearly the contribution of the electron spin relaxation depending on the assumed set of the ZFS parameters.

One can conclude that a proper description of the dipolar spin relaxation often meets with difficulties, because the perturbation descriptions of the S spin dynamics have limited validity ranges. In this section we have seen so far that the relative timescale of the rotational motion and the relaxation process of the electron spin resulting from the rotation is crucial for the nuclear spin dynamics *and that an analogous statement can be formulated regarding the relationship between the quadrupolar spin relaxation caused by the averaged (static) quadrupolar coupling and the motion, which leads to fluctuations of the averaged electric field gradient sensed momentarily by the quadrupolar spin. In the next section we shall deal with other types of difficulties.*

IX.2. Relaxation in spin systems of high symmetry at low magnetic field

A common factor of all approaches presented so far is that the static components of the second order interactions (the static ZFS or the static quadrupolar coupling) describe the energy level structure of the spin S at low magnetic field, while their transient counterparts induce transitions between these energy levels. The description has been established in agreement with the perturbation theory requiring that the perturbing Hamiltonian $H_1(t)$ must fulfill the condition $\omega_1 \tau_c \ll 1$, where the correlation time τ_c gives the timescale of the fluctuations of the perturbing interaction with respect to the principal axis system of the main Hamiltonian H_0, which fulfills the condition: $\omega_1 \tau_c \ll \frac{\omega_0}{\omega_1}$. It means that in the case of an electron spin system the transient and the static parts of the static ZFS must satisfy the relation: $\omega_{ZFS}^T \tau_D \ll \frac{\omega_{ZFS}^S}{\omega_{ZFS}^T}$. A corresponding relationship has to be satisfied by the quadrupolar interactions: $\omega_Q^T \tau_Q \ll \frac{\omega_Q^S}{\omega_Q^T}$, as well. The conditions are not always fulfilled. In this chapter I shall deal with the problem of a proper choice of the main and perturbing interactions for the spin S in the low field regime. I shall focus attention mainly on an electron spin system; however the results can be adapted in straightforward manner to systems containing quadrupolar spins, as usually.

If the static ZFS is weak relative to the transient part, the transient ZFS tends to become the main Hamiltonian for the electron spin [16]. The principal axis systems of the static and the transient parts of the ZFS tensor fluctuate with respect to each other with the correlation time τ_D (Fig.9.3).

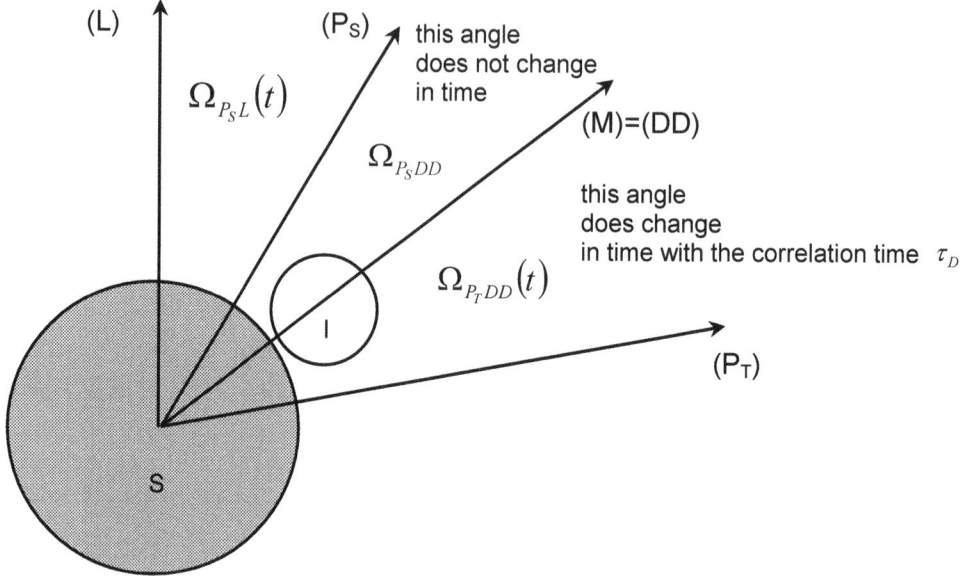

Fig.9.3
The principal axis system of the transient ZFS fluctuates in time with respect to the dipole-dipole axis and the principal axis system of the static ZFS due to the distortional (vibrational) motion characterized by the correlation time τ_D.

Therefore, the transient ZFS can be treated as the main Hamiltonian under the condition $\omega_{ZFS}^S \tau_D \ll \dfrac{\omega_{ZFS}^T}{\omega_{ZFS}^S}$. Comparing this relationship with the previous condition, $\omega_{ZFS}^T \tau_D \ll \dfrac{\omega_{ZFS}^S}{\omega_{ZFS}^T}$, one can conclude that in fact the ratio between the amplitudes of the static and the transient parts of the ZFS tensor determines the roles of the two interactions. If $\omega_{ZFS}^S \gg \omega_{ZFS}^T$ the static ZFS becomes the main interaction while the transient ZFS is the perturbing one (if, in addition, the

Redfield condition is fulfilled, *i.e.* $\omega_{ZFS}^T \tau_D \ll 1$). In the opposite case, when $\omega_{ZFS}^T \gg \omega_{ZFS}^S$ (and $\omega_{ZFS}^S \tau_D \ll 1$) the roles of the static and transient interactions reverse. Thus, when the magnitude of the transient ZFS approaches or surpasses the magnitude of the static ZFS the models presented so far break down and another approach must be adopted to achieve a correct description of the electron spin relaxation. The proper description must be formulated in terms of the energy level structure determined by the transient ZFS with the static ZFS, fluctuating with respect to the P_T frame, as the electron spin relaxation mechanism. One can modify in a straightforward manner the low field expressions for the individual electron spin relaxation rates derived when the static ZFS has been considered as the main interaction, making them appropriate for the present case. Since the roles of the static and transient parts of the ZFS are now inverted the spectral densities $J_m^S(\omega)$ occurring in the expressions for the electron spin relaxation, derived for example in Section IV.5, are proportional to Δ_S^2 instead of Δ_T^2. The appropriate transition frequencies contain now the terms $\omega_{D_T}, \omega_{E_T}$ and $\omega_{D_T \pm E_T}$ (the energy splitting D_T, E_T and $D_T \pm E_T$ resulting from the transient ZFS, expressed in the angular frequency units). They replace the static ZFS counterparts (ω_D, ω_E and $\omega_{D \pm E}$) in the corresponding expressions for the electron spin relaxation derived for the case when the static ZFS dominates over the transient component [15].

Thus, after this discussion we are in principle ready for working out the problem of the nuclear spin relaxation at low field, under the condition $\omega_{ZFS}^T \gg \omega_{ZFS}^S$. However, before we present the mathematical formulation, following the general philosophy of the earlier approaches, it is of primary importance to notice the difference between the present case and the earlier ones. The principal axis system of the static ZFS does not need to coincide with the dipole – dipole axis (even though for simplicity reasons we assumed always that it does), but their relative orientation remains unchanged in time. In the present case the principal axis system of the main Hamiltonian for the electron spin is not fixed relative to the direction of the dipole-dipole interaction. The fact that the electron spin relaxation is described in a frame which fluctuates relative to the

$I-S$ dipole-dipole axis, influences significantly the validity regimes of the present considerations. Now, the electron spin dynamics contributes to the modulations of the $I-S$ dipole-dipole coupling not only by its relaxation (provided by the static ZFS) but also by the stochastic fluctuations of its quantization axis (*i.e.* the P_T frame) with respect to the dipole-dipole axis. The expression for the nuclear spin relaxation, which we shall present in the second part of this chapter, includes the two elements. However, the entire treatment, and in consequence the result, does not reflect the fact that the electron spin dynamics is now considered in the fluctuating frame. In this context one can ask under what conditions the effective modulations of the dipole-dipole coupling can be described as a superposition of the relevant motional processes including, in particular, the electron spin relaxation in the P_T frame (not fixed relative to the dipole-dipole axis).

As it has been discussed in Chapter II, the very important step of the perturbation solution of the Liouville – von Neumann equation is the transformation of the density operator to the interaction representation. The transformation introduces oscillations of the density matrix elements with frequencies determined by the amplitude of the main Hamiltonian. From the perspective of the dipole-dipole frame, the transient ZFS can be treated as an interaction providing a stationary basis set for the electron spin only if the relative fluctuations of the P_T frame are slow compared to the oscillations of the electron spin density matrix elements due to the transformation to the interaction representation (generated in this case by the transient ZFS Hamiltonian). This condition has a simple mathematical formulation. Since the timescale of the fluctuations of the P_T frame relative to the dipole-dipole axis is given by the correlation time τ_D, while the timescale of the oscillations of the density operator is determined by $\left(\omega_{ZFS}^T\right)^{-1}$, the condition yields:

$$\tau_D >> \left(\omega_{ZFS}^T\right)^{-1}, \text{ i.e. } \omega_{ZFS}^T \tau_D >> 1.$$

The present case gives us opportunity to demonstrate at the same time the flexibility of the concept of the composite lattice se well as its danger traps. We start, as usual, from the spectral density $K_{1,1}^{DD}(-\omega_I)$ of Eq.4.21. Most of the essential elements of the general framework have been discussed earlier.

However, to explain better the present problem we shall comment on some important steps of the calculations in detail. First, we express the electron spin tensor components $S_p^{1(L)}$ in the principal axis system of the main Hamiltonian, *i.e.* in the P_T frame according to the transformation rule of Eq.3.3, assuming that the P_S and DD frames coincide:

$$S_p^{1(L)} = \sum_{m,k=-1}^{1} S_m^{1(P_T)} D_{m,k}^1\left(\Omega_{P_T P_S}(\tau)\right) D_{k,p}^1\left(\Omega_{P_S L}(\tau)\right) \tag{9.5}$$

After substituting the above representation of the tensor components in Eq.4.21, and contracting the Wigner matrices over the angle $\Omega_{P_S L} = \Omega_{DDL}$, the spectral density $K_{1,1}^{DD}(-\omega_I)$ yields the form [16]:

$$K_{1,1}^{DD}(-\omega_I) = 30\left(\frac{\mu_0}{4\pi}\gamma_I\gamma_S\hbar\right)^2 \sum_{p,q=-1}^{1} \begin{pmatrix} 2 & 1 & 1 \\ 1-q & q & -1 \end{pmatrix}\begin{pmatrix} 2 & 1 & 1 \\ 1-p & p & -1 \end{pmatrix} \times$$

$$\sum_{m,k=-1}^{1} \sum_{J,M,N=1}^{3} \sum_{m',K'=-1}^{1} \sum_{J',M',N'=1}^{3} (2J+1)(2J'+1)(-1)^{M'-N'+k-1} \times$$

$$\begin{pmatrix} 1 & 2 & J \\ -k & 0 & M \end{pmatrix}\begin{pmatrix} 1 & 2 & J \\ -q & q-1 & N \end{pmatrix}\begin{pmatrix} 1 & 2 & J' \\ k & 0 & M' \end{pmatrix}\begin{pmatrix} 1 & 2 & J' \\ p & 1-p & N' \end{pmatrix} \times \tag{9.6}$$

$$\int_0^\infty Tr_L\left\{\begin{matrix} \left[S_m^{1(P_T)+}D_{m,k}^{1*}\left(\Omega_{P_T DD}(\tau)\right)D_{M,N}^{J*}\left(\Omega_{DDL}(\tau)\right)\exp\left(-i\hat{L}_L\tau\right)\right] \\ \left[S_{m'}^{1(P_T)}D_{m',k'}^{1}\left(\Omega_{P_T DD}(0)\right)D_{-M',-N'}^{J'}\left(\Omega_{DDL}(0)\right)\rho_L^{eq}\right] \end{matrix}\right\} \exp(-i\omega_I\tau)d\tau$$

The quite complicated constants in front of the integral, in fact, do not make the calculations more cumbersome. Nevertheless, at this stage we focus attention on the 'pure' spectral density function, given just by the integral. Since the orientation of the dipole-dipole axis with respect to the P_T frame is modulated by a combined effect of the molecular tumbling and the distortional motion, both the angles $\Omega_{P_T DD}(\tau) = \Omega_{P_T P_S}(\tau)$, $\Omega_{DDL}(\tau)$ appear in Eq.9.6 explicitly. In the present case the lattice Liouvilian $\hat{L}_L$ contains the transient and the static ZFS superoperators and the distortional and the rotational Liouvilians: $\hat{L}_L = \hat{L}_{ZFS}^T + \hat{L}_{ZFS}^S(t) + \hat{L}_R + \hat{L}_D$. Since we aim for an explicit form of the spectral density $K_{1,1}^{DD}(-\omega_I)$ it is very important to consider correlations between all dynamic processes relevant for the nuclear spin relaxation. We have already concluded that the distortional and the rotational motion can be treated always as uncorrelated, because of their different physical mechanisms: the overall molecular reorientation does not affect any internal degrees of freedom of the molecule, while the distortional motion leads to fluctuations in the geometry of the molecular system. This assumption is essential for the present considerations. If the transient ZFS is the main Hamiltonian the static ZFS provides the mechanism of the electron spin relaxation. Since there is no correlation between the rotation and the distortional (vibrational) motion, the rotation just does not influence the electron spin relaxation. Thus the rotational correlation function can be separated from the remaining components of the lattice dynamics:

$$Tr_L \left\{ S_m^{1(P_T)^*} D_{m,k}^{1*}(\Omega_{P_T DD}(\tau)) D_{M,N}^{J*}(\Omega_{DDL}(\tau)) \left[\exp(-i\hat{L}_L \tau) S_{m'}^{1(P_T)} D_{m',k'}^{1}(\Omega_{P_T DD}(0)) D_{-M',-N'}^{J'}(\Omega_{DDL}(0)) \right] \rho_L^{eq} \right\} =$$

$$\frac{1}{2S+1} Tr_{(L-R)} \left\{ D_{m,k}^{1*}(\Omega_{P_T DD}(\tau)) S_m^{1(P_T)+} \left[\exp(-i\hat{L}_{L-R}\tau) S_{m'}^{1(P_T)} D_{m',k'}^{1}(\Omega_{P_T DD}(0)) \right] \right\} \times$$

$$\left\langle D_{M,N}^{J*}(\Omega_{DDL}(\tau)) D_{-M',-N'}^{J'}(\Omega_{DDL}(0)) \right\rangle$$

(9.7)

The next question concerns the correlation between the distortional motion and the electron spin relaxation, which is relevant for further evaluation of the correlation function: $Tr_{(L-R)} \left\{ D_{m,k}^{1*}(\Omega_{P_T DD}(\tau)) S_m^{1(P_T)+} \left[\exp(-i\hat{L}_{L-R}\tau) S_{m'}^{1(P_T)} D_{m',k'}^{1}(\Omega_{P_T DD}(0)) \right] \right\}$. This problem is fully equivalent to the case of the rotationally modulated static ZFS decoupled from the rotational modulation of the dipole-dipole coupling. We have concluded that the two processes are stochastically uncorrelated because, as a consequence of the Redfield condition $\omega_{ZFS}^S \tau_R \ll 1$, the electron spin relaxation associated with the rotational motion is much slower than the rotation by itself. Analogous arguments can be invoked in the present case. Since we assume that the static ZFS fulfills the condition: $\omega_{ZFS}^S \tau_D \ll 1$, the resulting electron spin relaxation is significantly slower than the distortional motion. Therefore one can write:

$$Tr_{(L-R)} \left\{ D_{m,k}^{1*}(\Omega_{P_T DD}(\tau)) S_m^{1(P_T)^*} \left[\exp(-i\hat{L}_{L-R}\tau) S_{m'}^{1(P_T)} D_{m',k'}^{1}(\Omega_{P_T DD}(0)) \right] \right\} =$$

$$\left\langle D_{m,k}^{1*}(\Omega_{P_T DD}(\tau)) D_{m',k'}^{1}(\Omega_{P_T DD}(0)) \right\rangle \times Tr_S \left\{ S_m^{1(P_T)+} \exp(-i\hat{L}_S \tau) S_{m'}^{1(P_T)} \right\}$$

(9.8)

where the electron spin Liouville operator, $\hat{\hat{L}}_S$, contains the transient ZFS and the relaxation superoperator: $\hat{\hat{L}}_S = \hat{\hat{L}}_{ZFS}^T + i\hat{\hat{R}}^{S(S)}$. The idea of the factorization of the entire correlation function is presented in Fig.9.4.

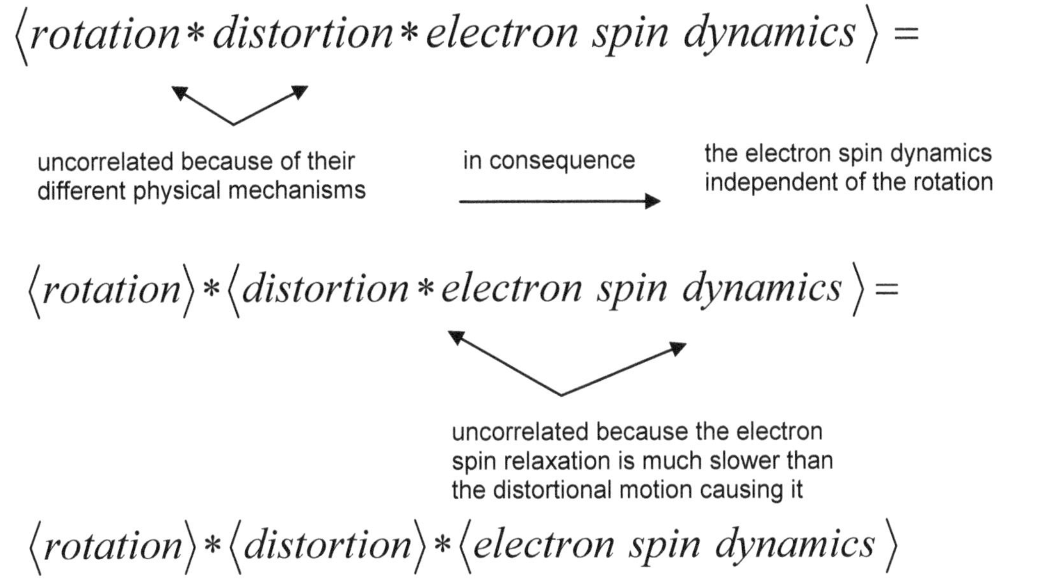

$$\langle rotation * distortion * electron\ spin\ dynamics \rangle =$$

uncorrelated because of their different physical mechanisms → in consequence → the electron spin dynamics independent of the rotation

$$\langle rotation \rangle * \langle distortion * electron\ spin\ dynamics \rangle =$$

uncorrelated because the electron spin relaxation is much slower than the distortional motion causing it

$$\langle rotation \rangle * \langle distortion \rangle * \langle electron\ spin\ dynamics \rangle$$

Fig.9.4
Factorization of the entire correlation function into the rotation, distortional and electron spin parts, performed for evaluating the nuclear spin relaxation at low field if the transient ZFS dominates over the static ZFS.

By employing the explicit forms of the rotational and distortional correlation functions $(\langle D_{m,k}^{1*}(\Omega(\tau))D_{m',k'}^{1}(\Omega(0))\rangle = \delta_{mm'}\delta_{kk'}\frac{1}{3}\exp\left(-\frac{\tau}{3\tau_c}\right))$ one can simplify Eq.9.6 to the form containing only the elements with $m = m', k = k', M = -M'$ and $N = -N'$. In addition, because of the symmetry properties of the 3j symbols [17-19] the terms with $J = 2$ and $J = 3$ vanish, and only the term for $J = 1$ contribute to the spectral density $K_{1,1}^{DD}(-\omega_I)$. Detailed considerations leading to the final formulation for the nuclear spin relaxation rate are presented in [16]. After substituting the single exponential correlation functions, characterized by the correlation times $3\tau_R$ and $3\tau_D$, for the rotational and the translational motion, respectively, (the factor 3 comes the fact that the we deal with the first order Wigner matrices), the explicit evaluations of the spectral density becomes straightforward. The final expression for the nuclear spin relaxation rate yields the form:

$$R_{1I} = \frac{4}{3}\left(\frac{\mu_0}{4\pi}\gamma_I\gamma_S\hbar\right)^2 \text{Re}\left\{s_{-1,-1}^{LF(T)} + s_{0,0}^{LF(T)} + s_{1,1}^{LF(T)}\right\} \qquad (9.9)$$

The labeling of the spectral densities indicates that they are evaluated at low field under the condition that $\omega_{ZFS}^T \ll \omega_{ZFS}^S$. They are defined as:

$$s_{m,m}^{LF(T)} = \frac{1}{2S+1}\left[S_m^{1(P_T)}\right]^+ \left[-i\hat{L}_{ZFS}^T + \hat{R}^{S(T)} + i\left(\frac{1}{3\tau_D}\right)\hat{1} + i\left(\omega_I + \frac{1}{3\tau_R}\right)\right]^{-1}\left[S_m^{1(P_T)}\right] \qquad (9.10)$$

where the vectors $\left[S_m^{1(P_T)}\right]$ include the expansion coefficients of the electron spin tensor components S_m^1 in the Liouville space constructed from the eigenfunctions of the transient ZFS Hamiltonian $H_{ZFS}^{T(P_T)}$. Eq.9.9 can be treated as a counterpart of the low field limit expression obtained under the opposite condition $\omega_{ZFS}^S \gg \omega_{ZFS}^T$. One can see immediately from the form of the spectral densities $s_{m,m}^{LF(T)}$ that the effective modulations of the dipole-dipole interaction are described as a superposition of the rotational motion, the distortional motion and the electron spin relaxation: $\tau_{c,i}^{-1} = (3\tau_R)^{-1} + (3\tau_D)^{-1} + R_{e,i}^{(P_T)}$. The first term reflects the 'direct' rotational modulation of the dipole-dipole axis split with the effects of the rotational contribution to the modulations of the P_T frame with respect to the laboratory axis. The second term originates from the fluctuations of the quantization axis for the electron spin with respect to the $I-S$ dipole-dipole axis due to the distortional motion, while the last term describes the electron spin relaxation in the P_T frame. This formulation does not reflect the fact that the electron spin correlation functions decaying multiexponentialy in the P_T frame, from the perspective of the dipole-dipole frame oscillate in time due to the relative movement of the two reference systems. One must be aware that the present case is valid if the condition $\omega_{ZFS}^T \tau_D \gg 1$, which 'fixes' the P_T frame to the dipole-dipole axis is fulfilled, otherwise it breaks down. The equivalent condition for the static ZFS is fulfilled always because the P_S frame is just fixed in the molecule. Finishing the discussion it is worthwhile to notice that the two relations: $\left|H_{ZFS}^S \tau_D\right| \ll 1$ and $\left|H_{ZFS}^T \tau_D\right| \gg 1$ imply that the secular approximation condition $\left(\omega_{ZFS}^S \tau_D\right)^2 \ll \omega_{ZFS}^T \tau_D$ is fulfilled as well.

IX.3. Validity regimes of relaxation theories based on perturbation treatment

We have treated in some detail in Chapters V and VI the electron (quadrupolar) spin and the dipolar spin relaxation, respectively, under the condition that the orientation of the principal axis system of the static interactions (the ZFS or the quadrupolar coupling) relative to the laboratory frame does not change in time. In contrary, in the present chapter we have allowed for very fast fluctuations of the molecular frame relative to the laboratory one, but assuming that the amplitudes of the static interactions dominate over the amplitudes of their transient counterparts. Being encouraged by the fact that at low magnetic field the perturbation approach gives a proper description of the dipolar spin relaxation independently of the timescale of these fluctuations we have put some effort to formulate an analogous description under the opposite condition, i.e. $\omega_{ZFS(Q)}^T \gg \omega_{ZFS(Q)}^S$ [16]. In fact, this description is much less successful. One can see from the examples that a rigorous treatment of the field dependent S and I spin relaxation has to overcome many difficulties. I shall not attempt to introduce such a treatment at this stage; I shall devote Chapter X to this issue. However, in this section I would like to discuss carefully validity regimes of the perturbation treatment applied to the $I - S$ spin systems, depending on the motional conditions and relative strengths of the spin interactions. Presenting, in this book, examples of various spin systems I always put attention on the validity criteria of the perturbation approach. Nevertheless, I believe that a compact overview of the mathematical formulations for the validity criteria of the perturbation relaxation theory based on the concept of the composite lattice and some comments how motional conditions combined with amplitudes of the Zeeman coupling and the second order interactions alter the computations can be interesting and useful for the reader.

If the orientation of the molecular frame with respect to the laboratory axis is fixed, the static parts of the corresponding second order interactions (the static ZFS or the static quadrupolar coupling) contribute, together with the Zeeman coupling, to the energy level structure of the electron (quadrupolar) spin. In this case the relaxation mechanism for the spin S is provided, independently of the applied magnetic field, by the transient ZFS or the transient part of the quadrupolar interaction, if they fulfill the Redfield condition $\omega_{ZFS}^T \tau_D \ll 1$ or $H_Q^T \tau_Q \ll 1$ for the electron and the quadrupolar spin, respectively. At the same time the main Hamiltonian must satisfy the relation $\dfrac{\omega_0}{\omega_1} \gg \omega_1 \tau_c$. The secular approximation condition is particularly important in the low field regime. We have seen that it leads to the conclusion that the static ZFS or the quadrupolar coupling, play the role of the main interactions only if they are large compared to the corresponding transient interactions: $\omega_{ZFS(Q)}^S \gg \omega_{ZFS(Q)}^T$. The validity criteria of the relaxation theory applied to the spin S have been formulated in Chapter V; we have repeated them at this stage for completeness of the discussion. They are depicted in Fig.9.5a,b.

a)

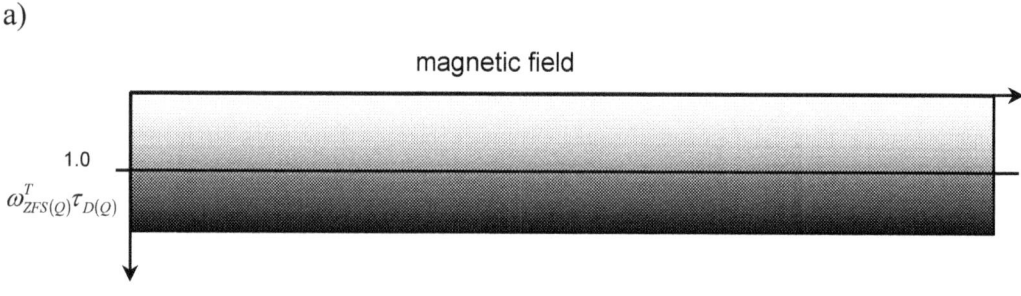

b)

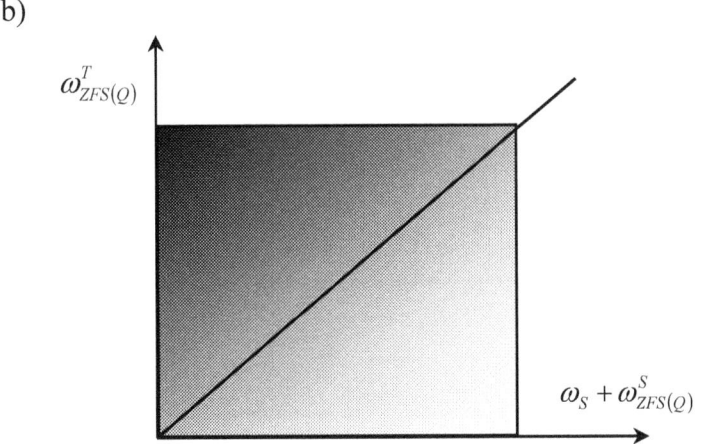

Fig.9.5
Validity criteria of the relaxation theory applied to the spin S under the condition that the principal axis system of the static ZFS (or the quadrupolar coupling) does not fluctuate relative to the laboratory frame. The transient ZFS (the transient quadrupolar coupling) provides, in this cases, the relaxation mechanism and therefore it must fulfill the Redfield condition $\omega_{ZFS(Q)}^T \tau_{D(Q)} << 1$ (a). The superposition of the Zeeman coupling and the static ZFS (the static quadrupolar coupling) determines the energy level structure of the electron (quadrupolar) spin if $\left(\omega_S + \omega_{ZFS(Q)}^S\right) >> \left(\omega_{ZFS(Q)}^T\right)$ (b); in particular at low field limit it is required that $\omega_{ZFS(Q)}^S >> \omega_{ZFS(Q)}^T$. The Redfield relaxation theory can be applied under the conditions illustrated by white areas while it breaks down for dark areas.

Since for the presently discussed case the S spin relaxation is in fact the only one, effective source of the modulation of the dipole-dipole interaction, describing, in the next step the I spin relaxation by the spectral density $K_{1,1}^{DD}(-\omega_I)$ (Chapter VI), one must consider with caution whether the relaxation of the spin S is efficient enough to fulfill the Redfield condition for the spin I: $\omega_{DD}^{IS} \tau_c << 1$. This problem is particularly important for systems containing slowly relaxing quadrupolar spins [5]; the electron spin relaxation is usually very fast.

One should be also aware that the Zeeman coupling can be treated as the main interaction for the spin I if it is strong enough to fulfill the secular approximation condition $\frac{\omega_I}{\omega_{DD}^{IS}} \gg \omega_{DD}^{IS}\tau_c$ (ω_{DD}^{IS} denotes here the magnitude of the $I-S$ dipole-dipole coupling in angular frequency units). The conditions for the spin I are illustrated in Fig.9.6a,b.

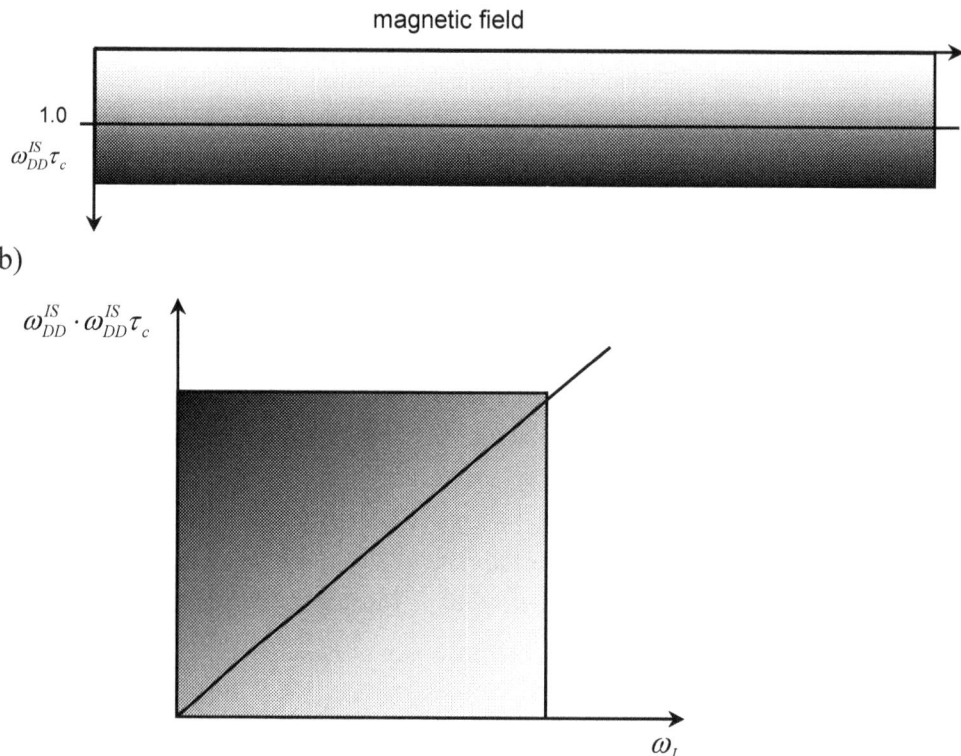

Fig.9.6
Validity conditions of the perturbation treatment applied to the I spin. The effective correlation time τ_c (including, in particular, the S spin relaxation) must be short enough to fulfill the Redfield condition, $\omega_{DD}^{IS}\tau_c \ll 1$ a). In addition, at low field one should check with caution on the secular approximation condition, $\frac{\omega_I}{\omega_{DD}^{IS}} \gg \omega_{DD}^{IS}\tau_c$. White areas indicate that the perturbation treatment is allowed while it is forbidden for dark areas.

This treatment can be extended to a certain degree to systems with faster modulations of the Ω_{DDL} angle. One can deal with motion fast enough to have a non-negligible effect on the dipole – dipole interaction, but still much slower than the S spin relaxation. In this motional regime the S spin possesses the same energy level structure resulting from the Zeeman coupling and the static ZFS (the static quadrupolar coupling), which from the point of view of the spin S still do not change in time their relative orientation. The motion modulating the orientation of the molecular frame and the S spin relaxation can be still treated as uncorrelated, because of timescale separation. Thus, the correlation function describing the whole system is still separable into a product of correlation functions, one for the S spin system and one for the 'direct' modulations of the dipole-dipole coupling. This motional regime has been discussed in the context of electron spin relaxation and rotational modulations of the $I-S$ dipole-dipole interaction in [20] and called the 'moderately slow rotation'. The reorientational correlation function, modeled in the simplest case as isotropic rotational diffusion, decays exponentially, contributing to the correlation function for the electron-nuclear dipole-dipole interaction. Since the electron spin relaxation is multiexponential and depends in addition on the molecular orientation it is very difficult to establish a strict mathematical condition for the rotational motion to be still significantly slower than the relaxation processes. The treatment breaks down starting from the high field limit since the S spin relaxation is here slower than at low field and therefore more close to the timescale of the motion responsible for the momentary orientation of the molecular and the laboratory frames. If this motion becomes faster the perturbation approach is not longer valid for the spin S. One cannot decompose the correlation function related to the modulations of the Ω_{DDL} angle from the S spin dynamics, because both the processes occur on a similar timescale [7, 20, 21]. From the perspective of the spin S one is not able to establish the role of the static component of the ZFS or the quadrupolar coupling, respectively. One may not include it into the main Hamiltonian, because its superposition with the Zeeman coupling does not lead to a well defined, time independent energy level structure any more. We cannot retain in this way any more the concept of stationary energy levels. One may not treat it as the relaxation mechanism, either, because the motion is not fast enough to fulfill the Redfield condition. Thus, the decomposition breaks down at the same time as the Redfield equation of motion is no longer valid for the electron spin. It implies of course that the relaxation of the spin I cannot be described in a manner which requires an explicit and clear definition of the S spin relaxation rates. If, because of high symmetry of the molecular environment, there is no static ZFS (no static quadrupolar coupling) the decomposition problem does not exist at all: the $I-S$ dipole-dipole axis changes its orientation due to, for example, molecular tumbling, while the electron (quadrupolar) spin relaxation is caused by a different motional process (for example lattice vibrations) affecting the corresponding transient interaction. Finally, we approach the motional limit when the static interactions are modulated fast enough, so that they fulfill the Redfield condition: for example $\omega^S_{ZFS(Q)}\tau_R \ll 1$. Under this condition, the static interactions provide a relaxation mechanism for the spin S. The analogous condition for the transient interactions is $\omega^T_{ZFS(Q)}\tau_{D(Q)} \ll 1$. Nevertheless, in the intermediate range of the magnetic field, if the Zeeman and the static second order interactions are comparable, it is impossible to decompose the total Hamiltonian into a main and a perturbing part, so that one cannot define the S spin relaxation. The validity regimes of the presented description of field dependent relaxation processes depending on the timescale of the motion affecting the $\Omega_{P_S(P_S^Q)L}$ angle are illustrated in Fig.9.7.

As has been mentioned above, in the next chapter I shall present an approach based on a full solution of the Liouville – von Neumann equation and therefore valid for arbitrary motional conditions and interaction strengths. Thus the discussed restrictions will not bother us any more. Nevertheless the price to pay for this great advantage is a high computational complexity of the general treatment, as we shall see.

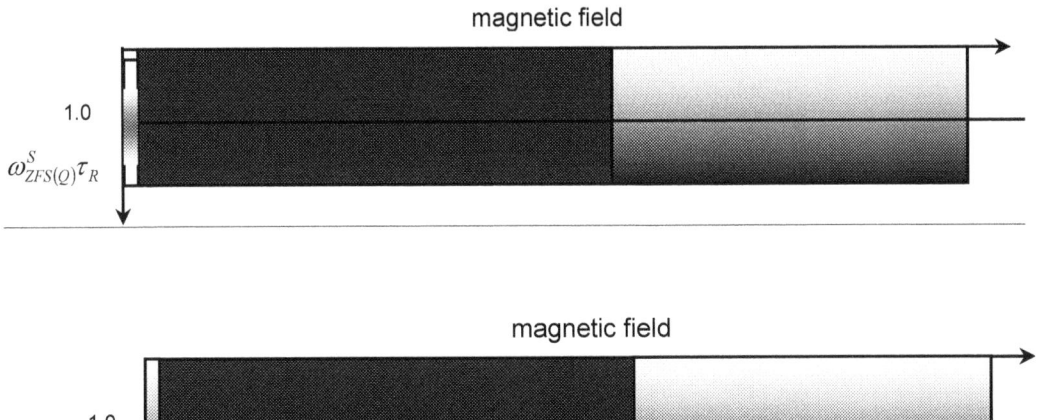

Fig.9.7
Validity regimes of the perturbation description of the electron (quadrupolar) spin relaxation processes depending on the timescale of the motion modulating the orientation of the principal axis system of the static ZFS (or of the static quadrupolar coupling, respectively) relative to the laboratory frame. It is assumed here that the static ZFS dominates over the transient counterpart (and analogously for the quadrupolar interaction). Then, at low field the static ZFS (the static quadrupolar coupling) is the main Hamiltonian and the electron (quadrupolar) spin is quantized in the molecule-fixed static ZFS principal frame (or the static quadrupolar interaction frame). Independently of the rate of the motion modulating the orientation of the molecular frame (the molecular tumbling or any other type of motion like the jump diffusion) the electron (quadrupolar) spin relaxation is caused by the transient ZFS (the transient quadrupolar coupling) modulated by the distortional (vibrational) motion. In the intermediate regime when the amplitudes of the static interaction and of the Zeeman couplings are comparable one cannot decompose the entire Hamiltonian into main and perturbing parts, except the case when the motion modulating the relative orientation of the principal axis systems of these two interactions is much slower than the electron (quadrupolar) spin relaxation, as illustrated already in Fig.9.5. A similar timescale of the modulations and the electron (quadrupolar) spin relaxation variables prevents one from variables prevent one from separating the motion responsible for the modulations or explicitly defining electron (quadrupolar) spin relaxation rates in terms of spectral densities of the Redfield-

like theory. In high field there are two relaxation channels for the electron (quadrupolar) spin relaxation; it is caused by the static and transient parts of the ZFS (quadrupolar) interaction, as described in the text. Therefore both the components have to fulfill the Redfield condition.

References:

1. D. Kruk, T. Nilsson, J. Kowalewski, Nuclear spin relaxation in paramagnetic systems with zero-field splitting and arbitrary electron spin, *Phys. Chem. Chem. Phys.* **3** (2001) 4907-4917

2. I. Bertini, J. Kowalewski, C. Luchinat, T. Nilsson, G. Parigi, Nuclear spin relaxation in paramagnetic complexes of S=1: electron spin relaxation effects, *J. Chem. Phys.* **111** (1999) 5795-5807

3. J. Kowalewski, D. Kruk, G. Parigi, NMR relaxation in solution of paramagnetic complexes: Recent theoretical progress for S >= 1, *Advances in Inorganic Chemistry*, **57** (2005) 41-104

4. R. Sharp, L. Lohr, J. Miller, Paramagnetic NMR relaxation enhancement: recent advances in theory, *Prog. NMR Spectr.* **38** (2001) 115-158

5. D. Kruk, O. Lips, Field-dependent nuclear relaxation of spins ½ induced by dipole –dipole couplings to quadrupolar spins: LaF_3 crystals as an example, *J. Magn. Reson.* **179** (2006) 250 – 262

6. D. Kruk, O. Lips, Evolution of solid state systems containing mutually coupled dipolar and quadrupolar spins: a perturbation treatment, *Solid State Nucl. Magn. Reson.* **28 (2-4)** (2005) 180-192

7. D. Kruk, J. Kowalewski, Nuclear spin relaxation in paramagnetic systems (S >= 1) under fast rotation conditions, *J. Magn. Reson.* **162 (2)** (2003) 229-240

8. S. Rast, P.H. Fries, E. Belorizky, A. Borel, L. Helm, A.E. Merbach, A general approach to the electronic spin relaxation of Gd(III) complexes in solutions. Monte Carlo simulations beyond the Redfield limit, *J. Chem. Phys.* **115** (2001) 7554-7563

9. S. Rast, P.H. Fries, E. Belorizky, Static zero field splitting effects on the electronic relaxation of paramagnetic metal ion complexes in solution, *J. Chem. Phys.* **113** (2000) 8724-8735

10. S. Rast, A. Borel, L. Helm, A. E. Belorizky, H. P. Fries, A. E. Merbach, EPR Spectroscopy of MRI-related Gd(III) complexes: simultaneous analysis of multiple frequency and temperature spectra, including static and transient crystal field effects, *J. Am. Chem. Soc.* **123** (2001) 2637-2644

11. A. Abragam, The principles of nuclear magnetism, *Oxford University Press, Oxford* (1961)

12. C.P. Slichter, Principles of magnetic resonance, *Springer - Verlag, Berlin* (1990)
13. M. Goldman, Formal theory of spin – lattice relaxation, *J. Magn. Reson.* **149** (2001) 160-187
14. J. Kowalewski, C. Luchinat, T. Nilsson, G. Parigi, Nuclear spin relaxation in paramagnetic systems: electron spin relaxation effects under near-Redfield limit conditions and beyond, *J. Phys. Chem A* **106** (2002) 7376-7382
15. T. Nilsson, J. Kowalewski, Low-field theory of nuclear spin relaxation in paramagnetic low – symmetry complexes for electron spin systems of S=1,3/2, 2, 5/2, 3 and 7/ 2, *Mol. Phys*. **98** (2000) 1617-1638, Erratum: *Mol. Phys*. **99** (2001) 369-370
16. D. Kruk, J. Kowalewski, Nuclear spin relaxation in solution of paramagnetic complexes with large transient zero-field splitting, *Mol. Physc*. **101 (18)** (2003) 2861-2874
17. M. Rotenberg, R. Bivins, N. Metropolis, J.K. Wooten, The 3-J and 6-J symbols, *Technology Press, Cambridge* (1959)
18. D. M. Brink, G.R. Satchler, Angular momentum, *Clarendon Press, Oxford* (1979)
19. D.A. Varshalovich, A.N. Moskalev, V.K. Khersonkii, Quantum theory of angular momentum, *Word Scientific Publishing, Singapore* (1988)
20. D. Kruk, J. Kowalewski, Field-dependent proton relaxation in aqueous solutions of some manganese (II) complexes: a new interpretation, *J. Biol. Inorg. Chem*. **8 (5)** (2003) 512-518
21. S.M. Abernathy, R. Sharp, Spin dynamics calculations of electron and nuclear spin relaxation times in paramagnetic solutions, *J. Chem. Phys*. **106** (1997) 9032-904

CHAPTER X

Dynamics of spin systems beyond validity regimes of the second order perturbation theory

So far we have considered in detail spin systems fulfilling the validity conditions of the second order perturbation theory. As it has been demonstrated in several examples, one can solve within this theoretical framework complicated problems of spin dynamics resulting from interplay between various spin interactions. In the previous chapters, devoted to different aspects of the spin evolution, we have discussed very cautiously validity criteria of the proposed approaches. That is why we become quite well acquainted with various restrictions and warnings. We learned a lot about the limited applicability of the perturbation solutions, however so far we have not managed to go beyond the limits. Therefore several important and interesting problems remained unsolved. We needed to stop our considerations, with the frustrating statement that it is not allowed to proceed further since the perturbation treatment breaks down.

In this chapter I shall present and discuss in detail a much more general treatment, valid far beyond the applicability of any perturbation approaches. It sounds like providing a perfect tool for describing the spin dynamics for arbitrary conditions. Actually it is so. The tool, which I shall discuss, is nothing else than a full solution of the Liouville von - Neumann equation based on the multipole representation of tensor operators introduced in Chapter II and the Wigner-Eckart theorem [1, 2]. In principle, one can ask at this stage why one puts so much effort on dealing with perturbation methods if we have the general treatment at our disposal. There are two reasons for that. The first one is a high conceptual and computational complexity of this general approach. In fact, the computational aspect becomes less important due to fast progress in computer technology, but there is no way to escape from the necessary conceptual effort. The second reason is even more important. Limiting approaches, based on the perturbation theory are very valuable. They end up with closed form expressions providing deep insight into the essential physical mechanisms of the spin dynamics, which are much less apparent otherwise.

I shall demonstrate the general approach applying it first to the S spin system. Since I intend to discuss mainly the case when the S spin does not fulfill the conditions of the relaxation perturbation theory and therefore its dynamics cannot be described in terms of relaxation rates, we shall focus attention on the S spin spectral density $s_{-1,-1}(\omega) = \int_0^\infty \langle S_{-1}^+(\tau) S_{-1}(0) \rangle \exp(-i\omega\tau) d\tau$. In the case of an electron spin S this quantity determines its ESR lineshape [3-9]. Next, I shall extend the spin system and apply the general theory to spin – lattice relaxation of a spin I coupled by a dipole-dipole interaction to the S spin. Thus, in the following a complete and general theory is presented where all the principal stages of the derivation of the S spin lineshape and the I spin relaxation are discussed in detail. Even though we shall refer to an electron spin S, it is worthwhile to remember that the case of a quadrupolar spin system can be treated in an analogous manner.

X.1. General description of ESR and quadrupolar spectra

As explained above, the complete ESR lineshape function $L(\omega)$ is determined by the spectral density $s_{-1,-1}(\omega)$ ($L(\omega) \propto s_{-1,-1}(\omega)$) corresponding to the single-quantum transitions of the electron spin. Thus, derivations of the ESR lineshapes have the same procedure as that in the part of the electron spin-spin spectral density, described in Chapters IV and V. The lineshape function is given as:

$$L(\omega_S - \omega) = \int_0^\infty Tr_S \left\{ S_{-1}^{1+} \left[\exp(-i\hat{\hat{L}}_S \tau) S_{-1}^1 \right] \rho_S^{eq} \right\} \exp(-i\omega\tau) d\tau \propto \left[S_{-1}^1 \right]^+ \left[\hat{\hat{M}}_{ESR} \right]^{-1} \left[S_{-1}^1 \right] \quad (10.1)$$

where the real and imaginary parts represent the absorption and dispersion spectra, respectively. The superoperator $\hat{\hat{M}}_{ESR}$ results from the same spin Hamiltonian model as the one employed in Chapters IV, V and IX to describe the electron spin relaxation under various motional conditions. It consists of the Zeeman Liouville operator taken at the frequency ω_S, $\hat{\hat{L}}_Z(\omega_S)$, the static and transient ZFS operators, the rotational and translational Liouvilians, and the operator $\omega\hat{\hat{1}}$ containing the frequency ω resulting from the Fourier transform, *i.e.*:

$$\hat{\hat{M}}_{ESR} = -i\left[\hat{\hat{L}}_Z(\omega_S) + \hat{\hat{L}}_{ZFS}^S + \hat{\hat{L}}_{ZFS}^T + \hat{\hat{L}}_R + \hat{\hat{L}}_D + \hat{\hat{1}}\omega \right] \quad (10.2)$$

One can provide an analytical description of the ESR lineshape only if the electron spin is within the Redfield limit. The requirements of the perturbation approach have been discussed in detail in the previous chapter. One could see from this discussion that the product of the amplitude of the static ZFS and the rotational correlation time, $\omega_{ZFS}^S \tau_R$, is crucial for the explicit, clear definition of the electron spin relaxation. ESR spectra are collected at rather high magnetic fields. The so called X band corresponds to the magnetic field of 0.3-0.35T, while the Q and W bands are measured at 1.1-1.3T and 3.3-3.5T, respectively. Therefore, if the condition $\omega_{ZFS}^S \tau_R \ll 1$ is fulfilled, the static ZFS provides, beside the transient ZFS (if $\omega_{ZFS}^T \tau_D \ll 1$) a relaxation channel for the electron spin. In this case the operator $\hat{\hat{M}}_{ESR}$ takes the form: $\hat{\hat{M}}_{ESR} = -i\hat{\hat{L}}_Z(\omega_S) + \hat{\hat{R}}_{ZFS}^{S(S)}(\omega_S) + \hat{\hat{R}}_{ZFS}^{S(T)}(\omega_S) - i\omega\hat{\hat{1}}$; the individual spectral densities encoded in the relaxation superoperators correspond to the frequency ω_S. In the opposite motional regime, when the molecular frame (the principal axis system of the static ZFS tensor) is fixed relative to the laboratory one, the static ZFS modifies the energy level structure of the electron spin. One can argue at this moment that typically the amplitude of the static ZFS is rather small comparing to the amplitude of the Zeeman coupling for the magnetic field at which the ESR spectra are measured and therefore it can be neglected altogether. This way of thinking is quite dangerous. It is not enough to compare just the amplitudes of the ZFS and Zeeman couplings. To be sure that the effect of the static ZFS is negligible one has to take into account how these two interactions influence the electron spin energy levels. To illustrate this statement let us consider the case of $S = \frac{7}{2}$. To simplify the problem, we assume that the molecular and laboratory frames coincide. The basis appropriate for the spin quantum number $S = \frac{7}{2}$ consists of $(2S+1) = 8$ functions formed by the Zeeman states $\{|m_S\rangle\}$ (m_S is the magnetic spin quantum number). The matrix element of the static Hamiltonian H_0^S including the Zeeman coupling as well as the static ZFS, $H_0^S = H_Z(S) + H_{ZFS}^S$ taken between the states of $m_S = \frac{1}{2}$ is equal to:

$\left\langle \frac{1}{2} \middle| H_0^S \middle| \frac{1}{2} \right\rangle = -5D_S + \frac{1}{2}\omega_S$. One can see from this expression that the real effect of the static ZFS is higher by a factor of 10 than one could expect comparing only the strengths of the two couplings. This example should be treated as a warning that a relatively weak interaction can alter significantly the energy level structure and such effects have to be always treated with high caution. However, we have seen in Chapter V how to describe the electron spin relaxation in the presence of the static ZFS tensor being fixed with respect to the laboratory frame and we are able to manage this for an arbitrary magnetic field. In this case the operator $\hat{M}_{ESR}$ must be defined for every orientation of the static ZFS tensor relative to the laboratory frame. It includes the main Liouvilian $\hat{L}_0^S = \hat{L}_Z^S(\omega_S) + \hat{L}_{ZFS}^S$ corresponding to the main Hamiltonian H_0^S and the relaxation operator $\hat{\hat{R}}_{ZFS}^{S(T)}$ representing the transient ZFS relaxation mechanism (if $\omega_{ZFS}^T \tau_D \ll 1$):

$\hat{\hat{M}}_{ESR} = -i\hat{\hat{L}}_Z(\omega_S) - i\hat{\hat{L}}_{ZFS}^S + \hat{\hat{R}}_{ZFS}^{S(T)}(\omega_S) - i\omega\hat{\hat{1}}$. Therefore, the evaluation of the ESR lineshape should be quite straightforward, even though we have to include the static ZFS. Indeed, we can calculate, following the procedure described in Chapters IV and V, the electron spin spectral densities $s_{m,n}$, in particular $s_{-1,-1}$. We have seen that, for this purpose, one has to calculate the projection vector $[S_{-1}^1]$ containing the expansion coefficients of the electron spin operator S_- into the Liouville basis constructed from the eigenvectors of the main Hamiltonian H_0^S. The eigenvectors are given as combinations of the Zeeman functions $\{|m_S\rangle\}$ mixed up due to the static ZFS contributing to the main Hamiltonian. In consequence, the representations of the operators S_- and S_z can contain some common elements, $|m_S\rangle\langle m_S'|$. This implies that the spin-spin and spin-lattice relaxation processes cannot be treated as independent processes. Thus, one can expect some effects of the spin-lattice electron spin relaxation on the ESR spectra, and this causes considerable complications. We know also from Chapter IX that despite the limiting cases of very slow molecular tumbling (when the relative orientation of the static ZFS tensor and the laboratory frame is fixed) and of fast molecular tumbling (that $\omega_{ZFS}^S \tau_R << 1$) one cannot describe the electron spin dynamics by a relaxation operator if the static ZFS is present. It brings us to the conclusion that, in fact, we are able to describe analytically the ESR spectra for few limiting cases: when the fluctuations of the orientation of the static ZFS tensor are very fast or when the static ZFS is very weak because of high molecular symmetry. It is somewhat difficult to accept such strong theoretical restrictions regarding such fundamental experimental results like ESR spectra. These restrictions do not concern the treatment presented below.

In order to evaluate the electron spin spectral density we have to set up a matrix representation of the operator $\hat{\hat{M}}_{ESR}$. This can be achieved even though the perturbation description of the electron spin dynamics breaks down. In principle this is a matter of constructing an appropriate basis, which includes all relevant degrees of freedom, represented by the Liouville operators in Eq.10.2. The main reasons for computational complexity are the classical stochastic processes, *i.e.* the rotational and distortional degrees of freedom. Their classical nature requires, in principle, setting up an infinite basis representing the continuum of rotational and distortional states. In this context it is much easier to deal with the spin variables; the number of the spin states is always finite and well defined ($(2S+1)$ in the Hilbert space and $(2S+1)^2$ within the Liouville representation). In fact, by introducing the relaxation superoperators $\hat{R}_{ZFS}^{S(T)}$ and $\hat{R}_{ZFS}^{S(S)}$ associated with the distortional and rotational motion, we avoid an explicit treatment of the infinite sets of the rotational and distortional states. If the electron spin is beyond the validity regimes of the Redfield description, one may not define the relaxation superoperator any more. One can attempt a solution of this problem by defining an infinite and complete, orthonormal basis $\{|O_i)\}$ [10-21] as direct product of three types of orthonormal basis operators, $|O_i) = |ABC) \otimes |LKM) \otimes |\Sigma\sigma)$. The first one, $|ABC)$, is determined by the quantum numbers A, B, C associated with the Wigner matrices $D_{B,C}^A$ representing the distortional motion. The second operator $|LKM)$ is defined in full analogy to the first one; it is related to the Wigner matrices $D_{K,M}^L$ and is associated with the rotation. The explicit definitions are as follows:

$$|ABC) = |ABC\rangle\langle ABC| = |\Psi_{BC}^A) = \sqrt{\frac{2A+1}{8\pi^2}} D_{BC}^A(\Omega_{P_T P_S}) \quad (10.3)$$

$$|LKM) = |LKM\rangle\langle LKM| = |\Psi_{KM}^L) = \sqrt{\frac{2L+1}{8\pi^2}} D_{KM}^L(\Omega_{P_S L}) \quad (10.4)$$

The operator $|\Sigma\sigma)$ is related to the basis vectors $|S, m_S\rangle\langle S, m_S'|$ forming the Liouville space for the S spin. It has been already defined in Chapter II; here, we repeat the definition for convenience of the reader:

$$|\Sigma\sigma) = \sum_m (-1)^{S-m-\sigma} \sqrt{2\Sigma+1} \begin{pmatrix} S & S & \Sigma \\ m+\sigma & -m & -\sigma \end{pmatrix} |S, m+\sigma\rangle\langle S, m| \quad (10.5)$$

where Σ ranges from 1 to $2S$. The orthonormal tensor operators $|\Sigma\sigma)$ differ from the spin tensor operators S_σ^Σ of the first and second ranks S_σ^1 and S_σ^2 by constant

Chapter X - Dynamics of spin systems beyond validity regimes

factors, according to Eq.2.28a,b. Due to the appropriate normalization introduced in Eqs.10.3, 10.4, 10.6 and 10.7 the operators $|O_i)$ fulfill the condition: $(O_i|O_j) \equiv \int d\Omega_{P_T P_S} d\Omega_{P_S L} Tr(O_i^+ O_j) = \delta_{ij}$. In the next step, we have to evaluate the matrix elements of the entire operator $\hat{\hat{M}}_{ESR}$ within the basis $\{|O_i)\}$: $[\hat{\hat{M}}_{ESR}]_{ij} = (O_i|\hat{\hat{M}}_{ESR}|O_j)$. The evaluations can be performed by employing the Wigner-Eckart theorem [1, 2]. Such an approach is very general and flexible, though tedious. It can be applied for any spin system evolving in time under an arbitrary Hamiltonian H. However, depending on the nature of the degrees of freedom relevant for the evolution of the spin system under interest and applied motional models, one has to adjust in an appropriate manner the generalized basis $\{|O_i)\}$. In the present case the basis includes, beside the spin functions, the vectors related to the Wigner rotational matrices. It is well suited to the applied models of the static and transient ZFS interactions. If one wishes to consider different motional models linking, for example, the fluctuations of the transient ZFS to temporary molecular geometry described in terms of normal coordinates, as it has been done in [22, 23], one has to include into the basis $\{|O_i)\}$ appropriate vectors related to the normal modes. Therefore, to profit fully from the generality of the present treatment it is particularly important to get some experience with calculations based on the Wigner-Eckart theorem. This issue is treated in the literature relatively less frequently compared to the perturbation approaches to spin dynamics. Therefore, even though final expressions for the matrix elements of the individual Liouville operators contributing to the operator $\hat{\hat{M}}_{ESR}$ have been presented in [19,20], we shall present here some important steps of the calculations. Before proceeding, it is worthwhile to repeat that all Hamiltonians generating the Liouvillians involved in the operator $\hat{\hat{M}}_{ESR}$ have to be expressed in the same reference frame. In addition, to avoid needless complications, it is highly advisable to choose the laboratory frame as the reference one. Let us begin from the static ZFS. The form of the Hamiltonian of Eq.3.10 is complicated enough to demonstrate how to apply the Wigner-Eckart theorem to nontrivial cases and, on the other hand, simple enough to obtain a closed form result avoiding long calculations. It is very useful for the evaluations to rewrite the static ZFS Hamiltonian in terms of the normalized electron spin tensor operators, $|\Sigma\sigma) \equiv |2n) \equiv Q_n^2$:

$$H_{ZFS}^S = \sqrt{\frac{(2S+3)(2S+1)(S+1)(2S-1)}{30}} \sum_{n,m=-2}^{2} (-1)^n \widetilde{V}_m^{2S(P_S)} D_{m,n}^2(\Omega_{P_SL}) Q_n^2 \qquad (10.6)$$

According to the Wigner-Eckart theorem we can separate in the evaluations of the matrix element, $\left(O_i | \hat{L}_{ZFS}^S | O_j\right) \equiv \left(O' | \hat{L}_{ZFS}^S | O\right)$, the spin and rotational variables:

$$\left(A'B'C' \left| L'K'M' \right| \left(\Sigma'\sigma' | \hat{L}_{ZFS}^S | \Sigma\sigma\right) | LKM \right| ABC\right) =$$

$$\delta_{A'A}\delta_{B'B}\delta_{C'C} \sqrt{\frac{(2S+3)(2S+1)(S+1)(2S-1)}{30}} \times \qquad (10.7)$$

$$\sum_{n,=-2}^{2} (-1)^n Tr_S\left\{Q_{\sigma'}^{\Sigma'+}\left[Q_{-n}^2, Q_\sigma^\Sigma\right]\right\} \times \left\{\sum_{m=-2}^{2} \widetilde{V}_m^{2S(P_S)} \int \Psi_{K'M'}^{L'*}(\Omega_{ML}) D_{m,n}^2(\Omega_{ML}) \Psi_{KM}^L(\Omega_{ML}) d\Omega_{ML}\right\}$$

Matters are greatly simplified if we make use of the two relations; the first one concerns the spin variables:

$$Tr\left\{Q_\alpha^{A+}\left[Q_\beta^B, Q_\gamma^C\right]\right\} =$$

$$\sqrt{(2A+1)(2B+1)(2C+1)} (-1)^{-\alpha} \begin{pmatrix} A & B & C \\ -\alpha & \beta & \gamma \end{pmatrix} \begin{Bmatrix} A & B & C \\ S & S & S \end{Bmatrix} \left[(-1)^{A+B+C} - 1\right] \qquad (10.8)$$

while the second one results from the properties of the Wigner matrices and gives us a closed form expression for the integral over the rotational variables:

$$\int \Psi_{K'M'}^{L'*}(\Omega_{ML}) D_{m,n}^2(\Omega_{ML}) \Psi_{KM}^L(\Omega_{ML}) d\Omega_{ML} =$$

$$\sqrt{(2L'+1)(2L+1)} (-1)^{K'-M'} \begin{pmatrix} L' & 2 & L \\ -K' & m & K \end{pmatrix} \begin{pmatrix} L' & 2 & L \\ -M' & n & M \end{pmatrix} \qquad (10.9)$$

Introducing Eq.10.8 and Eq.10.9 into the expression of Eq.10.7 one obtains:

$$\left(A'B'C'\left|\left(L'K'M'\left|\left(\Sigma'\sigma'\left|\hat{L}^S_{ZFS}\right|\Sigma\sigma\right)LKM\right)\right|ABC\right)=$$

$$\delta_{A'A}\delta_{B'B}\delta_{C'C}\sqrt{\frac{(2S+3)(2S+1)(S+1)(2S-1)}{30}}\times \qquad (10.10)$$

$$\sqrt{5}\sqrt{(2L'+1)(2L+1)(2\Sigma+1)(2\Sigma'+1)}\begin{Bmatrix}\Sigma' & 2 & \Sigma \\ S & S & S\end{Bmatrix}\times$$

$$\sum_{n,m=-2}^{2}(-1)^{n-\sigma'+K'-M'}\left[(-1)^{\Sigma+\Sigma'}-1\right]\widetilde{V}_m^{2S(P_S)}\begin{pmatrix}\Sigma' & 2 & \Sigma \\ -\sigma' & -n & \sigma\end{pmatrix}\begin{pmatrix}L' & 2 & L \\ -K' & m & K\end{pmatrix}\begin{pmatrix}L' & 2 & L \\ -M' & n & M\end{pmatrix}$$

Taking into account properties of the 3j coefficients [24 - 26] discussed in Chapter II, we can considerably simplify the above equations. Thus, the final expression for the matrix element $\left(O_i\left|\hat{L}^S_{ZFS}\right|O_j\right)$ of the static ZFS Liouville operator yields the form:

$$\left(O_i\left|\hat{L}^S_{ZFS}\right|O_j\right)=\left(A'B'C'\left|\left(L'K'M'\left|\left(\Sigma'\sigma'\left|\hat{L}^S_{ZFS}\right|ABC\right)LKM\right)\Sigma\sigma\right)\right.=\delta_{AA'}\delta_{BB'}\delta_{CC'}\times$$

$$\frac{1}{\sqrt{6}}(-1)^{\sigma+B'-C'}\widetilde{V}_{|K-K'|}^{2S(P_S)}\left[(-1)^{\Sigma'+\Sigma}-1\right]\times \qquad (10.11)$$

$$\sqrt{(2S+3)(2S+1)(S+1)S(2S-1)(2L'+1)(2L+1)(2\Sigma'+1)(2\Sigma+1)}\times$$

$$\begin{pmatrix}L' & 2 & L \\ -K' & K'-K & K\end{pmatrix}\begin{pmatrix}L' & 2 & L \\ -M' & M'-M & M\end{pmatrix}\begin{pmatrix}\Sigma' & 2 & \Sigma \\ -\sigma' & B'-B & \sigma\end{pmatrix}\begin{Bmatrix}\Sigma' & 2 & \Sigma \\ S & S & S\end{Bmatrix}$$

where the symbol { } denotes 6-j coefficients. Assuming, according to Eq.3.12, that the distortional motion has a dominant contribution to the modulations of the transient ZFS, one can obtain in a straightforward manner an analogous expression for the transient ZFS Liouvilian. It can be managed just by replacing the quantities $\widetilde{V}_m^{2S(P_S)}$ describing the static ZFS tensor by their counterparts $\widetilde{V}_m^{2T(P_T)}$ and the rotational quantum numbers LKM ($L'K'M'$) by the distortional ones ABC ($A'B'C'$). If one wishes to include the effects of the rotational motion on the transient ZFS present in Eq.3.13 the computations of the matrix elements require some more effort. I do not present here the computational details since they do not contain in principle new elements. One has to deal with two sets of Wigner rotation matrices instead of one and this makes the calculations more lengthy and cumbersome, however the computational procedure is essentially the same. At last the matrix elements of the transient ZFS Liouvilian including the distortional as well as rotational degrees of freedom take the form:

$$\left(O_i \middle| \hat{L}_{ZFS}^T \middle| O_j\right) =$$

$$\left(A'B'C'\middle|(\Sigma'\sigma'\middle|\hat{L}_{ZFS}^T\middle|ABC)\Sigma\sigma\right) = (-1)^{\sigma+B'+K'-C'-M'} \widetilde{V}_{|B-B'|}^{2T(P_T)} \frac{1}{\sqrt{6}} \left[(-1)^{\Sigma'+2+\Sigma}-1\right] \times$$

$$\sqrt{(2S+3)(2S+1)(S+1)S(2S-1)(2A'+1)(2A+1)(2L'+1)(2L+1)(2\Sigma'+1)(2\Sigma+1)} \times$$

$$\begin{pmatrix} A' & 2 & A \\ -B' & B'-B & B \end{pmatrix} \begin{pmatrix} A' & 2 & A \\ -C' & C'-C & C \end{pmatrix} \begin{pmatrix} L' & 2 & L \\ -K' & C'-C & K \end{pmatrix} \times$$

$$\begin{pmatrix} A' & 2 & A \\ -M' & M'-M & M \end{pmatrix} \begin{pmatrix} \Sigma' & 2 & \Sigma \\ -\sigma' & M-M' & \sigma \end{pmatrix} \begin{Bmatrix} \Sigma' & 2 & \Sigma \\ S' & S & S \end{Bmatrix}$$

(10.12)

It can be of some interest to demonstrate how one can derive the matrix elements of the Liouville operators representing the rotational and distortional motion. Since these operators describe isotropic rotational diffusion, their forms result from the rotational diffusion equation, *i.e.*: $\hat{L}_R = -iD_R \nabla^2_{\Omega_{PSL}}$ and $\hat{L}_D = -iD_D \nabla^2_{\Omega_{P_T P_S}}$; D_R and D_D are the corresponding diffusion constants. The derivations are confined, in principle, to the matrix elements taken between the rotational and distortional states, respectively. Since the eigenvalue equation for the diffusion operator is $\nabla^2_\Omega D^L_{MK} = -L(L+1)D^L_{MK}$ one gets for the rotational operator:

$$\left(A'B'C'\middle|(L'K'M'\middle|(\Sigma'\sigma'\middle|\hat{L}_R\middle|ABC)LKM)\Sigma\sigma\right) =$$

$$\delta_{AA'}\delta_{BB'}\delta_{CC'}\delta_{LL'}\delta_{KK'}\delta_{MM'}\delta_{\Sigma\Sigma'}\delta_{\sigma\sigma'} iD_R L(L+1)$$

(10.13)

The matrix elements of the distortional Liouville operator $\hat{L}_D$ have an analogous form:

$$\left(A'B'C'\middle|(L'K'M'\middle|(\Sigma'\sigma'\middle|\hat{L}_R\middle|ABC)LKM)\Sigma\sigma\right) =$$

$$\delta_{AA'}\delta_{BB'}\delta_{CC'}\delta_{LL'}\delta_{KK'}\delta_{MM'}\delta_{\Sigma\Sigma'}\delta_{\sigma\sigma'} iD_D A(A+1)$$

(10.14)

The diffusion constants are related to the correlation times (corresponding to second order Wigner matrices as: $D_{R(D)} = \left(6\tau_{R(D)}\right)^{-1}$). Evaluations of the quantities

$\left(O_i|\hat{L}_z|O_j\right)$ and $\left(O_i|\hat{1}|O_j\right)$ do not require any further explanations; we just write down the final expressions:

$$\left(A'B'C'|(L'K'M'|(\Sigma'\sigma'|\hat{L}_z|ABC)LKM)\Sigma\sigma\right) = \\ \delta_{AA'}\delta_{BB'}\delta_{CC'}\delta_{LL'}\delta_{KK'}\delta_{MM'}\delta_{\Sigma\Sigma'}\delta_{\sigma\sigma'}\omega_S\sigma \quad (10.15)$$

$$\left(A'B'C'|(L'K'M'|(\Sigma'\sigma'|\hat{1}|ABC)LKM)\Sigma\sigma\right) = \\ \delta_{AA'}\delta_{BB'}\delta_{CC'}\delta_{LL'}\delta_{KK'}\delta_{MM'}\delta_{\Sigma\Sigma'}\delta_{\sigma\sigma'} \quad (10.16)$$

We have completed in this way the matrix representation of the entire operator $\hat{M}_{ESR}$. It is of particular interest to notice that the non-diagonal elements are created by the ZFS term, while the diagonal part of the matrix represents the distortional and rotational motions as well as the Zeeman interaction. As it has been pointed out above, the matrix is in principle infinitely large, due to the incorporation of the classical distortional and rotational degrees of freedom; the quantum numbers A and L range from zero to infinity. To be able to perform the computations the basis set must be truncated. In practice, the size of the matrix is increased step by step, until convergence of the desired accuracy is accomplished. The convergence should be tested for every new set of parameter values. The projection vector $[S^1_{-1}]$ contains the expansion coefficients of the tensor operator S^1_{-1} in the basis $|O_i\rangle$. In fact, there is just one non-zero coefficient, namely the one associated with the basis vector $|ABC\rangle|LKM\rangle|\Sigma\sigma\rangle = |000\rangle|000\rangle|1-1\rangle$. This considerably simplifies the computational task, since we are interested only in one element of the inverted matrix $\left[\hat{M}_{ESR}\right]^{-1}$.

Fig.10.1 shows an example of ESR spectrum for the electron spin quantum number $S = \frac{7}{2}$ calculated assuming that there is no static ZFS. In this case, the rotational motion does not affect the lineshapes. The reader may wish to compare, for this particular case, the predictions of the perturbation treatment and the general one. Therefore, we present in the figures theoretical lineshapes evaluated in two ways: by introducing the relaxation superoperator $\hat{\hat{R}}_{ZFS}^{T}$ and by dealing with the large set of quantum states $\{|ABC\rangle|LKM\rangle|\Sigma\sigma\rangle\}$ available for the electron spin. The general treatment is not necessary in this case; however, we have performed such calculations to convince the reader that the two approaches lead to quite similar results.

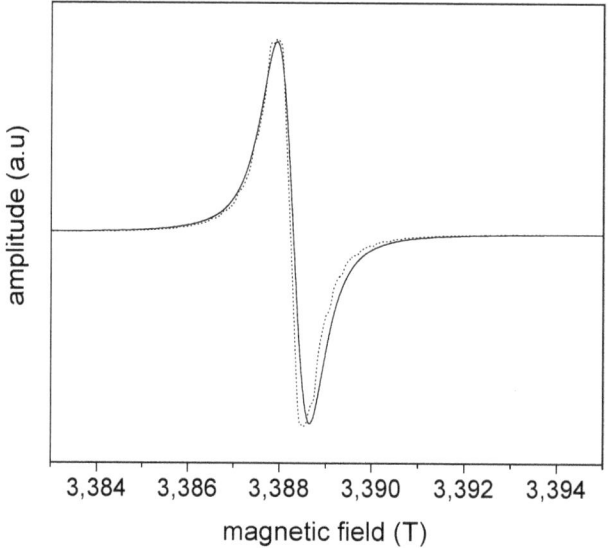

Fig.10.1

ESR spectrum – W band (3.39T) calculated within the general approach (solid line) and applying the perturbation treatment (dashed line); $D_T = 0.04 cm^{-1}$, $\tau_D = 30 ps$, no static ZFS.

Fig.10.2a,b and Fig.10.3-10.5 show next examples of ESR spectra ($S = \frac{7}{2}$) evaluated within the present theoretical framework. It is worthwhile to put some attention on the effects of the static ZFS and the rotational motion. Since experimental ESR spectra are collected as a derivative of absorption lines, the theoretical lineshapes are presented here in the same manner.

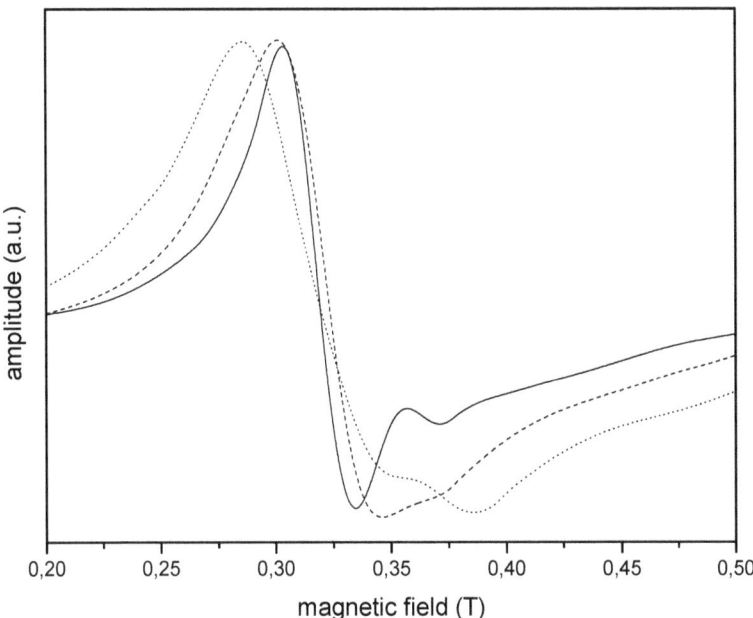

Fig.10. 2a

X band ESR spectra (the magnetic field of 0.33T), calculated for the following parameters: Solid line: static ZFS $D_S = 0.035 cm^{-1}$, rotational correlation time $\tau_R = 300 ps$, transient ZFS $D_T = 0.02 cm^{-1}$, distortional correlation time $\tau_D = 30 ps$. Here one can see clearly effects of the static ZFS not averaged by the rotational motion; the product $\omega_S \tau_R \cong 1$. Since the amplitude of the transient ZFS is rather small, the electron spin relaxation is not efficient enough to mask these effects. The spectrum presented by dashed line has been calculated for higher amplitude of the transient ZFS, $D_T = 0.03 cm^{-1}$; other parameters remained unchanged. More efficient electron spin relaxation diminishes the static ZFS effects, as one could expect. However if we increase the amplitude of the static ZFS to $D_S = 0.045 cm^{-1}$ ($\tau_R = 300 ps$ remains unchanged) the faster electron spin relaxation ($D_T = 0.03 cm^{-1}$, $\tau_D = 30 ps$) turns out to be not efficient enough, once again. This spectrum is presented as dotted line.

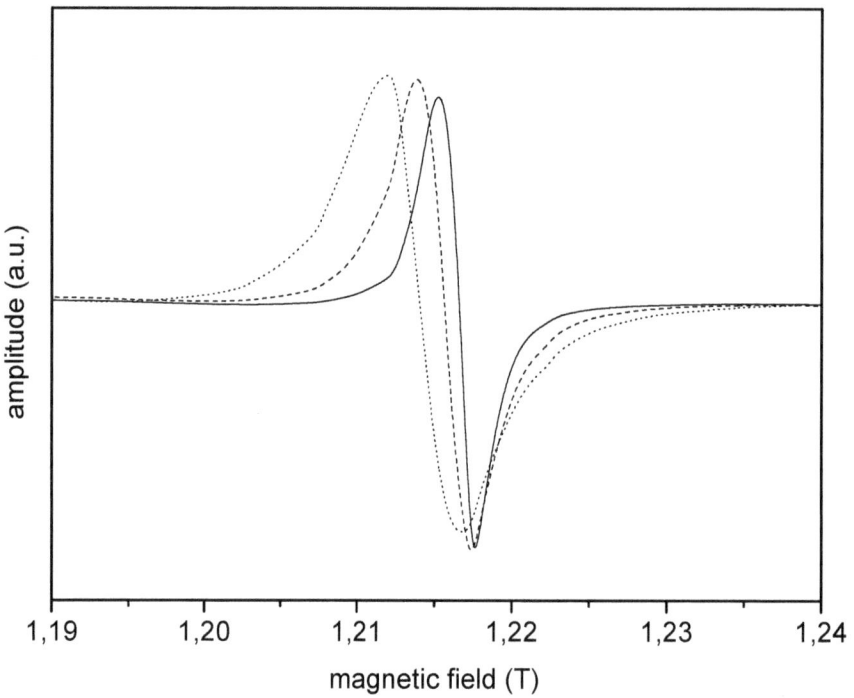

Fig.10.2b

Q band spectra (1.21T) corresponding to the X band spectra presented in Fig.12.2a. For higher magnetic fields the effects of the static ZFS are less pronounced.

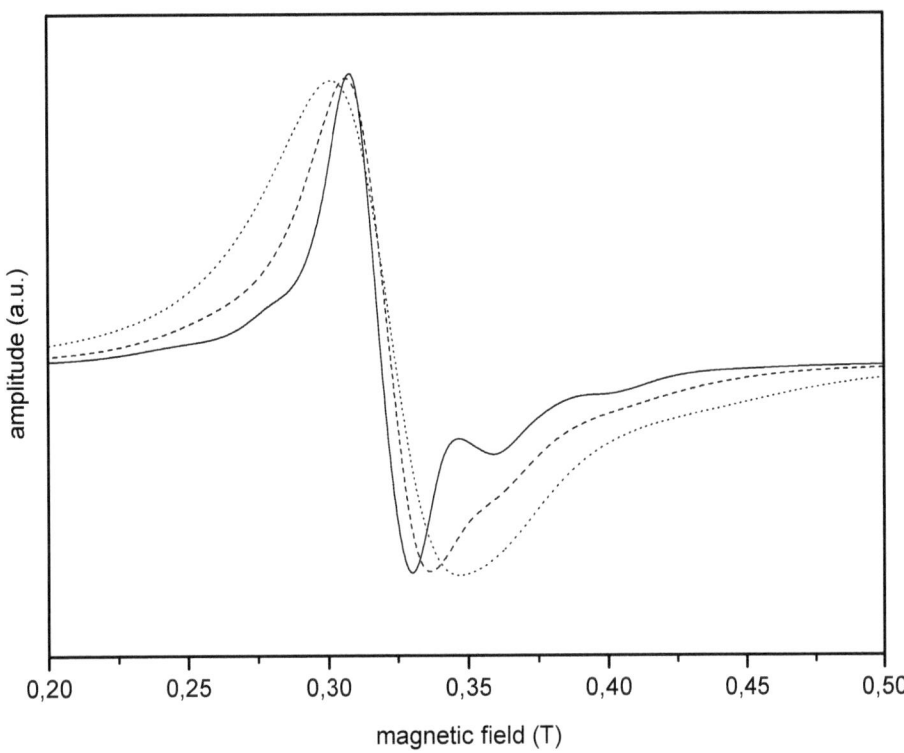

Fig.10.3

Examples of X band ESR spectra calculated for relatively slowly rotating systems. Solid line: $D_S = 0.02 cm^{-1}$, $D_T = 0.02 cm^{-1}$, $\tau_D = 40 ps$, $\tau_R = 10 ns$
Dashed line: $D_T = 0.03 cm^{-1}$, Dotted line: $D_T = 0.04 cm^{-1}$, other parameters remain unchanged.

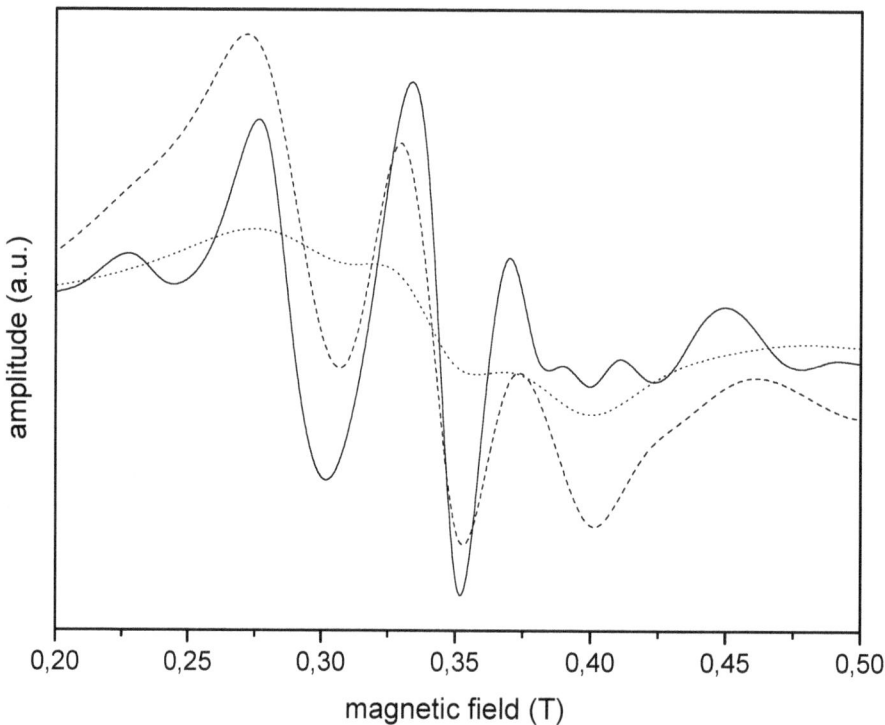

Fig.10.4

X band ESR spectra calculated for the same parameters as in Fig.12.3, except the static ZFS witch is now larger, i.e:

Solid line: $D_S = 0.05 cm^{-1}$, $D_T = 0.02 cm^{-1}$, $\tau_D = 40 ps$, $\tau_R = 10 ns$

Dashed line: $D_T = 0.03 cm^{-1}$, *Dotted line:* $D_T = 0.04 cm^{-1}$, *other parameters remain unchanged. The larger static ZFS affects the spectra significantly.*

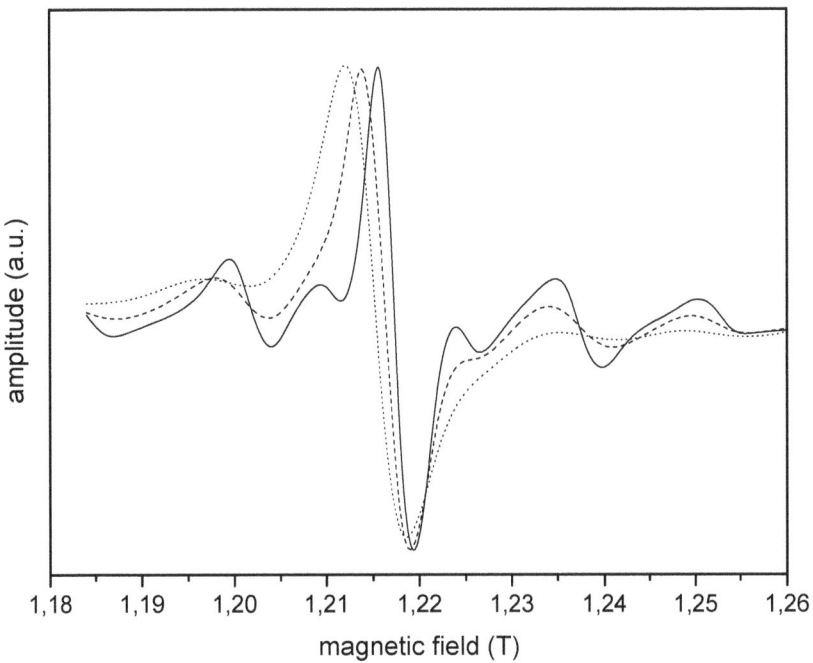

Fig.10.5
Examples of Q band ESR spectra (1.21T). Since the electron spin relaxation is not very efficient one can see clearly the effects of the static ZFS.
Solid line: $D_S = 0.02 cm^{-1}$, $D_T = 0.02 cm^{-1}$, $\tau_D = 20 ps$, $\tau_R = 10 ns$; Dashed line $D_T = 0.03 cm^{-1}$, Dotted line $D_T = 0.04 cm^{-1}$, other parameters remain unchanged.

IX.2. Dipolar spin relaxation treated within the general formalism

Let us now turn attention to the $I-S$ spin system and the spin-lattice relaxation of the spin I. As it has been demonstrated in Chapters IV and VI, the way of computing the dipolar spectral density, $K_{1,1}^{DD}(-\omega_I)$, is setting up and inverting a supermatrix $\hat{\hat{M}}$ (in the previous chapters we have used different labeling, depending on the magnetic field regime). Let us compare the expression for the dipolar spectral density, $K_{1,1}^{DD}(-\omega_I)$, of Eq.4.19 with the electron spin spectral density $s_{-1,-1}(-\omega)$ of Eq.10.1. To make the comparison more straightforward, one can rewrite Eq.4.16 as follows:

$$K_{1,1}^{DD}(-\omega_I) = \int_0^\infty Tr_L\left\{T_1^{1(DD)+}\left[\exp(-i\hat{\hat{L}}_L\tau)T_1^{1(DD)}\right]\rho_L^{eq}\right\}\exp(-i\omega_I\tau)d\tau =$$
$$\frac{1}{2S+1}[T_1^{1(DD)}]^+\left[\hat{\hat{M}}_{REL}\right]^{-1}[T_1^{1(DD)}]$$
(10.17)

Essentially, there is no difference between the operator $\hat{\hat{M}}_{REL} = -i\hat{\hat{L}}_L - i\omega_I\hat{\hat{1}}$ determining the spin-lattice relaxation of the dipolar spin I at the frequency ω_I and the $\hat{\hat{M}}_{ESR}$ operator. Since the lattice for the nuclear spin is formed by the electron spin subsystem coupled by the ZFS interaction to the distortional and rotational degrees of freedom, the lattice Liouvilian $\hat{\hat{L}}_L$ contains the same components which are relevant for the electron spin dynamics: $\hat{\hat{L}}_L = \hat{\hat{L}}_Z(S) + \hat{\hat{L}}_{ZFS}^S + \hat{\hat{L}}_{ZFS}^T + \hat{\hat{L}}_D + \hat{\hat{L}}_R$ (see Eq.4.20 where the static ZFS term is not included due to the assumption that it vanishes). Therefore, the already set up matrix $\left[\hat{\hat{M}}_{ESR}\right]$ is appropriate also for the calculations of the nuclear spin relaxation after a small modification. It is just enough to replace in Eq.10.2 the frequency ω (in the term $\omega\hat{\hat{1}}$) at which the ESR spectrum is evaluated by the frequency ω_I (the Larmor frequency of the spin I) at which one wishes to evaluate the spin-lattice relaxation rate R_{1I}. The projection vector $[T_1^{1(DD)}]$ describes not only the representation of the 'pure' S spin operators in the Liouville basis $\{O_i\}$, but the entire lattice tensor operators. They result from the form of the spin-lattice coupling, that is in the present case the $I-S$ dipole-dipole interaction, and from physical origins of eventual modulations of the orientation of the $I-S$ axis with respect to the molecular frame. Since at the moment we consider the case of isotropic tumbling of the whole molecule containing the spins I and S, the $I-S$ axis is fixed relative to the principal frame of the static ZFS tensor. We shall see in Chapter XII that relative translational diffusion of the I and S spins alters the $T_1^{1(DD)}$ operators. Even though this problem will bother us later, at this instant we can easily extract from Eq. 4.18 the form of the operator $T_1^{1(DD)}$ appropriate for the present case:

$$T_1^{1(DD)} = \sqrt{5} a_{DD}^{IS} \sum_{q=-1}^{1} \begin{pmatrix} 2 & 1 & 1 \\ 1-q & q & -1 \end{pmatrix} S_q^1 D_{0,1-q}^2(\Omega_{ML}) =$$

$$a_{DD}^{IS} \sqrt{\frac{5(2S+1)(S+1)S}{3}} \sum_{q=-1}^{1} \begin{pmatrix} 2 & 1 & 1 \\ 1-q & q & -1 \end{pmatrix} Q_q^1 D_{0,1-q}^2(\Omega_{ML}) \quad (10.18)$$

In the last equality we have made use of the relation between the operators S_q^1 and Q_q^1 given by Eq.2.28a. We have included also the dipole-dipole constant a_{DD}^{IS} into the expression for $T_1^{1(DD)}$, since the form of the spectral density $K_{1,1}^{DD}(-\omega_I)$ of Eq.4.19 has been derived assuming that the spin-lattice Hamiltonian is given by the equation: $H_{IL} = \sum_{n=-1}^{1}(-1)^n I_n^1 T_{-n}^1$ (see Chapter II). Closer inspection of Eq.10.18 brings us to the conclusion that there are three non-zero expansion coefficients of the tensor operators T_1^{1DD} in the basis $\{O_i\}$. The coefficients, corresponding to the basis vectors $|000\rangle|202\rangle|1-1\rangle$, $|000\rangle|201\rangle|10\rangle$ and $|000\rangle|200\rangle|11\rangle$, are equal to $\begin{pmatrix} 1 & 2 & 1 \\ -1 & 2 & -1 \end{pmatrix}$, $\begin{pmatrix} 1 & 2 & 1 \\ 0 & 1 & -1 \end{pmatrix}$ and $\begin{pmatrix} 1 & 2 & 1 \\ 1 & 0 & -1 \end{pmatrix}$, respectively. To avoid any confusion, let us explain this in more detail. The operator $T_1^{1(DD)}$ contains elements, which are associated with the 'direct' sources of the modulations of the spin I - lattice coupling, i.e. the rotational motion and the S spin dynamics. The distortional motion affects the dipole-dipole interaction in an indirect way, through the S spin relaxation and therefore it is not present in Eq.10.18. In fact this statement is quite obvious if one realizes that the form of the operator $T_1^{1(DD)}$ comes from the transformation of the dipole-dipole Hamiltonian from the molecular frame to the laboratory frame. Since $D_{0,0}^0(\Omega)=1$, one can treat the $T_1^{1(DD)}$ operator as related to the distortional states $|ABC\rangle=|000\rangle$. We see from Eq.10.18 that three spin functions $|\Sigma\sigma\rangle=|1-1\rangle,|10\rangle,|1,1\rangle$ are of importance. This is not a surprise since we have discussed in Chapters IV and VI that there are three types of the S spin transitions involved in the I spin relaxation; they are characterized by $\Delta m_S=-1,0,1$. The spin functions $|1-1\rangle,|10\rangle$ and $|1,1\rangle$ are associated with the Wigner matrices corresponding to the rotational states $|202\rangle$, $|201\rangle$ and $|200\rangle$, respectively. Combining Eq.10.17 and Eq.10.18 and taking into account that $R_{1I}=2\operatorname{Re}\{K_{11}^{DD}(-\omega_I)\}$ we can write a closed expression for the R_{1I} quantity:

$$R_{1I} = \frac{10}{3}\left(a_{DD}^{IS}\right)^2 S(S+1)\operatorname{Re}\left\{[c_1]^+\left[\hat{M}_{REL}\right]^{-1}[c_1]\right\} \quad (10.19)$$

The vector $[c_1]$ is of the dimension of the matrix $\hat{M}_{REL}$ (determined in turn by the number of states $|ABC\rangle|LKM\rangle|\Sigma\sigma\rangle$ involved into the calculations). It contains at appropriate positions depending on the ordering of the states in the direct-product basis, $\{|O_i\rangle\}$, the three numbers: $\begin{pmatrix}1 & 2 & 1\\-1 & 2 & -1\end{pmatrix}$, $\begin{pmatrix}1 & 2 & 1\\0 & 1 & -1\end{pmatrix}$ and $\begin{pmatrix}1 & 2 & 1\\1 & 0 & -1\end{pmatrix}$. I remind at this moment that the dipole-dipole constant has been defined as $a_{DD}^{IS}=\sqrt{6}\dfrac{\mu_0}{4\pi}\dfrac{\gamma_I\gamma_S\hbar}{r_{IS}^3}$.

The general derivations, performed within the superoperator formalism, are closely related to the perturbation treatment and reveal the relation of the S spin lineshape and the I spin relaxation to some well defined spectral densities taken at appropriate frequencies. However, the evaluation of the spectral density in the general theory becomes more cumbersome, due to averaging of the in principle infinitely many classical (distortional and rotational) degrees of freedom. To understand the great advantages of this treatment one should realize that to calculate the S spin spectral densities, one does not need to define the relaxation times (rates) for this spin. They are not directly involved in calculations, nor do they appear in any intermediate step. Therefore, one does not need to care whether it is possible to define these quantities, fulfilling the conditions of the perturbation theory.

Fig.10.6 und Fig.10.7 show proton relaxation profiles which can be properly evaluated only by applying the general treatment. Choosing the examples, I have put emphasis on the effects of a large transient ZFS and of the rotational motion. More interesting examples one can find in [20, 21].

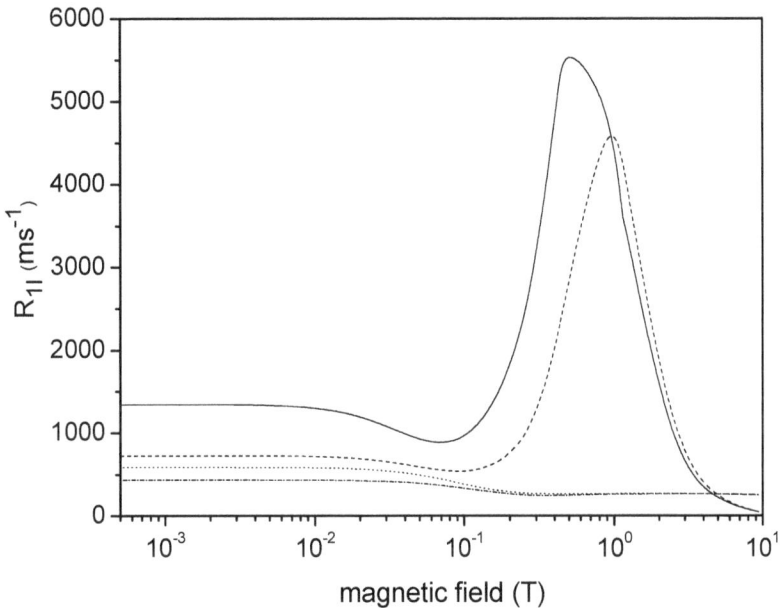

Fig.10.6
Proton spin – lattice relaxation profiles for the electron spin quantum number $S = \frac{7}{2}$.
Solid line: $D_S = 0.02 cm^{-1}$, $D_T = 0.03 cm^{-1}$, $\tau_D = 30 ps$, $\tau_R = 10 ns$
Dotted line: $D_S = 0.02 cm^{-1}$, $D_T = 0.03 cm^{-1}$, $\tau_D = 30 ps$, $\tau_R = 100 ps$
Dashed line: $D_S = 0.02 cm^{-1}$, $D_T = 0.05 cm^{-1}$, $\tau_D = 30 ps$, $\tau_R = 10 ns$
Dashed – dotted line: $D_S = 0.02 cm^{-1}$, $D_T = 0.05 cm^{-1}$, $\tau_D = 30 ps$, $\tau_R = 100 ps$
The interspin distance has been set to 300 pm.

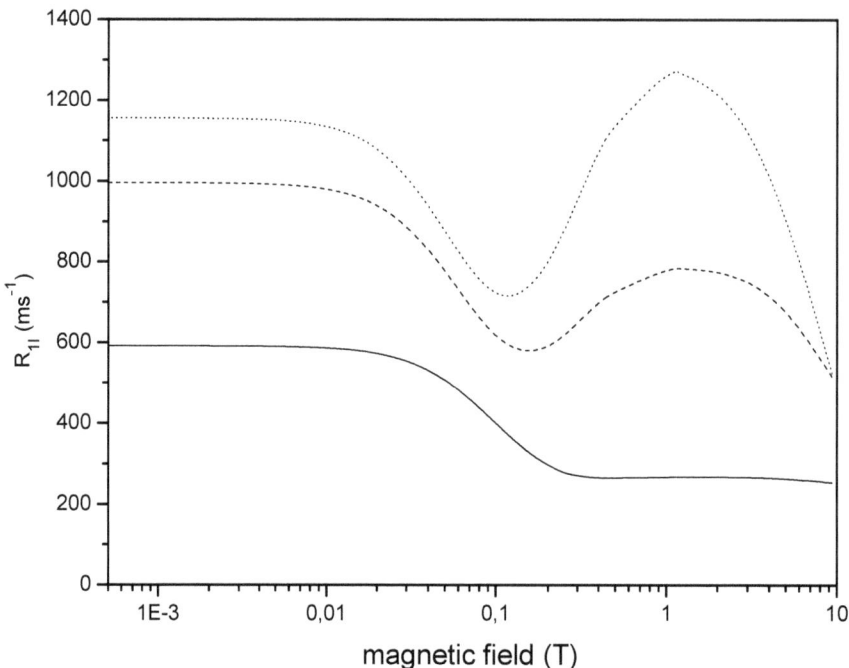

Fig.10.7

Proton spin – lattice relaxation profiles for $S = \frac{7}{2}$.
Solid line: $D_S = 0.03 cm^{-1}$, $D_T = 0.03 cm^{-1}$, $\tau_D = 30 ps$, $\tau_R = 100 ps$
Dashed line: $\tau_R = 300 ps$, Dotted line: $\tau_R = 500 ps$; other parameters remain unchanged
The interspin distance has been set to 300 pm.

The approach presented here originates from the relatively early works [10-15]. It has been developed and modified depending on the applied Hamiltonian models. In principle, the development follows the direction of increasing complexity of the considered systems by allowing for more motional degrees of freedom and more complex (and more realistic at the same time) models of the spin interactions [16-23]. This approach is known in the literature as the 'slow motion theory' or the 'general theory' [10-23, 27]. It can be interesting to understand the origin of this terminology because it comes from the type of problems to which the theory is dedicated. The name 'slow motion' originates from the situation when the motion modulating the relevant interaction is too slow compared to the timescale of the spin dynamics; in other words: it is not true that the motion is much faster than the resulting spin dynamics and therefore the perturbation theory breaks down. In fact, one does not need to explain why this theory is called 'general'. It is just so, because it is valid for general cases and not only the limiting ones. There are no limitations of this approach as far as the S spin is concerned. Within this theoretical framework one can properly evaluate ESR or quadrupolar spectra for arbitrary motional conditions and arbitrary strengths of the spin interactions. In particular, the general treatment accounts for a correlation between the spin and reorientational degrees of freedom. However, if one aims for a description of the I spin relaxation rate, obviously one must be sure that this quantity exists. Therefore one should check carefully the condition $\omega_{DD}^{IS}\tau_c \ll 1$. This problem exists always, independently on other restrictions of applied theories and models. One should also pay attention if the secular approximation condition $\omega_{DD}^{IS}\tau_c \cdot \omega_{DD}^{IS} \ll \omega_I$ is fulfilled at low magnetic field, *i.e.* if the Zeeman coupling determines in fact the energy levels of the I spin. There is one more restriction of this approach when applied to the I spin relaxation. It is not related to the methods used for evaluations of the spectral density $K_{1,1}^{DD}(-\omega_I)$, but to the problem whether the spectral density really describes the I spin relaxation. This issue has been also discussed previously, in particular in Chapter VI. We have concluded that it is so if the predominant relaxation mechanisms of the spin S is provided by the ZFS (quadrupolar) interaction and therefore it is independent of the presence of the spin I (in other words: the contribution to the S spin relaxation due to fluctuations of the mutual $I-S$ dipole-dipole coupling is negligible) or if the spin S relaxes slowly compared to other motional processes modulating the $I-S$ dipole-dipole axis.

I finish this chapter stating that properties of the general theory make it possible to adapt it to various systems, when any simplified treatments break down, and therefore it is worthwhile to put some effort to become more familiar with this type of evaluations. It is also highly appropriate to turn attention to other works devoted to spin systems beyond the Redfield limit and the formalism proposed therein [28, 29].

References:

1. D.A. Varshalovich, A.N. Moskalev, V.K. Khersonkii, Quantum theory of angular momentum, *Word Scientific Publishing, Singapore* (1988)
2. D. M. Brink, G.R. Satchler, Angular momentum, *Clarendon Press, Oxford* (1979)
3. J. H. Freed, G.V. Bruno, C.F. Polnaszek, ESR line shapes for triplets undergoing slow rotational reorientation, *J. Phys. Chem.* **75** (1971) 3385-3389
4. J. H. Freed, G.V. Bruno, C.F. Polnaszek, Electron spin resonance line shapes and saturation in the slow motional region, *J. Phys. Chem.* **75** (1971) 3385-3389
5. J.H. Freed, Spin labelling: theory and applications, L.J. Berliner (Ed.) *Academic Press, New York* (1976) 53-132
6. R.M. Lynden-Bell, Density matrix formulation of theory of magnetic-resonance spectra in slowly reorienting systems, *Mol. Phys.* **22** (1971) 837-851
7. X. Z. Zhou, P. Caravan, R. B. Clarkson, P.O. Westlund, On the philosophy of optimizing contrast agents. An analysis of H-1 NMRD profiles and ESR lineshapes of the Gd (III) complexes MS-325+HAS, *J. Magn. Reson.* **167 (1)** (2004) 147-160
8. X. Z. Zhou, P.-O. Westlund, H-1 NMRD profiles and ESR lineshapes of Gd(III) complexes: a comparison between the generalized SBM and the stochastic Liouville approach, *J. Magn.Reson.* **173 (1)** (2005) 75-83
9. A. Abragam, The principles of nuclear magnetism, *Oxford University Press, Oxford* (1961)
10. N. Benetis, J. Kowalewski, L. Nordenskiöld, H. Wennerström, P.-O. Westlund, Nuclear spin relaxation in paramagnetic systems. The slow motion problem for electron spin relaxation, *Mol. Phys.* **48** (1983) 329-346
11. N. Benetis, J. Kowalewski, L. Nordenskiöld, H. Wennerström, P.-O. Westlund, Nuclear spin relaxation in paramagnetic systems (S=1) in the slow motion regime for the electron spin: II. The dipolar T_2 and the role of scalar interaction, *J. Magn. Reson.* **58** (1984) 261-281
12. P.-O. Westlund, H. Wennerström, L. Nordenskiöld, J. Kowalewski, N. Benetis, Nuclear spin – lattice and spin – spin relaxation in paramagnetic

systems in the slow motion regime for electron spin. III. Dipole – dipole and scalar spin – spin interaction for $S = \frac{3}{2}$ and $\frac{5}{2}$, *J. Magn. Reson.* **59** (1984) 91-109

13. J. Kowalewski, L. Nordenskiöld, N. Benetis, P.-O. Westlund, Theory of nuclear spin relaxation in paramagnetic systems, *Progr. NMR. Spectr.* **17** (1985) 141 – 185

14. P.-O. Westlund, N. Benetis, H. Wennerström, Paramagnetic proton nuclear magnetic-relaxation in the Ni^{2+} hexa-aquo complex – a theoretical study, *Mol. Phys.* **61** (1987) 177-194

15. J. Kowalewski, T. Larrson, P.-O. Westlund, Proton spin – lattice relaxation in aqueous – solution of the nickel (II) ion, *J. Magn. Reson.* **74** (1987) 56-65

16. P.-O. Westlund, T.P. Larsson, O. Teleman, Paramagnetic enhanced proton spin – lattice relaxation in the Ni^{2+} hexa-aquo complex: A theoretical and molecular dynamics simulation study of the Bloembergen- Morgan decomposition approach, *Mol. Phys.* **78** (1993) 1365-1384

17. T. Larsson, P.-O. Westlund, J. Kowalewski, S.H. Koening, Nuclear-spin relaxation in paramagnetic complexes in the slow-motion regime for the electron spin: the anisotropic pseudorotational model for S=1 and the interpretation of nuclear magnetic relaxation dispersion results for a low-symmetry Ni (II) complex, *J. Chem. Phys.* **101** (1994) 1116-1128

18. P.-O. Westlund, Nuclear paramagnetic spin relaxation theory: Paramagnetic spin probes in homogenous and microheterogenous solutions, in Dynamics of solutions and fluid mixtures by NMR, J. J. Delpuech (Ed.), *Wiley, Chichester* (1995) 173-229

19. T. Nilsson, J. Svoboda, P.-O. Westlund, J. Kowalewski, Slow-motion theory of nuclear spin relaxation in paramagnetic complexes (S=1) of arbitrary symmetry, *J. Chem. Phys.* **109** (1998) 6364-6375

20. T. Nilsson, J. Kowalewski, Slow-motion theory of nuclear spin relaxation in paramagnetic low-symmetry complexes: a generalization to high electron spin, *J. Magn. Reson.* **146** (2000) 345-358

21. T. Nilsson, G. Parigi, J. Kowalewski, Experimental NMRD profiles for some low-symmetry Ni (II) complexes (S=1) in solution and their interpretation using slow-motion theory, *J. Phys. Chem.* **106** (2002) 4476-4488

22. D. Kruk, J. Kowalewski, Vibrational motions and nuclear spin relaxation in paramagnetic complexes: Hexaaquonickel (II) as an example, *J. Chem. Phys.* **116** (2002) 4079-4086

23. D. Kruk, J. Kowalewski, P.-O. Westlund, Nuclear and electron spin relaxation in paramagnetic complexes in solution: Effects of the quantum nature of molecular vibrations, *J. Chem. Phys.* **121 (5)** (2004) 2215-2227

24. M. Rotenberg, R. Bivins, N. Metropolis, J.K. Wooten, The 3-J and 6-J symbols, *Technology Press, Cambridge* (1959)

25. M. E. Rose, Elementary theory of angular momentum, *Wiley, New York* (1957)

26. A. R. Edmunds, Angular momentum in quantum mechanics, *Princeton University Press, Princeton* (1974)

27. J. Kowalewski, D. Kruk, G. Parigi, NMR relaxation in solution of paramagnetic complexes: Recent theoretical progress for S >= 1, *Advances in Inorganic Chemistry*, **57** (2005) 41-104

28. A. A. Nevzorov, J.H. Freed, Spin relaxation by dipolar coupling: from motional narrowing to the rigid limit, *J. Chem. Phys.* **112 (3)** (2000) 1413-1424

29. A. A. Nevzorov, J.H. Freed, Direct-product formalism for calculating magnetic resonance signals in many body systems of interacting spins, *J. Chem. Phys.* **115 (6)** (2001) 2401-2415

CHAPTER XI

Effects of neighbouring dipolar spins on dynamics of quadrupolar spin

Talking, in the previous chapters, about the relaxation processes in the $I-S$ spin system we have emphasised the hierarchy of events: the S spin dynamics influences very significantly the I spin relaxation, being at the same time independent of the presence of the I spin. Under this condition the dipolar spectral density $K_{1,1}^{DD}(-\omega_I)$ determines the spin-lattice relaxation of the I spin. In Chapter VII we have discussed the case when, in fact, the motions which results from coupling with the dipolar spin on the S spin relaxation, however 'direct' modulations of the $I-S$ dipole – dipole axis are much faster than the dynamics of the spin S resulting from the dipole-dipole interaction. We have pointed out that this requirement is equivalent to the Redfield condition for the spin S. If the relation $\omega_{DD}^{IS}\tau_c \ll 1$ (where ω_{DD}^{IS} is the amplitude of the $I-S$ dipole – dipole interaction, represented by the Hamiltonian $H_{DD}(I,S)$, expressed in angular frequency units) is fulfilled we can safely assume that the relaxation of the spin S due to the $I-S$ dipole-dipole coupling is significantly slower than the motion causing this relaxation and is characterised by the correlation time τ_c (for example jump diffusion or exchange motion of the participating spins, or molecular tumbling). Therefore, to be still allowed to treat the spin S as a lattice, we neglect the contribution to the S spin dynamics related to the spin I. Such a treatment is definitely not satisfactory from the perspective of the spin S, which is involved in fact in two types of relaxation processes. The spin S relaxes due to its own relaxation mechanism (the ZFS or the quadrupolar interactions) and via the dipole-dipole relaxation channel. Since the electron spin relaxation is usually dominated by the ZFS mechanism, this problem concerns mainly the quadrupolar spins. Discussing the S spin dynamics in this chapter I shall attach considerable importance to the way in which the dipolar spin I contributes to the relaxation of the spin S.

XI.1. Quadrupolar spin relaxation due to dipole-dipole couplings to dipolar spins

In this section I am concerned mainly with the relaxation pathway of the quadrupolar spin S provided by the $I-S$ dipole-dipole interaction. The spins I and S are mutually coupled and therefore they evolve in time as one, combined system [1-7]. The total Hamiltonian H including all interactions which determine the evolution of this entire spin system has been already discussed in this book. It contains the single spin interactions, *i.e.* the Zeeman couplings for the spins I and S, and the quadrupolar interaction decomposed into the averaged and fluctuating parts (the static and transient parts, respectively), as well as the two spin interaction, *i.e.* the mutual dipole-dipole coupling:

$$H = H_Z(I) + H_Z(S) + H_Q^S(S) + H_Q^T(S) + H_{DD}(I,S) \qquad (11.1)$$

It is very important to realise that even though we intend to focus attention on the S spin it is necessary to consider the full Hamiltonian, containing in particular the single I spin term, $H_Z(I)$. This statement is not surprising, it is even obvious. Since we intend to discuss effects of the $I-S$ coupling on the S spin dynamics, all interactions relevant for the I spin become relevant also for the spin S. We got used to think that it is easier to describe the dynamics of the spin S than of the spin I, because in the first case we can restrict the calculations to the spin S subsystem, while the second case requires to consider the entire $I-S$ spin system. In this chapter we leave for the first time the assumption that the spin S

can be treated independently. We shall see soon that, in consequence, the description of the spin S becomes much more difficult.

In order to describe the S spin dynamics within the framework of the second order perturbation theory, we have to split unambiguously the total Hamiltonian H into main and perturbing parts: $H(t) = H_0 + H_1(t)$. We deal now with situations when there are some kinds of motional processes modulating the $I-S$ dipole-dipole interaction for which the condition $\omega_{DD}^{IS}\tau_c \ll 1$ is fulfilled. In the opposite motional regime, $\omega_{DD}^{IS}\tau_c \gg 1$, one observes polarization transfer effects described in Chapter VIII. To specify better the problem let us take up as an example a crystal lattice formed by dipolar spins I and quadrupolar spins S with the $I-S$ dipole – dipole interactions is modulated by jump diffusion of the spins I within the crystal sites. Obviously the dipole-dipole interaction contributes to the relaxation (perturbing) Hamiltonian for the $I-S$ spin system, independently of the origin of the time fluctuations. Also the role of the transient (fluctuating) component of the quadrupolar interaction is well defined; it is responsible for the S spin relaxation and therefore has to be included into the perturbing Hamiltonian as well. Special caution is needed regarding the static (averaged) part of the quadrupolar interaction. If the quadrupolar spin remains fixed on its position (as for LaF$_3$) or jump between equivalent lattice sites (this implies the same electric field gradient tensor at all positions available for the quadrupolar spin) the averaged quadrupolar coupling contributes to the energy level structure of the spin S. Thus, in this case the main Hamiltonian contains the Zeeman interactions and the averaged (static) quadrupolar coupling:

$$H_0 = H(I) + H_Z(S) + H_Q^S \tag{11.2}$$

while the perturbing Hamiltonian consists of the fluctuating quadrupolar interaction and the dipole-dipole coupling:

$$H_1(t) = H_Q^T(t) + H_{DD}(I,S)(t) \tag{11.3}$$

The problem becomes essentially different when the quadrupolar spin can move between non-equivalent crystal sites. Then the spin S senses momentarily different local electric field gradients and the role of the static quadrupolar interaction becomes more complex. In particular, when the quadrupolar spin dynamics is fast enough, the averaged quadrupolar interaction can contribute to the S spin relaxation. This subject has been treated in some detail in Chapter VII.

In next steps of the considerations we shall treat the Hamiltonian H by the second order perturbation theory. The energy levels, E_r, for the entire system can be obtained diagonalising the matrix representation of the main Hamiltonian H_0 in the basis $\{|n\rangle = |m_I, m_S\rangle\}$. The energy levels are associated with corresponding eigenvectors $|\psi_p\rangle$. At this moment I would like to remind once again the reader that all interactions contributing to the main Hamiltonian have to be expressed in the same reference frame. In this chapter we shall refer to the laboratory representation of the Zeeman and quadrupolar couplings, given in Chapter III. It is worthwhile to notice that the eigenfunctions $|\psi_p\rangle$ are obtained as an outer product of the eigenfunctions $|\psi_\alpha^S\rangle$ of the spin S (derived in Chapter V) and the Zeeman states $|m_I\rangle$ of the spin I : $|\psi_p\rangle = |\psi_\alpha^S\rangle \otimes |m_I\rangle$ [7]. It is due to this fact that there is no time independent interaction between the spins I and S. The number of the eigenstates (eigenvectors) N for the entire $I-S$ spin system is given by a product of the numbers of eigenstates available for the I and S spins, separately, i.e. $N = (2I+1)(2S+1) = 2(2S+1)$ (since $I = \frac{1}{2}$). The eigenvectors $\{|\psi_p\rangle\}$ form the Liouville space $\{|\psi_p\rangle\langle\psi_{p'}|\}$ appropriate for the entire spin system. The vectors $|\psi_p\rangle\langle\psi_{p'}|$ can be expressed in terms of the Zeeman vectors $|n\rangle\langle n'|$, according to the relation:

$$\{|\psi_p\rangle\langle\psi_{p'}|\} = \left\{ \sum_{n,n'=1}^{N} a_{pn}^*(\Omega_{P_SL}) a_{p'n'}(\Omega_{P_SL}) |n\rangle\langle n'| \right\} \tag{11.4}$$

where the coefficients a_{pn} link the eigenvectors $|\psi_p\rangle$ to the Zeeman functions $|n\rangle$: $|\psi_p\rangle = \sum_{n=1}^{N} a_{pn}|n\rangle$, while the angle Ω_{P_SL} describes the orientation of the averaged electric field gradient tensor at the position of the quadrupolar spin relative to the laboratory axis. Actually, so far we have just extended the treatment of the spin S presented in Chapter V to the $I-S$ spin system; in particular Eq.11.4 is a straightforward extension of Eq. 5.8. Therefore, I am not going to comment more on this issue. Nevertheless, one should notice that the coefficients a_{pn} cannot be directly linked to the coefficients $c_{\alpha r}$ (see Eq.5.8) because the two sets of coefficients are determined by different Hamiltonians. We are also quite familiar with the evaluations of the relaxation rates $R_{pp'qq'}$ connecting the time evolution of the individual density matrix elements $\rho_{pp'}$ and $\rho_{qq'}$; they are given by the Redfield formula (Eq.2.14), which has been used many times in this book.

In the present case there are two independent relaxation mechanisms: the fluctuating part of the quadrupolar interaction and the dipole-dipole coupling. Therefore the relaxation rates are now given as a superposition of the spectral densities $J^Q(\omega)$ (Eq.4.59) associated with the quadrupolar relaxation pathway and the spectral densities $J^{DD}(\omega)$, corresponding to the dipolar relaxation channel. We have derived in Chapter V expressions for the S spin relaxation rates (Eq.5.11a,b), which are valid for an arbitrary magnetic field. The relaxation rates have been given in terms of the spectral densities resulting from the transient ZFS (quadrupolar) interaction. On this basis we can write analogous expressions for the relaxation rates $R_{pp'qq'}$, which determining the time evolution of the density matrix elements of the entire $I-S$ spin system, including the dipole-dipole relaxation mechanism. Thus, the relaxation rates $R_{pp'qq'}$ can be evaluated from the equations [7]:

$$R_{pqpq} = 2\left[\xi_{pq} J^Q(0) + \varsigma_{pq} J^{DD}(0)\right] - $$
$$\sum_{s=1}^{N}\left[\xi_{sp} J^Q(\omega_{sp}) + \varsigma_{sp} J^{DD}(\omega_{sp})\right] - \sum_{s=1}^{N}\left[\xi_{sq} J^Q(\omega_{sq}) + \varsigma_{sq} J^{DD}(\omega_{sq})\right] \quad (11.5a)$$

$$R_{ppqq} = \left[\xi_{pq} J^Q(\omega_{pq}) + \varsigma_{pq} J^{DD}(\omega_{pq})\right] \quad (11.5b)$$

The diagonal relaxation rates R_{pppp} are given as: $R_{pppp} = -\sum_{q=1}^{N} R_{ppqq}$. These formulas can be treated as a quite straightforward extension of the corresponding equations, Eq.5.11a,b. However, some further comments regarding the evaluation of the coefficients ξ_{pq} and ζ_{pq} can be useful. The coefficients result, respectively, from the relationship between the representations of the perturbing Hamiltonians, $H_Q^{T(L)}(S)$ and $H_{DD}^{(L)}(I,S)$ in the two bases: $|n\rangle$ and $|\psi_p\rangle$. According to Chapter III the Hamiltonians are given by: $H_Q^{T(L)}(S)(t) = \sum_{m=-2}^{2}(-1)^m \tilde{A}_{-m}^{2T(L)}(t)T_m^2(S)$ (Eq.3.4) and

$H_{DD}^{(L)}(I,S)(t) = \sum_{m=-2}^{2}(-1)^m \tilde{F}_{-m}^{2(L)}(t)T_m^2(I,S)$. As has been explained in detail in Chapter V the coefficients come directly from the form of the Hamiltonian matrix elements in the eigenbasis $\{|\psi_p\rangle\}$, for example for the dipole-dipole interaction one gets: $\langle \psi_p | H_{DD}^{(L)}(I,S) | \psi_q \rangle = \sum_{m=-2}^{2}(-1)^m \tilde{F}_{-m}^{2(L)} \sum_{n,n'=1}^{N}\left(a_{pn}^* a_{qn'}\langle n|T_m^2(I,S)|n'\rangle\right)$. Thus the coefficients ξ_{pq} and ζ_{pq} are of the form [7]:

$$\xi_{pq} = \sum_{m=-2}^{2}\left(\sum_{n,n'}^{N}|a_{pn}^* a_{qn'}|^2 |\langle n|T_m^2(S)|n'\rangle|^2\right) \quad (11.6)$$

$$\zeta_{pq} = \sum_{m=-2}^{2}\left(\sum_{n,n'}^{N}|a_{pn}^* a_{qn'}|^2 |\langle n|T_m^2(I,S)|n'\rangle|^2\right) \quad (11.7)$$

Even though Eq.11.6 and Eq.5.12 are very similar one should keep in mind that now we deal with the entire spin system and the vectors $|n\rangle$ correspond to the quantum states characterised by the m_S and m_I quantum numbers. The matrix elements $\langle n|T_m^2(S)|n'\rangle$ of the S spin tensor components $T_{2,m}(S)$ have a non-zero value only if the magnetic quantum numbers m_I characterising the states $|n\rangle$ and $|n'\rangle$ are the same, i.e. $\Delta m_I = m_{I|n\rangle} - m_{I|n'\rangle} = 0$. The spectral densities $J^Q(\omega)$ and $J^{DD}(\omega)$ are defined as: $J^Q(\omega) = J_m^Q(\omega) = \int_0^\infty \langle \tilde{A}_m^{2T(L)*}(\tau)\tilde{A}_m^{2T(L)}(0)\rangle \exp(-i\omega\tau)d\tau$

and $J^{DD}(\omega) = J_m^{DD}(\omega) = \int_0^\infty \langle \tilde{F}_m^{2(L)*}(\tau)\tilde{F}_m^{2(L)}(0)\rangle \exp(-i\omega\tau)d\tau$; I have assumed here as earlier in this book that the spectral densities are independent of m.

XI.2. Influence of dipole-dipole relaxation mechanism on quadrupolar spectra

Having seen how to evaluate the relaxation rates for the $I-S$ spin system, let us discuss in more detail effects of the dipolar relaxation of the spectrum of the spin S. We shall focus attention on the simplest case of the quadrupolar spin quantum number $S=1$. In fact, by this example one can illustrate all relevant aspects of the dipolar relaxation avoiding at the same time cumbersome calculations. We shall simplify this problem even more assuming that the contribution of the static (averaged) quadrupolar interaction to the energy level structure of the spin S is really negligible. This assumption should be always carefully verified. One should keep in mind Section X.1 of this book, in particular the warning regarding the real effects of the ZFS (quadrupolar) interaction on the energy levels of the electron (quadrupolar) spin.

The Zeeman basis consists in this case of the six functions $|n\rangle = |m_I, m_S\rangle$ labelled as follows:

$$|1\rangle = \left|\frac{1}{2},1\right\rangle, |2\rangle = \left|\frac{1}{2},0\right\rangle, |3\rangle = \left|\frac{1}{2},-1\right\rangle, |4\rangle = \left|-\frac{1}{2},1\right\rangle, |5\rangle = \left|-\frac{1}{2},0\right\rangle, |6\rangle = \left|-\frac{1}{2},-1\right\rangle.$$

Thus, the $\langle S_+ \rangle$ quantity can be represented as:

$$\langle S_+ \rangle \cong \left|\frac{1}{2},1\right\rangle\left\langle\frac{1}{2},0\right| + \left|\frac{1}{2},0\right\rangle\left\langle\frac{1}{2},-1\right| + \left|-\frac{1}{2},1\right\rangle\left\langle-\frac{1}{2},0\right| + \left|-\frac{1}{2},0\right\rangle\left\langle-\frac{1}{2},-1\right| \quad (11.8)$$

$$\cong \rho_{12} + \rho_{23} + \rho_{45} + \rho_{56}$$

The assumption that the main Hamiltonian H_0 is diagonal in the Zeeman basis, $\{|n\rangle\}$, (the Hamiltonian includes actually the Zeeman interactions) greatly simplifies our problem. In consequence, the evaluation of the spectral density $s_{-1-1}(\omega)$, describing the quadrupolar spectrum (according to Eq.10.1) [8-11], can be restricted to the single-quantum block of the relaxation matrix of the dimension 4×4:

$$L(\omega) \cong s_{-1-1}(\omega) \cong \int_0^\infty Tr_S \left\{ S_1^{1+} \exp\left[\left(-i\hat{L}_Z(S) - \hat{\hat{R}}\right)\tau\right] S_1^1 \right\} \exp(-i\omega\tau) d\tau \cong$$

$$[1 \ 1 \ 1 \ 1] \times \begin{bmatrix} i(\omega_s - \omega) + R_{1212} & R_{1223} & R_{1245} & R_{1256} \\ R_{1223} & i(\omega_s - \omega) + R_{2323} & R_{2345} & R_{2356} \\ R_{1245} & R_{2345} & i(\omega_s - \omega) + R_{4545} & R_{4556} \\ R_{1256} & R_{2356} & R_{4556} & i(\omega_s - \omega) + R_{5656} \end{bmatrix}^{-1} \times \begin{bmatrix} 1 \\ 1 \\ 1 \\ 1 \end{bmatrix}$$

(11.9)

According to the secular approximation the relevant coherences ($\rho_{12}, \rho_{23}, \rho_{45}$ and ρ_{56}) are mutually coupled, because all of them are characterised by the same frequency ω_S. It can be of some interest to see explicitly how the fluctuating (transient) part of the quadrupolar coupling and the $I - S$ dipole-dipole interaction contribute to the relaxation rates of Eq.11.9. As already explained they can be evaluated from the Redfield formula applied to the matrix representation of the perturbing Hamiltonian

$$H_1(t) = H_Q^{T(L)} + H_{DD}^{(L)} = \sum_{m=-2}^{2}(-1)^m \tilde{A}_{-m}^{2T(L)} T_m^2(S) + \sum_{m=-2}^{2}(-1)^m \tilde{F}_{-m}^{2(L)} T_m^2(I,S):$$

$$[H_1] = \begin{bmatrix}
 & |1\rangle & |2\rangle & |3\rangle & |4\rangle & |5\rangle & |6\rangle \\
\langle 1| & \frac{1}{\sqrt{6}}\tilde{A}_0^{2(L)} + \frac{1}{\sqrt{6}}\tilde{F}_0^{2(L)} & \frac{1}{\sqrt{2}}\tilde{A}_{-1}^{2(L)} + \frac{1}{2\sqrt{2}}\tilde{F}_{-1}^{2(L)} & \tilde{A}_{-2}^{2(L)} & \frac{1}{2}\tilde{F}_{-1}^{2(L)} & \frac{1}{\sqrt{2}}\tilde{F}_{-2}^{2(L)} & 0 \\
\langle 2| & & -\frac{2}{\sqrt{6}}\tilde{A}_0^{2(L)} & -\frac{1}{\sqrt{2}}\tilde{A}_{-1}^{2(L)} + \frac{1}{2\sqrt{2}}\tilde{F}_{-1}^{2(L)} & -\frac{1}{2\sqrt{3}}\tilde{F}_0^{2(L)} & 0 & \frac{1}{\sqrt{2}}\tilde{F}_{-2}^{2(L)} \\
\langle 3| & & & \frac{1}{\sqrt{6}}\tilde{A}_0^{2(L)} - \frac{1}{\sqrt{6}}\tilde{F}_0^{2(L)} & 0 & -\frac{1}{2\sqrt{3}}\tilde{F}_0^{2(L)} & -\frac{1}{2}\tilde{F}_{-1}^{2(L)} \\
\langle 4| & & & & \frac{1}{\sqrt{6}}\tilde{A}_0^{2(L)} - \frac{1}{\sqrt{6}}\tilde{F}_0^{2(L)} & \frac{1}{\sqrt{2}}\tilde{A}_{-1}^{2(L)} - \frac{1}{2\sqrt{2}}\tilde{F}_{-1}^{2(L)} & \tilde{A}_{-2}^{2(L)} \\
\langle 5| & & & & & -\frac{2}{\sqrt{6}}\tilde{A}_0^{2(L)} & -\frac{1}{\sqrt{2}}\tilde{A}_{-1}^{2(L)} - \frac{1}{2\sqrt{2}}\tilde{F}_{-1}^{2(L)} \\
\langle 6| & & & & & & \frac{1}{\sqrt{6}}\tilde{A}_0^{2(L)} + \frac{1}{\sqrt{6}}\tilde{F}_0^{2(L)}
\end{bmatrix}$$

(11.10)

I have provided here the explicit form of the Hamiltonian matrix for convenience of the reader. The individual relaxation matrix elements are as follows:

$$R_{1212} = R_{4545} = -\frac{3}{2}J_0^Q(0) - \frac{3}{2}J_1^Q(\omega_S) - J_2^Q(2\omega_S) - $$
$$\frac{1}{6}J_0^{DD}(0) - \frac{3}{8}J_1^{DD}(\omega_I) - \frac{3}{4}J_1^{DD}(\omega_S) - \frac{1}{12}J_0^{DD}(\omega_I - \omega_S) - \frac{1}{2}J_2^{DD}(\omega_I + \omega_S)$$

(11.11a)

$$R_{2323} = R_{5656} = -\frac{3}{2}J_0^Q(0) - \frac{3}{2}J_1^Q(\omega_S) - J_2^Q(2\omega_S) -$$

$$\frac{1}{6}J_0^{DD}(0) - \frac{1}{2}J_1^{DD}(\omega_I) - \frac{3}{8}J_1^{DD}(\omega_S) - \frac{1}{6}J_0^{DD}(\omega_I - \omega_S) - \frac{1}{2}J_2^{DD}(\omega_I + \omega_S) \quad (11.11b)$$

$$R_{1223} = R_{4556} = -\frac{1}{2}J_1^Q(\omega_S) + \frac{1}{8}J_1^{DD}(\omega_S) \quad (11.11c)$$

$$R_{1256} = \frac{1}{2}J_2^{DD}(\omega_I + \omega_S) \quad (11.11d)$$

$$R_{2345} = \frac{1}{12}J_0^{DD}(\omega_I - \omega_S) \quad (11.11e)$$

I have introduced here the index of the individual spectral densities, $J_m^Q(\omega)$ and $J_m^{DD}(\omega)$ just for completeness and generality of these expressions.

Evaluating the relaxation matrix elements we have treated the two perturbing interactions as completely uncorrelated. In fact, it is quite difficult to imagine a spin system for which there is a correlation between the motion responsible for the fluctuations of the electric field gradient tensor around its averaged value at the position of the quadrupolar spin and the motion modulating the $I - S$ dipole-dipole coupling. In particular, it is obvious in the case of crystal solid state systems. One can safely treat the lattice vibrations leading to the fluctuations of the transient quadrupolar interaction as stochastically independent of the jump diffusion of the dipolar or quadrupolar spins (as long as they jump between equivalent lattice sites). Nevertheless, the description can be adapted to some other cases, for which the problem of correlation is important. It can happen that despite the relaxation mechanism associated with the fluctuating part of the quadrupolar interaction, the static (averaged) part of the quadrupolar coupling contributes also to the relaxation processes of the spin S. This can take place if, for example, the quadrupolar spin jumps between non-equivalent crystal sites, as has been already mentioned. This jump diffusion contributes also to the modulations of the $I - S$ dipole-dipole interaction and the two relaxation pathways (via the averaged quadrupolar coupling and via the dipole-dipole coupling) become correlated. In consequence one has to include into the combinations of spectral densities, resulting from the Redfield formula, appropriate cross-correlation terms defined as:

$$K_{pqp'q'}^{m,m'}(\omega) =$$

$$(-1)^{m+m'} \langle \psi_p | T_m^2(S) | \psi_q \rangle \langle \psi_{p'} | T_{m'}^2(I,S) | \psi_{q'} \rangle \int_0^\infty \langle \tilde{A}_m^{2T(L)}(\tau) \tilde{F}_{m'}^{2(L)}(0) \rangle \exp(-i\omega\tau) d\tau \qquad (11.12)$$

This can be done relatively easily, but the really difficult aspect of this problem is to establish a strict mathematical form of the correlation function $\langle \tilde{A}_m^{2T(L)}(\tau) \tilde{F}_{m'}^{2(L)}(0) \rangle$. I do not want to go into details of such calculations in this book. I leave this difficult subject and illustrate the effects of the dipole-dipole relaxation mechanism on the S spin spectra in figures. Some illustrative calculations performed for the Lorentzian forms of the quadrupolar and dipolar spectral densities, i.e.: $J_m^Q(\omega) = \left[\frac{1}{2}\sqrt{\frac{3}{2}} \frac{\Delta_Q^T}{S(2S+1)}\right]^2 \frac{1}{5} \frac{\tau_Q}{1+\omega^2 \tau_Q^2}$, and

$J_m^{DD}(\omega) = (a_{DD}^{IS})^2 \frac{1}{5} \frac{\tau_c}{1+\omega^2 \tau_c^2}$; they are presented in Fig.11.1-11.4.

Finishing this section it is of some interest to write down a set of equations describing the coupled evolution of the $\langle I_z \rangle$ and $\langle S_z \rangle$ quantities. This can be achieved by linking the observables of interest to the density matrix elements: $\langle I_z \rangle \cong \rho_{11} + \rho_{22} + \rho_{33} - \rho_{44} - \rho_{55} - \rho_{66}$ and $\langle S_z \rangle \cong \rho_{11} + \rho_{44} - \rho_{33} - \rho_{66}$ [1,12], and calculating relaxation rates relevant for the individual populations, which appear in this representation. As a result one obtains:

$$\frac{d}{dt}\langle I_z \rangle = \qquad (11.13a)$$
$$-\left[\frac{1}{12}J_0^{DD}(\omega_I - \omega_S) + \frac{1}{4}J_1^{DD}(\omega_I) + \frac{1}{2}J_2^{DD}(\omega_I + \omega_S)\right]\langle I_z \rangle + \left[\frac{1}{2}J_2^{DD}(\omega_I + \omega_S) - \frac{1}{12}J_0^{DD}(\omega_I - \omega_S)\right]\langle S_z \rangle$$

$$\frac{d}{dt}\langle S_z \rangle =$$
$$-\left[\frac{1}{12}J_0^{DD}(\omega_I - \omega_S) + \frac{1}{4}J_1^{DD}(\omega_S) + \frac{1}{2}J_2^{DD}(\omega_I + \omega_S) + \frac{1}{2}J_1^Q(\omega_S) + 2J_2^Q(2\omega_S)\right]\langle S_z \rangle + \qquad (11.13b)$$
$$\left[\frac{1}{2}J_2^{DD}(\omega_I + \omega_S) - \frac{1}{12}J_0^{DD}(\omega_I - \omega_S)\right]\langle I_z \rangle$$

This set of equations can be treated as a modified Solomon – Bloembergen - Morgan description [13-16]. It reflects the essential elements of the spin-lattice relaxation in the $I-S$ spin system: the coupled evolution of the spin magnetisations caused by the mutual dipole-dipole interaction and the additional, independent relaxation channel for the quadrupolar spin.

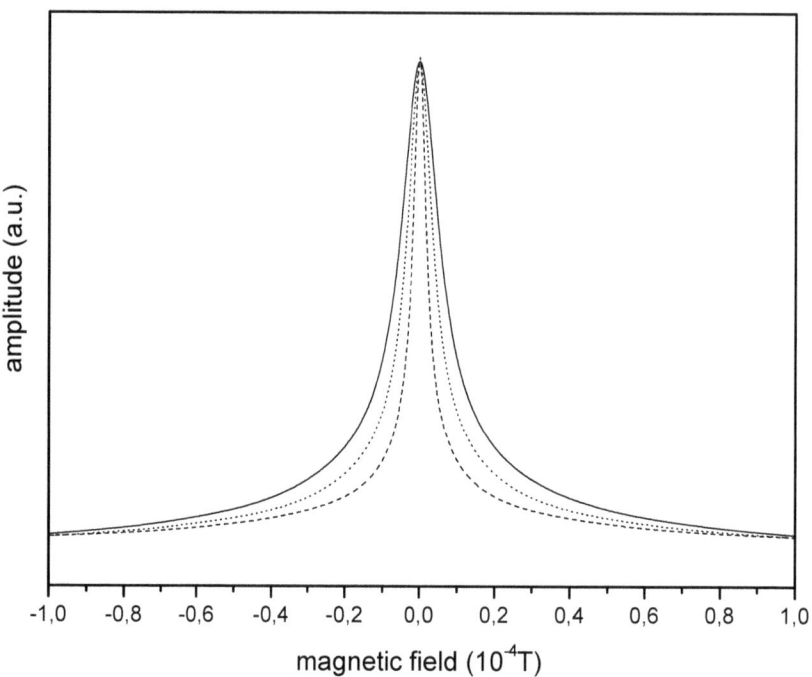

Fig.11.1
A series of ^{14}N spectra at 1T. Dotted line: the relaxation of the quadrupolar spin results from a dipole-dipole coupling to a proton spin; the interspin distance is 300 pm, the correlation time τ_c describing fluctuations of the dipole-dipole interaction is $\tau_c = 10^{-5} s$ (no contribution from the quadrupolar relaxation mechanism). Dashed line: the lineshape is determined by the quadrupolar relaxation mechanism: $a_Q^T = 3 MHz$, $\tau_Q = 10^{-10} s$ (no contribution from the dipole-dipole coupling). Solid line: the lineshape results from both relaxation mechanisms.

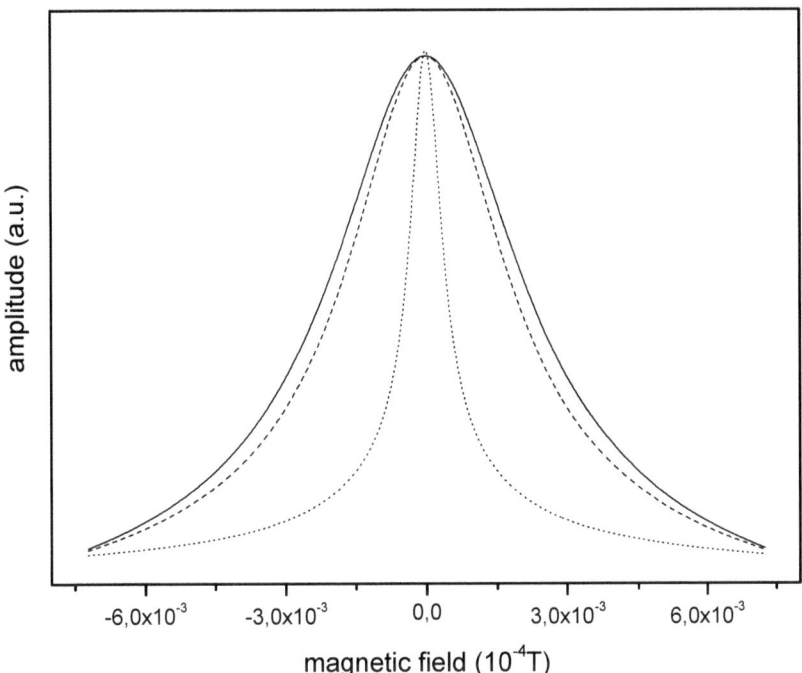

Fig.11.2
^{14}N spectra at 1T. Dotted line: the relaxation of the quadrupolar spin results from a dipole-dipole coupling to a proton spin; the interspin distance is still 300 pm, but the correlation time τ_c is shorter, $\tau_c = 10^{-6} s$ (no contribution from the quadrupolar relaxation mechanism). Dashed line: the lineshape is determined by the quadrupolar relaxation mechanism: $a_Q^T = 3MHz$, $\tau_Q = 10^{-10} s$ (the same curve as in Fig.11.2, no contribution from the dipole-dipole coupling). Solid line: the lineshape results from both the relaxation mechanisms.

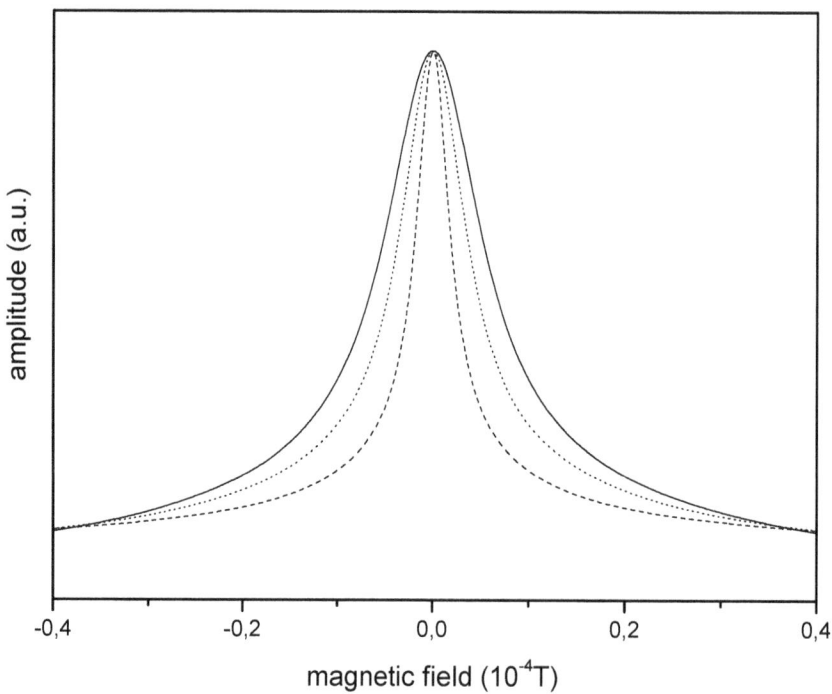

Fig.11.3

This figure corresponds to Fig.11.1, but the spectra have been simulated for the magnetic field of 7T.

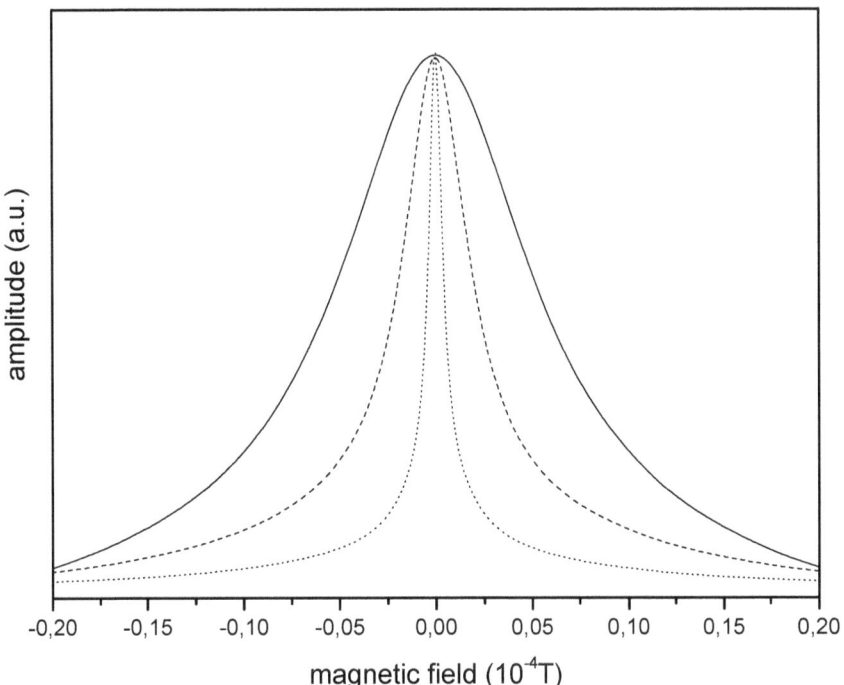

Fig.11.4
It corresponds to Fig.11.2 for the magnetic field of 7T.

In the next section I shall consider time development of an arbitrary spin observable using still as an example the case of mutually coupled quadrupolar and dipolar spins.

XI.3. Time evolution of spin observables in the presence of relaxation processes

We know from Chapter II that if the spin transitions in each multi-quantum manifold are well separated, such that $\left|\omega_{pq} - \omega_{p'q'}\right| \gg R_{pqp'q'}$, then each element of the density matrix ρ_{pq} ($p \neq q$) evolves in time independently of all others, according to the differential equations:

$$\frac{d\rho_{pq}(t)}{dt} = -(i\omega_{pq} - R_{pqpq})\rho_{pq}(t) \qquad (11.14)$$

In consequence, the density matrix element oscillates with the frequency ω_{pq} decaying exponentially at the same time with the rate constant $R_{pqp'q'}$:

$$\rho_{pq}(t) = \rho_{pq}(0)\exp(-i\omega_{pq}t)\exp(-R_{pqpq}t) \qquad (11.15)$$

The zero quantum coherences (populations) ρ_{pp} remain unchanged if there is no relaxation. Otherwise they evolve in time in a complicated, multi-exponential manner. The evolution of the individual populations ρ_{pp} is mutually coupled by the coefficients R_{ppqq}: $\frac{d\rho_{pp}}{dt} = \sum_{q=1}^{N} R_{ppqq}\rho_{qq}$. The spin-lattice relaxation is also usually multi-exponential, especially for higher spin quantum numbers. Now, I shall say more about consequences of these facts. In Chapter VIII we have discussed in detail the evolution of spin observables under a time independent Hamiltonian, neglecting any relaxation effects. Now, we shall include relaxation processes into the considerations.

We should be able to describe, for example, time evolution of the S spin magnetisation, *i.e.* the $\langle S_z \rangle$ observable, depending on the applied magnetic field. Such problems are, in fact, very fundamental and closely related to what can be observed experimentally. Nevertheless, they are quite demanding from the theoretical point of view. Actually, Eqs.11.13a,b do not confirm this statement. It is very easy to solve this set of equations and obtain the well know bi-exponential evolution of the I and S spin magnetisations (see Chapter IV). It is important to realise that, at this stage, we profit from the fact that the main Hamiltonians of both the participating spins ($H_z(I)$ and $H_z(S)$, respectively) are diagonal in the Zeeman basis $\{|n\rangle = |m_I, m_S\rangle\}$. Let us try to set up an analogous set of equations for an arbitrary magnetic field, when the main Hamiltonian for the quadrupolar spin contains the Zeeman as well as the static quadrupolar contributions:

$H_0(S) = H_Z(S) + H_Q^S$. For simplicity reasons, let us take up once again the case of $S = 1$. First of all, we have to express the $\langle S_z \rangle$ quantity in the Liouville basis $\{|\psi_p\rangle\langle\psi_q|\}$, generated by the eigenstates of the Hamiltonian H_0 (Eq.11.2). Employing Eq.11.4 one can set up the relationship between the Liouville vectors $|n\rangle\langle n'|$ and $|\psi_p\rangle\langle\psi_q|$: $|n\rangle\langle n'| = \sum_{p,q=1}^{N} a_{np}^{-1*} a_{n'q}^{-1} |\psi_p\rangle\langle\psi_q|$. The coefficients a_{np}^{-1} are given as the corresponding elements of the inverted matrix a: $a_{np}^{-1} \equiv (a^{-1})_{np}$. In consequence on obtains:

$$\langle S_z \rangle \cong \rho_{11}^Z + \rho_{22}^Z + \rho_{33}^Z - \rho_{44}^Z - \rho_{55}^Z - \rho_{66}^Z \cong$$
$$\sum_{p,q=1}^{6} \left(a_{1p}^{-1} a_{1q}^{-1*} + a_{2p}^{-1} a_{2q}^{-1*} + a_{3p}^{-1} a_{3q}^{-1*} - a_{4p}^{-1} a_{4q}^{-1*} - a_{5p}^{-1} a_{5q}^{-1*} - a_{6p}^{-1} a_{6q}^{-1*} \right) \rho_{pq} \quad (11.16)$$

In the above expression I have followed the labelling of the Zeeman functions $|n\rangle$ from the previous section. The density matrix elements $\rho_{nn'}^Z \equiv |n\rangle\langle n'|$ refer to the Zeeman basis $\{|n\rangle\langle n'|\}$ (the index z points this out explicitly), while $\rho_{pq} \equiv |\psi_p\rangle\langle\psi_q|$. We can see that now the S spin magnetisation depends not only on the populations of the energy levels of the $I - S$ spin system, which are included into the quantities $\langle I_z \rangle$ and $\langle S_z \rangle$ (reducing in this way the number of coupled equations to 2). Generally speaking, the evolution of the $\langle S_z \rangle$ observable is coupled to various modes, encoded into the linear combination on the right hand side of Eq.11.16. The mixture of mutually coupled density matrix elements depends on the relative strengths of the Zeeman and quadrupolar interactions. Setting up Eqs.11.13a,b we have dealt with the simplest case when $a_{np}^{-1} = \delta_{np}$. This example brings us to the conclusion that in order to evaluate the function $\langle S_z \rangle(t)$ we have to solve, in fact, the whole set of the differential equations: $\frac{d\rho_{pp'}(t)}{dt} = -i\omega_{pp'} \rho_{pp'}(t) + \sum_{qq'} R_{pp'qq'} \rho_{qq'}(t)$. This applies to any other observable under interest. It is not surprising, especially in the context of Eq.8.4, which tells us clearly that time development of an arbitrary observable follows the evolution of individual density matrix elements. The computational complexity concerns, in principle, the evaluation of the populations $\rho_{pp}(t)$, because the multi-quantum coherences evolve in time independently according to Eq. 11.15. As long as the number of mutually coupled linear differential equation is reasonable, the computational methodology is well known. However, aiming for exact expressions for $\langle O(t) \rangle$ quantities (O denotes here an arbitrary observable) one should be aware that they are determined by the initial density operator, which depends, for example, on the experimental procedure (see Chapter VIII). The initial conditions must be incorporated in an appropriate manner into the evaluations. We shall work out this problem below.

Since we are interested in the time development of individual coherences, we have to diagonalize the zero-quantum block of the relaxation matrix. This procedure leads to a set of magnetisation modes (eigenvectors), $\tilde{\rho}_\alpha$. This means that one can write them as linear combinations of the populations: $\tilde{\rho}_\alpha = \sum_{p=1}^{N} \Lambda_{\alpha p} \rho_{pp}$, with the coefficients $\Lambda_{\alpha p}$ resulting from the diagonalisation. The magnetisation modes $\tilde{\rho}_\alpha$ develop exponentially with the characteristic rate constant $\tilde{R}_\alpha$ obtained as the corresponding eigenvalues: $\tilde{\rho}_\alpha(t) = \tilde{\rho}_\alpha(0)\exp(-\tilde{R}_\alpha t)$. Thus we can consider the evolution of the individual coherences ρ_{pp} as a superposition of exponential processes associated with the already obtained modes:

$$\rho_{pp}(t) = \sum_{\alpha=1}^{N} (\Lambda^{-1})_{p\alpha} \tilde{\rho}_\alpha(0)\exp(-\tilde{R}_\alpha t) \tag{11.17}$$

where the initial magnetisation modes $\tilde{\rho}_\alpha(0)$ reflect the initial state of the spin system; they are related to the initial density operators: $\tilde{\rho}_\alpha(0) = \sum_{q=1}^{N} \Lambda_{\alpha q} \rho_{qq}(0)$. Thus, the density matrix elements $\rho_{pp}(t)$ are related to the initial populations $\rho_{qq}(0)$ according to the expression:

$$\rho_{pp}(t) = \sum_{\alpha=1}^{N} (\Lambda^{-1})_{p\alpha} \left[\sum_{q=1}^{N} \Lambda_{\alpha q} \rho_{qq}(0) \right] \exp(-\tilde{R}_\alpha t) \tag{11.18}$$

We are now in a position to evaluate expectation values of any observables, $\langle Q(t) \rangle$. Following Eq.8.4 we can write:

$$\langle O(t) \rangle = Tr\{\rho(t)O\} =$$

$$\langle O(t) \rangle_{0Q} + \langle O(t) \rangle_{MQ} = \sum_{p=1}^{N} \langle p|\rho(t)|p\rangle\langle p|O|p\rangle + \sum_{p,q=1, p \neq q}^{N} \langle p|\rho(t)|q\rangle\langle q|O|p\rangle \tag{11.19}$$

The first term, $\langle O(t) \rangle_{0Q}$, represents the influence of the populations on the detected observable, while the second one, $\langle O(t) \rangle_{MQ}$, describes the effect of the multi-quantum transitions. The second term can be relatively easily calculated, by making use of Eq.11.15:

$$\langle O(t) \rangle_{MQ} = \sum_{p,q=1, p \neq q}^{N} \langle p|\rho(t)|q\rangle\langle q|O|p\rangle =$$

$$\sum_{p,q=1, p \neq q}^{N} \langle p|O|q\rangle\langle q|\rho(0)|p\rangle \exp(-i\omega_{pq}t) \exp(-R_{pqpq}t) \tag{11.20}$$

In fact, as far as the multi-quantum effects are concerned, the present calculations differ from the derivation of Chapter VIII by the exponentially decaying factor. A proper evaluation of the population part requires more effort. Employing Eq.11.18 one obtains:

$$\langle O(t) \rangle_{0Q} = \sum_{p=1}^{N} \langle p|O|p\rangle \left[\sum_{\alpha=1}^{N} (\Lambda^{-1})_{p\alpha} \tilde{\rho}_\alpha(0) \exp(-\tilde{R}_\alpha t) \right] =$$

$$\sum_{p=1}^{N} \langle p|O|p\rangle \left\{ \sum_{\alpha=1}^{N} (\Lambda^{-1})_{p\alpha} \left[\sum_{q=1}^{N} \Lambda_{\alpha q} \langle q|\rho(0)|q\rangle \right] \exp(-\tilde{R}_\alpha t) \right\} \tag{11.21}$$

This formula is quite complicated, especially if one takes into account that so far it has been written in the eigenbasis of the main Hamiltonian $|\psi_p\rangle$. It would be very useful to proceed one step further and express the quantity $\langle O(t) \rangle_{0Q}$ in the Zeeman basis $\{|n\rangle\}$. This concerns also the multi-quantum term, $\langle O(t) \rangle_{MQ}$.

Before I shall write a final, general expression for the expectation value $\langle O(t) \rangle$, I would like to present a simple example. This should make the further considerations easier to follow and closer to some well defined experimental results. Let us assume, referring to Chapter VIII, that the $I-S$ spin system has been initially polarized in a strong magnetic field, and therefore the initial density operator is given as: $\rho(0) \cong \gamma_I I_z + \gamma_S S_z$. We are interested to follow the time evolution of the spin S magnetisation, $\langle S_z \rangle$ under a Hamiltonian H. In this case, Eq.11.20 takes the form:

$$\langle O(t) \rangle_{MQ} \equiv \langle S_z(t) \rangle_{MQ} \cong$$
$$\sum_{p,q=1, p \neq q}^{6} \left[\gamma_S \langle p|S_z|q \rangle^2 + \gamma_I \langle p|I_z|q \rangle \langle p|S_z|q \rangle \right] \exp(-i\omega_{pq} t) \exp(-R_{pqpq} t) \quad (11.22)$$

We are well aquatinted with the problem how to obtain the matrix elements of an operator within different representations (bases). Therefore, we just make use once again of the relationship between the two basis $\{|\psi_p\rangle\}$ and $\{|n\rangle\}$, and write:

$$\langle S_z(t) \rangle_{MQ} \cong$$
$$\sum_{p,q=1, p \neq q}^{6} \sum_{n,n'=1}^{6} (a_{pn}^* a_{qn'})^2 \left[\gamma_S \langle n|S_z|n' \rangle^2 + \gamma_I \langle n|S_z|n' \rangle \langle n'|I_z|n \rangle \right] \exp(-i\omega_{pq} t) \exp(-R_{pqpq} t) \quad (11.23)$$

where, in fact, the summation is restricted to the terms with $n = n'$, since the matrix representation of the I_z as well as S_z operators in the Zeeman basis is diagonal. Matters are greatly simplified if we take into account that $I = \frac{1}{2}$ and $S = 1$; this implies that:

$$\langle S_z(t) \rangle_{MQ} \cong$$
$$\gamma_S \sum_{p,q=1, p \neq q}^{6} \left[(a_{p1}^* a_{q1})^2 + (a_{p3}^* a_{q3})^2 + (a_{p4}^* a_{q4})^2 + (a_{p6}^* a_{q6})^2 \right] \exp(-i\omega_{pq} t) \exp(-R_{pqpq} t) \quad (11.24)$$

Proceeding carefully we can evaluate in a similar manner the $\langle S_z(t) \rangle_{0Q}$ quantity. We begin from the expression in the brackets [ ] (Eq.11.21), describing the initial values of the magnetisation mode $\tilde{\rho}_\alpha(0)$:

$$\tilde{\rho}_\alpha(0) = \sum_{q=1}^{6} \Lambda_{\alpha q} \langle q|\rho(0)|q \rangle \cong \sum_{q=1}^{6} \Lambda_{\alpha q} \left[\sum_{n,n'=1}^{6} (a_{qn}^* a_{qn'}) \left(\gamma_S \langle n|S_z|n' \rangle + \gamma_I \langle n|I_z|n' \rangle \right) \right] \quad (11.25)$$

In the particular case considered here, we obtain:

$$\sum_{q=1}^{6} \Lambda_{\alpha q} \langle q|\rho(0)|q\rangle =$$

$$\sum_{q=1}^{6} \Lambda_{\alpha q} \left[\gamma_S \left(a_{q1}^* a_{q1} + a_{q3}^* a_{q3} - a_{q4}^* a_{q4} - a_{q6}^* a_{q6} \right) + \frac{\gamma_I}{2} \left(a_{q1}^* a_{q1} + a_{q2}^* a_{q2} + a_{q3}^* a_{q3} - a_{q4}^* a_{q4} - a_{q5}^* a_{q5} - a_{q6}^* a_{q6} \right) \right]$$

(11.26)

In consequence, the magnetisation mode $\tilde{\rho}_p$ at the time t, $\tilde{\rho}_p(t)$, represented by the brackets { } in Eq.11.20, can be written as:

$$\tilde{\rho}_p(t) = \sum_{\alpha=1}^{6} (\Lambda^{-1})_{p\alpha} \left\{ \sum_{q=1}^{6} \Lambda_{\alpha q} \left[\begin{array}{l} \gamma_S \left(a_{q1}^* a_{q1} + a_{q3}^* a_{q3} - a_{q4}^* a_{q4} - a_{q6}^* a_{q6} \right) + \\ \frac{\gamma_I}{2} \left(a_{q1}^* a_{q1} + a_{q2}^* a_{q2} + a_{q3}^* a_{q3} - a_{q4}^* a_{q4} - a_{q5}^* a_{q5} - a_{q6}^* a_{q6} \right) \end{array} \right] \right\} \exp(-\tilde{R}_\alpha t)$$

(11.27)

Thus, if the observable under interest is the magnetisation of the spin S, the final expressions take the form:

$$\langle O(t) \rangle_{0Q} \cong \sum_{p=1}^{6} \left(a_{p1}^* a_{p1} + a_{p3}^* a_{p3} - a_{p4}^* a_{p4} - a_{p6}^* a_{p6} \right) \tilde{\rho}_p(t) \quad (11.28)$$

where the magnetisation modes $\tilde{\rho}_p(t)$ have been already evaluated (Eq.10.26). We have used in the last equation the explicit form of the matrix element $\langle p|S_z|p\rangle$ (for $S=1$), written in the Zeeman basis $\{|n\rangle = |m_I, m_S\rangle\}$. The treatment presented here can be applied to any observable and any spin system. It is, in fact, an extension of the theory of Chapter VIII by including relaxation effects. This extension is unfortunately not quite straightforward because of computational and conceptual difficulties caused mainly by spin-lattice relaxation processes. Nevertheless, it is worthwhile to put some effort to develop a theoretical treatment allowing one to consider in great detail information encoded into experimental results and fully profit from them. This is particularly important when the experimental procedure is relatively simple, like for example a detection of the spin magnetisation versus time. In such cases an appropriate theoretical description including various aspects of the molecular dynamics and their effects on the measured quantities becomes especially valuable. Therefore, we shall complete our description by rewriting the general expressions of Eq.11.20 and Eq.11.21 in the Zeeman basis $\{|n\rangle\}$ [7]:

$$\langle O(t)\rangle_{MQ} = \sum_{p,q=1, p\neq q}^{N} \left(\sum_{n,n'=1}^{N} a_{pn}^* a_{qn'} \langle n|O|n'\rangle \times \sum_{n,n'=1}^{N} a_{qn'}^* a_{pn} \langle n'|\rho(0)|n\rangle \right) \exp(-i\omega_{pq}t)\exp(-R_{pqpq}t) \quad (11.29)$$

and

$$\langle O(t)\rangle_{0Q} = \sum_{p=1}^{N} \left\{ \sum_{n,n'=1}^{N} \left(a_{pn}^* a_{pn'} \right) \langle n|O|n'\rangle \right\} \left\{ \sum_{\alpha=1}^{N} \left(\Lambda^{-1} \right)_{p\alpha} \left[\sum_{q=1}^{N} \Lambda_{\alpha q} \left(\sum_{n,n'=1}^{N} \left(a_{qn}^* a_{qn'} \right) \langle n|\rho(0)|n'\rangle \right) \right] \exp(-\tilde{R}_\alpha t) \right\} \quad (11.30)$$

The general formulas seem to be rather cumbersome; however they can be considerably simplified when applied to given observables and experimental conditions, as shown above. Presenting the perturbation treatment of the spin observables, $\langle O(t)\rangle$, I have decomposed the effects related to the spin-lattice and spin-spin relaxation by evaluating separately the terms $\langle O(t)\rangle_{0Q}$ and $\langle O(t)\rangle_{MQ}$. The decomposition must be treated with high caution. Since the energy level structure of the spin S results from a superposition of the Zeeman and quadrupolar interactions, one can expect several energy level crossings. This implies that at certain magnetic fields some coherences ρ_{pq} ($p \neq q$) can be characterised by zero frequency, $\omega_{pq} \cong 0$, and for that reason the populations ρ_{pp} cannot be considered as decoupled from them. The way to solve this problem is to extend the set of mutually coupled equations (describing originally the populations) by including the relevant coherences.

The presented approach has been applied to follow time evolution of some relevant observables characterising a selected multi-spin system [7]. I wish to mention in this context two examples, considered in detail in [7]. The first example concerns a spin system composed of two non-equivalent dipolar spins I_1 and I_2, and one quadrupolar spin S. We have considered the situation when the spin I_1 has been involved simultaneously in two essentially different processes: the polarization transfer to the quadrupolar spin, caused by the time independent $I_1 - S$ dipole-dipole interaction and the relaxation, caused by the time fluctuating $I_1 - I_2$ coupling. We have followed the evolution of the entire system, focusing attention on a competition between the two 'activity channels' of the spin I_1. In the second example, which we wish to mention, we have dealt with the quadrupolar spin is coupled to two non-equivalent dipolar spins. It has been assumed that one of the dipole-dipole couplings is time-independent, while the second one fluctuates in time. By mentioning these examples I wish to point out that the present approach can be successfully applied to analyse various aspects of spin dynamics.

References:

1. L. Werbelow, Relaxation-Induced Transfer of Nuclear Spin Polarization as a Probe of Molecular Structure and Dynamics in Mobile Phases, in Nuclear Magnetic Resonance Probes of Molecular Dynamics, R. Tycko (Ed.) *Kluwer Academic Publishers* (1994) 223-263
2. R.E. London, D.M. Lemaster, L.G. Werbelow, Unusual NMR multiplet structures of spins-1/2 nuclei coupled to spin-1 nuclei, *J. Am. Chem. Soc.* **116** (1994) 8400-8401
3. S. Grzesiek, A, Bax, Interference between dipolar and quadrupolar interactions in the slow molecular tumbling limit: s source of line shift and relaxation in H-2-labeled compounds, *J. Am. Chem. Soc.* **116** (1994) 10196-10201
4. P. Bernatowicz, J. Kowalewski, D. Kruk, L.G. Werbelow , 'C-13 NMR line shapes in the study of dynamics of perdeuterated methyl groups', *J. Phys. Chem. A 108 42* (2004) 9018-9025
5. P. Bernatowicz, D. Kruk, J. Kowalewski, L.G. Werbelow, '^{13}C NMR Lineshapes for the $^{13}C^2H^2H'$ isotopomeric spin grouping', *Chem. Phys. Chem.* **3** (2002) 933-938
6. C.N. Banvell, H. Primas, On the analysis of high-resolution nuclear magnetic resonance spectra: I. Methods of calculating NMR spectra, *Mol. Phys.* **6** (1963) 225-256
7. D. Kruk, O. Lips, Evolution of solid state systems containing mutually coupled dipolar and quadrupolar spins: a perturbation treatment, *Solid State Nucl. Magn. Reson.* **28 (2-4)** (2005) 180-192
8. J.A. Pople, Effect of quadrupolar relaxation on nuclear magnetic resonance multiplets, *Mol. Phys.* **1** (1958) 168-174
9. A. Abragam, The principles of nuclear magnetism, *Oxford University Press, Oxford* (1961)
10. R. Kimmich, NMR - Tomography, Diffusometry, Relaxometry, *Springer – Verlag Berlin* (1997)
11. J. Kowalewski, L. Mäler, Nuclear spin relaxation in liquids: theory, experiments and applications, Series in Chemical Physics, *Taylor & Francis Group* (2006)

12. A. Kumar, R. C. R. Grace, P.K. Madhu, Cross-correlations in NMR, *Prog. Nucl. Magn. Reson. Spectr.* **37** (2000) 191-319
13. I. Solomon, Relaxation processes in a system of two spins, *Phys. Rev.* **99** (1955) 559-565
14. I. Solomon, N. Bloembergen, Nuclear magnetic interactions in the HF molecule, *J. Chem. Phys.* **25** (1956) 261-266
15. N. Bloembergen, Proton relaxation times in paramagnetic solutions, *J. Chem. Phys.* **27** (1957) 572-573
16. N. Bloembergen, L.O. Morgan, Proton relaxation times in paramagnetic solutions: Effects of electron spin relaxation, *J. Chem.Phys.* **34** (1961) 842-850

CHAPTER XII

Relaxation processes caused by translational diffusion of interacting spins

In this chapter I focus attention on molecular systems for which modulations of the $I-S$ dipole-dipole interaction arise from relative translational diffusion of the coupled spins and the S spin dynamics. Various molecular systems, especially liquids, exhibit translational degrees of freedom and it is therefore highly appropriate to include them into our considerations. Nevertheless, the fact that we allow for a relative translation movement of molecules carrying the spins under interest, does not exclude their rotational motion. This implies that to describe such systems we have to deal with more degrees of freedom than in the previous chapters. The necessary calculations are much more demanding and cumbersome. Even so it is worthwhile to undertake the effort, because of a wide applicability of the results. One could argue that we have considered systems with translational degrees of freedom, talking in Chapter VII about solid state systems and jump diffusion. Even though we really have allowed for relative translational motion of the dipolar and quadrupolar spins, we have described this motion by an exponential correlation function. Since the rotational diffusion is also characterised by an exponential correlation functions, we could, to a certain extent, employ in our calculations analogous mathematical formulations. Translational motion in liquids cannot be described by exponential correlation functions [1-8] and therefore we have to modify considerably our treatment. Before I start detailed considerations, I would like to point out the essential difference between the systems considered so far and the ones being under our current interest. In Chapters VI and IX we have dealt with spin systems when the $I-S$ dipole-dipole axis is fixed relative to the principal axis system of the static ZFS tensor (the molecular frame (P_s); we have even assumed that both the frames coincide), and the relative orientation of the molecular (and obviously the dipole-dipole) and laboratory frames is modulated by the molecular tumbling. When the relative translational motion of the interacting spins I-S is allowed, the molecular tumbling does not influence any more the direction of the dipole-dipole axis. The orientation of the I-S dipole-dipole axis with respect to the laboratory frame is mediated by the translational diffusion, while the rotational motion still affects the orientation of the molecular frame.

I shall discuss below systems containing dipolar and electron spins, however as always in this book one can adapt the considerations to quadrupolar spins.

XII.1. Analytical description of dipolar spin relaxation in the presence of translational diffusion

The general form of the quantum-mechanical spectral density, $K_{1,1}^{DD}(-\omega_I)$, which is a starting point for evaluating the I spin relaxation in the case when the distance between the interacting spins varies in time, has been given by Eq.7.1. This expression has been set up in Chapter VII to describe dipolar spin relaxation in solid state systems, for which the inter-spin distance r_{IS} as well as the orientation of the $I-S$ dipole-dipole axis with respect to the laboratory frame (described by the angle $\Omega_{DDL} \equiv \Omega_{IS}$) is modulated by jump diffusion of the participating spins within available lattice sites. In this chapter, we shall apply it to spin systems, for which the lattice Liouville operator $\hat{\hat{L}}_L$ consists of the following terms:

$$\hat{\hat{L}}_L = \hat{\hat{L}}_Z(S) + \hat{\hat{L}}^S_{ZFS} + \hat{\hat{L}}^T_{ZFS} + \hat{\hat{L}}_D + \hat{\hat{L}}_R + \hat{\hat{L}}_{Diff} \qquad (12.1)$$

Most of the operators, present in this expression, have been already defined and used many times in this book; they represent the Zeeman interaction of the electron spin, the static and transient parts of the ZFS interactions, and the distortional and rotational Liouvilians, respectively. Now, we have to define the operator $\hat{\hat{L}}_{Diff}$ associated with the translational motion. The translational diffusion is a classical Markov process, described by the operator:

$$\hat{\hat{L}}_{Diff} = -i D_{Diff} \nabla^2_{\vec{r}_{IS}} \qquad (12.2)$$

where the index $\vec{r}$ indicates that the differential operator $\nabla^2_{\vec{r}_{IS}}$ contains terms related to changes of the interspin distance as well as the orientation of the dipole-dipole axis (*i.e.* $\vec{r}_{IS} \equiv (r_{IS}, \Omega^L_{IS})$). At this stage it can be useful to comment more on the rotational motion of the molecule carrying the S spin. The role of the molecular tumbling for systems with translational degrees of freedom has been explained in the introduction to this chapter. This problem is however crucial for understanding the dynamics and relaxation mechanisms of such systems. To avoid confusions and questions why and when the molecular tumbling is important if we deal with systems, for which translational motion modulates mutual spin interactions, the systems of interest is shown graphically in Fig.12.1. One can easy understand from this picture that now the role of the rotational motion is restricted to the modulations of the orientation of the principal axis system of the static ZFS (or the principal axis system of the averaged electric field gradient tensor) with respect to the laboratory frame. When the molecular tumbling is slow, to evaluate the I spin relaxation we have to consider the effect of the S spin dynamics resulting from all possible orientations of the molecular frame. Nevertheless one cannot treat the present case in full analogy to the problem of slow molecular tumbling described in Chapter VI when the orientation of the $I-S$ dipole-dipole axis relative to the laboratory frame is fixed, since now it changes in time due to the translational motion. When the molecular tumbling is rather fast it influences the I spin relaxation in the low field regime by modulating the orientation of the quantization axis of the electron spin; I shall discuss this issue soon.

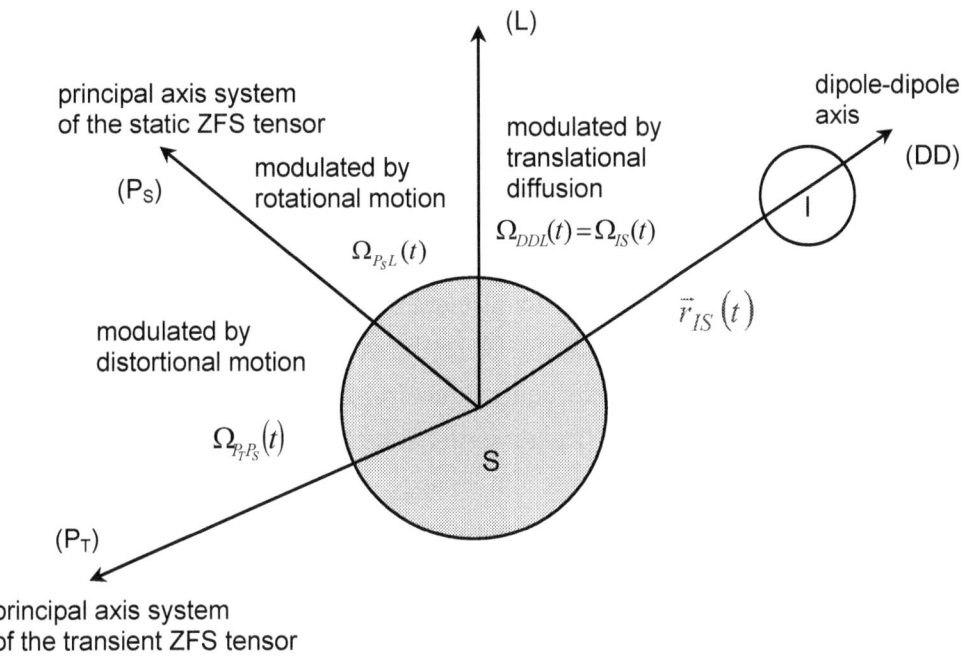

Fig.12.1
A schematic view of a nuclear spin – electron spin system in the presence of relative translational diffusion of the spins. The orientation of the I – S dipole – dipole axis with respect to the laboratory frame is modulated by the translational motion; it is independent of the tumbling of the molecule carrying the S spin. The molecular tumbling is responsible for the fluctuations of the principal axis system of the static ZFS with respect to the laboratory frame.

As we know from the previous considerations, to attempt analytical evaluations of the spectral density $K_{1,1}^{DD}(-\omega_I)$ we have to analyze carefully dependencies (correlations) between different processes contributing to the lattice dynamics. The crucial point is to decompose from the lattice correlation function the part $C_{Diff}(\tau)$ corresponding to the modulations of the dipole-dipole interaction by the translational diffusion $C_{Diff}(\tau) = \left\langle \dfrac{D^{2*}_{0,1-q}(\Omega_{DDL}(\tau)) D^{2}_{0,1-p}(\Omega_{DDL}(0))}{r^3_{IS}(\tau) \, r^3_{IS}(0)} \right\rangle$, i.e. to express the quantum-mechanical correlation function in the form of Eq.7.3. Formally, this decomposition is allowed if the remaining terms of the lattice Liouvilian are independent of the translational motion. An analogous problem has bothered us in Chapter VII, when the correlation function has been mediated by jump diffusions of the coupled spins. We have concluded then that except for the case when the quadrupolar spins jump between non-equivalent crystal sites and therefore they can sense different local electric field gradients at these sites, we can safely perform this factorisation. Here the situation seems to be simpler; all positions in a liquid are just equivalent in this sense. Unfortunately, this conclusion is drawn too fast. Relative translational diffusion of the molecules carrying the S and I spins does not change obviously the structure and the internal dynamics of the molecule containing the S spin and therefore it does not affect the transient ZFS (quadrupolar) interactions. Nevertheless, to separate the correlation function $C_{Diff}(\tau)$ from other degrees of freedom of the lattice it is required that the relative translational motion of the S and I spins is uncorrelated with the reorientational motion of the molecule containing the S spin, and this can create some uncertainties. Let us assume at this moment that the translational motion and the molecular tumbling are uncorrelated. I shall come back to this issue later by discussing deeper the physical simplifications introduced to our treatment by this assumption. Let us profit at this moment from Eq.7.3, which we can use under this condition, not bothering too much. To proceed further we need to specify the form of the correlation function $C_{Diff}(\tau)$. In Chapter VII we have assumed that the correlation function mediated by jump diffusion of ions within a crystal lattice can be modelled as an exponential decay. This model has simplified considerably the evaluations. In the present case we cannot keep this assumption. The form of the correlation function associated with translational diffusion in liquids depends on the assumed diffusional model but it is not exponential. For a force-free diffusion with a uniform distribution of the solvent molecules outside the distance of closest approach, d, and under the assumption of the reflecting wall boundary condition at $r_{IS} = d$, (see Fig.12.2) the correlation function for translational diffusion takes the well – known closed analytical form [5]:

$$\left\langle \frac{D_{0,1-q}^{2*}(\Omega_{DDL}(\tau)) D_{0,1-p}^{2}(\Omega_{DDL}(0))}{r_{IS}^{3}(\tau)} \frac{}{r_{IS}^{3}(0)} \right\rangle = \qquad (12.3)$$

$$\delta_{pq} \frac{72}{5} \frac{1}{d^3} N_S \int_0^\infty \frac{u^2}{81 + 9u^2 - 2u^4 + u^6} \exp\left(-\frac{D_{diff}}{d^2} u^2 \tau\right) du$$

where N_S is the number of spins S per unit volume, while D_{Diff} is the relative translational diffusion coefficient of the molecules carrying the spins I and S, and is defined as a sum of the diffusion coefficients of these molecules, i.e. $D_{Diff} = D_{Diff}^{I} + D_{Diff}^{S}$.

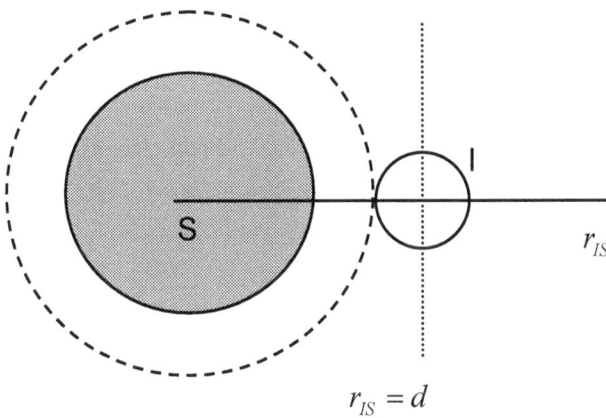

Fig.12.2

An illustration of the assumption of the reflecting wall (dashed sphere) boundary condition at the distance of closest approach $r_{IS} = d$.

In fact, the formula of Eq.12.3 seems to be quite complicated, especially if we realize that we suppose to insert it into Eq.7.3, which is complicated enough by its own. Nevertheless, this expression for the correlation function has a great advantage that the part including explicitly the time variable t has still an exponential form. We shall appreciate this soon (nevertheless, its spectral density function is highly non-Lorenzian due to the contributions at various interspin distances). By force free diffusion correlation function into Eq.7.3 one obtains for the spectral density $K_{1,1}^{DD}(-\omega_I)$:

$$K_{1,1}^{DD}(-\omega_I) = 30\left(\frac{\mu_0 \hbar \gamma_I \gamma_S}{4\pi}\right)^2 \frac{1}{2S+1}\frac{72}{5}\frac{1}{d^3} N_S \sum_{p=-1}^{1}\begin{pmatrix} 2 & 1 & 1 \\ 1-p & p & -1 \end{pmatrix}^2 \times$$

$$\int_0^\infty Tr_{(L-Diff)}\left\{S_p^{1(L)+}\left[\exp\left(-i\hat{L}_{(L-Diff)}\tau\right)S_p^{1(L)}\right]\right\} \times \qquad (12.4)$$

$$\left[\int_0^\infty \frac{u^2}{81+9u^2-2u^4+u^6}\exp\left(-\frac{D_{diff}}{d^2}u^2\tau\right)du\right]\exp(-i\omega_I\tau)d\tau$$

Writing down this expression we have taken into account that the equilibrium operator for translational motion is equal to one $\rho_T^{eq} = 1$. The first consideration by looking at this expression is that it is hardly manageable, because of the integral over the u variable. However a closer inspection brings us to the conclusion that we do not need to integrate first over u. We can bring together the exponential terms depending on τ and change the order of integration. When we proceed in this way the expression takes the form:

$$K_{1,1}^{DD}(-\omega_I) = 30\left(\frac{\mu_0 \hbar \gamma_I \gamma_S}{4\pi}\right)^2 \frac{1}{2S+1}\frac{72}{5}\frac{1}{d^3} N_S \sum_{p=-1}^{1}\begin{pmatrix} 2 & 1 & 1 \\ 1-p & p & -1 \end{pmatrix}^2 \times$$

$$\int_0^\infty \frac{u^2}{81+9u^2-2u^4+u^6}\left\{\int_0^\infty Tr_{(L-Diff)}\left[S_p^{1(L)+}\exp\left(-\left(i\hat{L}_{(L-Diff)}+\frac{D_{diff}}{d^2}u^2\hat{1}+\omega_I\hat{1}\right)\tau\right)S_p^{1(L)}\right]d\tau\right\}du \qquad (12.5)$$

This formula can be treated as a starting point for evaluating the dipolar spin relaxation resulting from the translational motion of the spins I and S, and the S spin dynamics [9, 10].

Next parts of this section are devoted to detailed treatments of the I spin relaxation depending on the applied magnetic field and motional conditions characterising the considered spin system, in particular the timescale of the molecular tumbling. We know very well from Chapter IX that the rotational motion can cause serious difficulties when one attempts to describe the S spin relaxation within the framework of the second order perturbation theory. The point is that except the limiting cases of very slow and very fast molecular tumbling we are not able to divide unambiguously the electron spin Hamiltonian into main and perturbing parts. There is no way to escape from this problem and we have to deal with it once again under even more complex circumstances than in Chapter IX, because now the system possesses in addition translational degrees of freedom. Thus, I shall proceed further with the discussion of relaxation processes in the presence of translational motion, increasing gradually the level of complexity.

We start with the simplest case when the static ZFS vanishes. It implies that the electron spin exhibits a Zeeman energy level structure, independently of the applied magnetic field, while its relaxation occurs due to the transient ZFS. Then we can just rewrite Eq.12.5 in the form:

$$R_{1I} = 2\operatorname{Re}\{K_{1,1}^{DD}(-\omega_I)\} = 864\left(\frac{\mu_0 \hbar \gamma_I \gamma_S}{4\pi}\right)\frac{1}{2S+1}\frac{1}{d^3}N_S \sum_{p=-1}^{1}\begin{pmatrix} 2 & 1 & 1 \\ 1-p & p & -1 \end{pmatrix}^2 \times$$
$$\int_0^\infty \frac{u^2}{81+9u^2-2u^4+u^6} s_{p,p}^{Diff}(u,\omega_I)\,du \qquad (12.6)$$

The spectral densities, $s_{p,p}^{Diff}(u,\omega_I)$ result from the integration over time and can be evaluated in the well known way as a product of three matrices:

$$s_{p,p}^{Diff}(u,\omega_I) =$$
$$\int_0^\infty Tr_{(L-Ddiff)}\left\{S_p^{1+(L)}\left[\exp\left(-\left(i\hat{L}_{(L-Diff)} + \frac{D_{Diff}}{d^2}u^2\hat{1} + i\omega_I\hat{1}\right)\tau\right)S_p^{1(L)}\right]\right\}d\tau = [S_p^1]^\dagger [\hat{M}(u)][S_p^1] \qquad (12.7)$$

The operator $\hat{M}(u)$, which is generally defined as:

$$\hat{M}(u) =$$
$$i\hat{L}_{(L-Diff)} + \frac{D_{Diff}}{d^2}u^2\hat{1} + i\omega_I\hat{1} = i\hat{L}_Z + i\hat{L}_{ZFS}^S + i\hat{L}_{ZFS}^T + i\hat{L}_R + i\hat{L}_D + \frac{D_{Diff}}{d^2}u^2\hat{1} + i\omega_I\hat{1} \qquad (12.8)$$

takes now a much simpler form, namely:

$$\hat{M}(u) = i\hat{L}_Z + \hat{R}_{ZFS}^T + \frac{D_{Diff}}{d^2}u^2\hat{1} + i\omega_I\hat{1} \qquad (12.9)$$

where the variable u ranges from zero to infinity (see Eq.12.3).

where $\hat{R}_{ZFS}^T$ is the relaxation operator corresponding to the transient ZFS relaxation mechanism. The form of Eq.12.9 indicates clearly that as long as we neglect the rotational contribution to the modulations of the transient ZFS with respect to the laboratory frame, the nuclear spin relaxation is not affected by the molecular tumbling, since it does not influence directly the dipole-dipole coupling and does not affect the electron spin relaxation, either. The matrices $[S_p^1]$ contain expansion coefficients of the tensor components (Eqs.6.1a,b,c) in the Liouville basis constructed from the Zeeman eigenvectors of the electron spin $\{|m_S\rangle\langle m_S'|\}$. As a result we get:

$$R_{1I} = \frac{144}{5}\left(\frac{\mu_0 \hbar \gamma_I \gamma_S}{4\pi}\right)\frac{1}{2S+1}\frac{1}{d^3} N_S \times$$

$$\int_0^\infty \frac{u^2}{81+9u^2-2u^4+u^6}\left[s_{11}^{Diff}(u,\omega_I)+3s_{00}^{Diff}(u,\omega_I)+6s_{-1-1}^{Diff}(u,\omega_I)\right]du \quad (12.10)$$

This expression can be treated as a counterpart of the SBM theory (Chapter IV) when the $I-S$ dipole-dipole coupling is modulated by a translational motion of the participating spins. Applying it to the simplest case of $S=1$ one obtains:

$$R_{1I} = \frac{48}{5}\left(\frac{\mu_0}{4\pi}\gamma_I\gamma_S\hbar\right)^2 \frac{1}{dD_{Diff}} N_S \times$$

$$\int_0^\infty \frac{u^2}{81+9u^2-2u^4+u^6}\left[\frac{y_2+u^2}{(y_2+u^2)^2+(d^2(\omega_I-\omega_S)/D_{Diff})^2} + 3\frac{y_1+u^2}{(y_1+u^2)^2+(d^2\omega_I/D_{Diff})^2} + 6\frac{y_2+u^2}{(y_2+u^2)^2+(d^2(\omega_I+\omega_S)/D_{Diff})^2}\right]du \quad (12.11)$$

where the quantities $y_i = \frac{d^2 R_{iS}}{D_{Diff}}$ include the electron spin relaxation rates R_{1S} and R_{2S} of Eqs.4.41a,b.

It is useful and interesting to formulate a description of the nuclear spin relaxation in the presence of translational motion for slowly rotating systems and compare it with the description presented in Chapter VI. A complete theory of the nuclear spin relaxation affected by translational motion of the participating spins (assuming slow molecular tumbling) for an arbitrary electron spin quantum number has been developed in [11]. It is highly appropriate at this point to turn attention of the reader to other works devoted to the problem of the nuclear spin relaxation caused by relative translational movement of the nuclear and electron spins [11-13]. For didactic reasons we shall recapitulate here briefly the main stages of the treatment presented in [11]. In fact, after reading Chapters VI and VII we are well prepared to perform the calculations and there is no need to consider this issue in detail. Let us just follow the already discussed procedure, adjusting it in an appropriate manner to the present case. We shall perform the calculations, describing the electron spin dynamics in the principal axis system of the static ZFS interaction. We begin with Eq.12.5. The Liouville operator, $\hat{L}_{L-Diff}$ depends now on the molecular orientation, described by the angle $\Omega_{P_S L}$, and is given as:

$$\hat{L}_{L-Diff} \equiv \hat{L}_S(\Omega_{P_S L}) = \hat{L}_Z(\Omega_{P_S L}) + \hat{L}_{ZFS}^S + \hat{R}_{ZFS}^T(\Omega_{P_S}) + \omega_I \hat{1}.$$

The field-dependent electron spin dynamics for slow molecular tumbling has been described in great detail in Chapter V; the energy level structure of the electron spin is given by a superposition of the Zeeman and ZFS interactions, while the transient ZFS causes the relaxation described by the operator, $\hat{R}^T(\Omega_{P_S L})$. At this stage one should not forget that the Redfield relaxation theory requires that $\omega_{ZFS}^T \tau_D \ll 1$ and $\omega_{ZFS}^T \tau_D \ll \frac{\omega_{ZFS}^T}{\omega_{ZFS}^S}$. Since we consider the electron spin relaxation in the (P_S) frame, we have to express the tensor components $S_p^{1(L)}$ in the same reference frame; we have discussed this subject in Chapter V. Therefore, the correlation function, $Tr_{(L-Diff)}\left\{S_p^{1(L)+} \exp\left[-\left(i\hat{L}_S(\Omega_{P_S L}) + \frac{D_{Diff}}{d^2}u^2\right)\tau\right]S_p^{1(L)}\right\}$, has to be calculated employing the transformation rule (Eq.3.3) for tensor operators:

$$Tr_{(L-Diff)}\left\{S_p^{1(L)+} \exp\left[-\left(i\hat{L}_S(\Omega_{P_S L}) + \frac{D_{Diff}}{d^2}u^2\right)\tau\right]S_p^{1(L)}\right\} =$$

$$\sum_{m,k=-1}^{1} Tr_S\left\{S_m^{1(P_S)+} \exp\left[-\left(i\hat{L}_S(\Omega_{P_S L}) + \frac{D_{Diff}}{d^2}u^2\right)\tau\right]S_k^{1(P_S)}\right\} D_{m,p}^{1*}(\Omega_{P_S L}) D_{k,p}^{1}(\Omega_{P_S L})$$

(12.12)

Substituting this expression into Eq.12.5 one obtains after a couple of obvious mathematical manipulations, that:

$$K_{11}^{DD}(-\omega_I) = 30\left(\frac{\mu_0 \hbar \gamma_I \gamma_S}{4\pi}\right)^2 \frac{1}{2S+1} \frac{72}{5} \frac{1}{d^3} N_S \times$$

$$\sum_{m,k=-1}^{1} \int_0^\infty \frac{u^2}{81+9u^2-2u^4+u^6} \left\langle g_{mk}^P(\Omega_{P_SL}) s_{m,k}^{Diff}(\Omega_{P_SL},u,\omega_I)\right\rangle du \quad (12.13)$$

where the brackets <> denote the averaging over molecular orientations. The spectral densities $s_{m,k}(\Omega_{P_SL},u,\omega_I)$ are now defined for individual molecular orientations:

$$s_{m,k}^{Diff}(\Omega_{P_SL},u,\omega_I) = \int_0^\infty Tr_S\left\{S_m^{1(P_S)+}\left[-\left(i\hat{L}_S(\Omega_{P_SL})+\frac{D_{Diff}}{d^2}u^2\hat{1}+i\omega_I\hat{1}\right)\tau\right]S_k^{1(P_S)}\right\}d\tau =$$

$$[S_m^1(\Omega_{P_SL})]^+\left[\hat{M}(\Omega_{P_SL},u)\right]^{-1}[S_k^1(\Omega_{P_SL})] \quad (12.14)$$

The corresponding weight factors $g_{mk}(\Omega_{P_SL})$ yield:

$$g_{mk}(\Omega_{P_SL}) = \sum_{p=-1}^{1}\begin{pmatrix}2 & 1 & 1\\ 1-p & p & -1\end{pmatrix}^2 D_{m,p}^{1*}(\Omega_{P_SL})D_{k,p}^1(\Omega_{P_SL}) \quad (12.15)$$

The way of evaluating the quantities $s_{m,k}^{Diff}(\Omega_{P_SL},u,\omega_I)$ does not need any other explanations, except the current definition of the superoperator $\hat{M}(\Omega_{P_SL},u)$:

$$\hat{M}(\Omega_{P_SL},u) = -i\hat{L}_Z(\Omega_{P_SL})-i\hat{L}_{ZFS}^S + \hat{R}^T(\Omega_{P_SL}) + \left(i\omega_I + \frac{D_{Diff}}{d^2}u^2\right)\hat{1} \quad (12.16)$$

and the statement that, as usual, the matrices $[S_m^1(\Omega_{P_SL})]$ and $[\hat{M}(\Omega_{P_SL},u)]$ are matrix representations of the tensor operators S_m^1 and $\hat{M}(\Omega_{P_SL},u)$ in the Liouville basis constructed from eigenstates of the main Hamiltonian: $H_Z(\Omega_{P_SL}) + H_{ZFS}^S$, respectively. Knowing how to evaluate the orientation-dependent spectral densities $s_{m,k}^{Diff}(\Omega_{P_SL},u,\omega_I)$ with the u variable as a parameter, we can obtain the nuclear spin relaxation rate R_{1I} from the expression:

$$R_{1I} = 864\left(\frac{\mu_0\hbar\gamma_I\gamma_S}{4\pi}\right)^2 \frac{1}{2S+1}\frac{1}{d^3}N_S \operatorname{Re}\left\{\sum_{m,k=-1}^{1} s_{mk}^{Diff}(\omega_I)\right\} \quad (12.17)$$

where the spectral densities $s_{m,k}^{Diff}(\omega_I)$ result from the integration over the u variable after the averaging over molecular orientations of the quantities $s_{m,k}^{Diff}(\Omega_{P_SL}, u, \omega_I)$ (see Eq.12.13):

$$s_{m,k}^{Diff}(\omega_I) = \int_0^\infty \frac{u^2}{81 + 9u^2 - 2u^4 + u^6} \langle g_{mk}(\Omega_{P_SL}) s_{m,k}^{Diff}(\Omega_{P_SL}, u, \omega_I) \rangle du \qquad (12.18)$$

Fig.12.3 shows examples of nuclear spin relaxation profiles for slowly rotating systems in the presence of relative translational motion of the I and S spins. More examples one can find in [10, 14]; the second work concerns fast rotating systems. One can see that field dependencies of the nuclear spin relaxation when the $I - S$ dipole-dipole interaction is affected by the translational diffusion differ significantly from the relaxation profiles evaluated in Chapter VI. It is worthwhile to notice that the nuclear spin relaxation rates obtained in the presence of translational motion are rather low compared to the relaxation rates obtained in Chapter VI.

a)

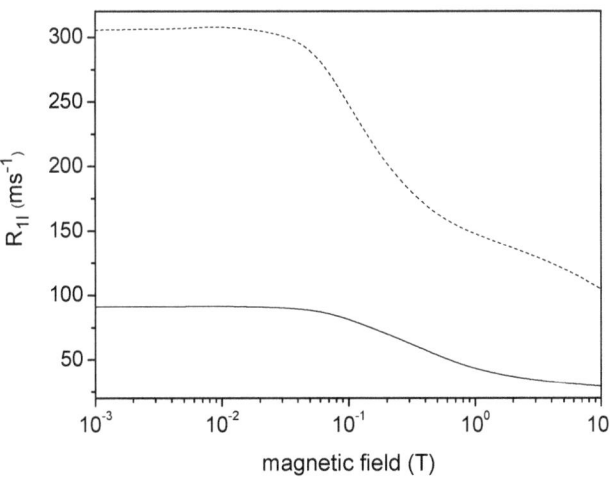

b)

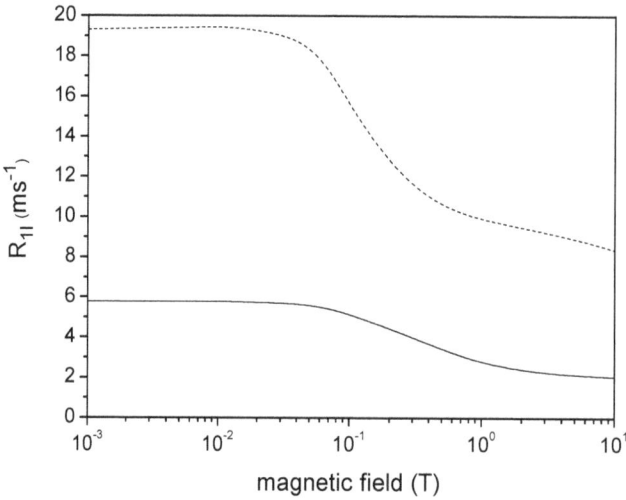

Fig.12.3

Proton a) and carbon b) spin – lattice relaxation rates in the presence of translational diffusion for the electron spin quantum number $S = 1$
Solid line: $D_{diff} = 2.3*10^{-9} m^2/s$, dashed line: $D_{diff} = 5.0*10^{-10} m^2/s$
The other parameters are:
$D_S = 0.05 cm^{-1}$, $D_T = 0.02 cm^{-1}$, $\tau_D = 10 ps$, $\tau_R = 1 \mu s$, $d = 300 pm$.

The modulations of the electron spin - nuclear spin dipole - dipole interaction result from the electron spin relaxation and the translational diffusion; if the second process is very fast it masks the electron spin relaxation. This can happen especially at high field limit when the electron spin relaxation becomes slower. The efficiency of the nuclear spin relaxation increases when the relative translational diffusion is slower.

We have already concluded that in the high field limit the rotational motion of the molecule carrying the S spin does not influence the nuclear spin relaxation. This is due to the fact that the molecular tumbling does not affect the electron spin dynamics and that the quantization axis of the electron spin does not fluctuate with respect to the laboratory axis; the electron spin is just locked in the laboratory frame. The second condition is not fulfilled at low field limit. Even though the rotational motion still does not influence either the energy level structure of the electron spin (which is in fact fully determined by the static ZFS) or the electron spin relaxation (the fluctuations of the transient ZFS with respect to the P_S frame are caused by the distortional motion), we have to take into account that now we describe the electron spin relaxation in a reference frame (the P_S frame) which fluctuates with respect to the laboratory axis. We can quite easily derive analytical expressions for the nuclear spin relaxation at the low field limit, when the $I-S$ dipole-dipole coupling is mediated by the translational motion. Such a description has been developed in [9] for an arbitrary electron spin quantum number S, but it can be worthwhile to comment at this stage on the most important steps of the calculations.

Even though, as we shall see soon, it is not necessary to begin our considerations from the beginning, starting from Eq.7.1 I wish to do so. When we transform the electron spin tensor components S_p^1 to the P_S frame, Eq.5.1 takes the form:

$$K_{1,1}^{IS}(-\omega_I) = 30\left(\frac{\mu_0 \hbar \gamma_I \gamma_S}{4\pi}\right)^2 \frac{1}{2S+1} \sum_{p,q=-1}^{1} \sum_{m,n=-1}^{1} \begin{pmatrix} 2 & 1 & 1 \\ 1-p & p & 1 \end{pmatrix} \begin{pmatrix} 2 & 1 & 1 \\ 1-q & q & 1 \end{pmatrix} \times$$

$$\int_0^\infty \left\{ \begin{array}{l} Tr_S\left\{S_m^{1(P_S)+}\left[\exp\left(-i\hat{L}_S\tau\right)S_n^{1(P_S)}\right]\right\} \times \left\langle D_{m,q}^{1*}(\Omega_{P_SL}(\tau))D_{n,p}^1(\Omega_{P_SL}(0))\right\rangle \times \\ \left\langle \frac{D_{0,1-q}^{2*}[\Omega_{DDL}(\tau)]}{r_{IS}^3(0)} \frac{D_{0,1-p}^2[\Omega_{DDL}(0)]}{r_{IS}^3(t)} \right\rangle \exp(-i\omega_I\tau) d\tau \end{array} \right\} \quad (12.19)$$

Generally, the transformation introduces into our calculations the rotational degrees of freedom; however for slowly rotating systems the angle Ω_{P_SL} remains, in fact, unchanged in time. It is noteworthy that in Eq.12.19 we have separated the correlation functions describing the electron spin dynamics, the molecular tumbling and the translational diffusion. This is actually the reason for starting the considerations from the main expression of Eq.7.1; I have intended to make apparent for the reader why the statement that the three dynamic processes are uncorrelated is crucial for the calculations. Instead of averaging over molecular orientations we need now to make use of the explicit form of the rotational correlation function: $\langle D_{m,q}^{1*}(\Omega_{P_S L}(\tau))D_{n,p}^{1}(\Omega_{P_S L}(0))\rangle = \delta_{m,n}\delta_{pq}\frac{1}{3}\exp\left(-\frac{\tau}{3\tau_R}\right)$, where τ_R corresponds to Wigner rotational matrices of the second rank. Introducing into Eq.12.19 this expression as well as Eq.12.3 we obtain a general formula for the nuclear spin relaxation rate R_{1I} at low field [9]:

$$R_{1I} = 96\left(\frac{\mu_0 \hbar \gamma_I \gamma_S}{4\pi}\right)^2 \frac{1}{2S+1}\frac{1}{d^3}N_S \times$$
$$\int_0^\infty \frac{u^2}{81+9u^2-2u^4+u^6}\left[s_{-1,-1}^{Diff,LF}(u,\omega_I)+s_{0,0}^{Diff,LF}(u,\omega_I)+s_{1,1}^{Diff,LF}(u,\omega_I)\right]du \quad (12.20)$$

with the electron spin spectral densities $s_{m,n}^{Diff,LF}(u,\omega_I) = [S_m^1]^+\left[\hat{\hat{M}}^{LF}(u)\right]^{-1}[S_m^1]$, which now do not depend on the molecular orientation. At low field limit one can neglect in Eq.12.16 the Zeeman term and therefore the operator $\hat{\hat{M}}^{LF}(u,\omega_I)$ can be defined as:

$$\hat{\hat{M}}^{LF}(u) = -i\hat{\hat{L}}_{ZFS}^S + \hat{\hat{R}}^T + \left(i\omega_I + \frac{1}{3\tau_R} + \frac{D_{Diff}}{d^2}u^2\right)\hat{\hat{1}} \quad (12.21)$$

It is important to notice that the Zeeman interaction does not contribute to the energy level structure of the electron independently of the timescale of the molecular tumbling. It implies that Eq.12.20 describes the nuclear spin relaxation at low field for an arbitrary rotational correlation time. Therefore we could obtain it just from Eq.12.13 noticing that at the low field limit $g_{mk}(\Omega_{P_S L}) \equiv g_{mk} = \frac{1}{3}\sum_{p=-1}^{1}\begin{pmatrix}2 & 1 & 1\\1-p & p & -1\end{pmatrix}$, modifying the operator $\hat{\hat{M}}(\Omega_{P_S L},u)$ of Eq.12.16 to the form $\hat{\hat{M}}^{LF}(u)$ of Eq.12.21 and skipping, because of the already explained and mathematically obvious reasons, the averaging over molecular orientations.

Making use of the fact that the matrices $[S_m^1]$ represent now the tensor components expanded in the Liouville basis constructed from eigenstates of the static ZFS one can write closed form expressions for R_{1I} depending on the spin quantum number S [9]. For example for $S = 1$, according to [10], one gets for an axially symmetric ZFS tensor:

$$R_{1I} = \frac{32\pi}{9}\left(\frac{\mu_0}{4\pi}\gamma_I\gamma_S\hbar\right)^2 \frac{1}{dD_{Diff}} N_A[M]10^3 \times$$

$$\left(\int_0^\infty \xi(u)\frac{y_1 + u^2}{(y_1 + u^2)^2 + (d^2\omega_I/D_{Diff})^2}du + \int_0^\infty \xi(u)\frac{y_2 + u^2}{(y_2 + u^2)^2 + (d^2(\omega_I + \omega_D)/D_{Diff})^2}du\right)$$

(12.22)

where $y_i = \frac{d^2}{D_{Diff}}\left(R_{iS} + \frac{1}{3\tau_R}\right)$ with the electron spin relaxation rates R_{1S} and R_{2S} given by Eqs.4.41a,b. The spin density N_S (in units of m^{-3}) has been here replaced by Avogadro number N_A times molar concentration [M] of the molecules possessing the S spin (the factor 10^3 accounts for the change of volume units). The function $\xi(u)$ has been defined in [9] as: $\xi(u) = \frac{18}{\pi}\frac{u^2}{81 + 9u^2 - 2u^4 + u^6}$ and therefore the constants in front of Eq.12.22 and 12.20 are different; we have decided here to follow the convention of this work.

XII.2. General treatment of relaxation mediated by translational diffusion

The purpose of this section is to show how to adapt and extend the treatment presented in Chapter X to molecular systems which exhibit translational motion. We intend to keep very close analogy to the considerations of Chapter X and follow the same way of thinking. Therefore, let us come back once again to Eq.7.1 which represents the quantum-mechanical spectral density, $K_{1,1}^{DD}(-\omega_I)$, in its initial form, before the translational correlation function $C_{Diff}(\tau)$ has been extracted. The expression results from the general definition of the quantum-mechanical spectral density, introduced in Chapter II and invoked in Eq.10.17. To make the further development more straightforward, let us remain with the original form of the spectral density $K_{1,1}^{DD}(-\omega_I)$ of Eq.4.19:

$$K_{11}^{DD}(-\omega_I) = \int_0^\infty Tr_L \left\{ T_1^{1(DD)+} \left[\exp(-i\hat{L}_L \tau) T_1^{1(DD)} \rho_L^{eq} \right] \right\} \exp(-i\omega_I \tau) d\tau, \text{ where } T_1^{1(DD)} \text{ is the}$$

tensor operator describing the dipole-dipole coupling between the I spin and the lattice which involves, in particular, the S spin. To adapt the general treatment discussed in Chapter X to the present case we have to write down the explicit form of the tensor $T_1^{1(DD)}$ in the presence of the translational diffusion. Comparing the definition with Eq.7.1 we can easily conclude that now the tensor $T_1^{1(DD)} = T_1^{1(DD,Diff)}$ has the form:

$$T_1^{1(DD,Diff)} = \sqrt{5}\left(\sqrt{6}\gamma_I\gamma_S\hbar\frac{\mu_0}{4\pi}\right)\sum_{q=-1}^{1}\begin{pmatrix} 2 & 1 & 1 \\ 1-q & q & -1 \end{pmatrix} S_q^1 \frac{D_{0,1-q}^2(\Omega_{DDL})}{r_{IS}^3} \qquad (12.23)$$

Actually, this expression is quite obvious if one compares it with Eq.10.18. Instead of the Wigner rotation matrices $D_{0,1-q}^2(\Omega_{ML})$ describing the orientation of the molecular frame (and at the same time the I-S dipole-dipole axis) with respect to the laboratory axis, modulated by the rotational motion, Eq.12.23 contains the Wigner matrices $D_{0,1-q}^2(\Omega_{DDL})$ which also determine the relative orientation of the I-S dipole-dipole and laboratory axes, however now the orientation is modulated by the translational diffusion. Since the translational motion changes also the interspin distance r_{IS} we have excluded it from the dipole-dipole constant and combined with the Wigner matrices $D_{0,1-q}^2(\Omega_{DDL})$. Employing the relationship between the tensor components S_q^1 and the normalised operators $Q_q^1 \equiv |1,q\rangle$, one obtains:

$$T_1^{1(DD,Diff)} = a_{DD}^T \sqrt{\frac{5(2S+1)(S+1)S}{3}} \sum_{q=-1}^{1}\begin{pmatrix} 2 & 1 & 1 \\ 1-q & q & -1 \end{pmatrix} Q_q^1 \frac{D_{0,1-q}^2(\Omega_{DDL})}{r_{IS}^3} \qquad (12.24)$$

where $a_{DD}^T = \sqrt{6}\gamma_I\gamma_S\hbar\frac{\mu_0}{4\pi}$. Let us now rewrite the expression for spectral density $K_{1,1}^{DD}(-\omega_I)$ (Eq.12.5) in a form similar to Eq.10.17:

$$K_{1,1}^{DD}(-\omega_I) = (a_{DD}^T)^2 \frac{1}{2S+1} \frac{72}{5} \frac{1}{d^3} N_S \frac{5(2S+1)(S+1)S}{3} \sum_{p,q=-1}^{1} \begin{pmatrix} 2 & 1 & 1 \\ 1-p & p & -1 \end{pmatrix} \begin{pmatrix} 2 & 1 & 1 \\ 1-q & q & -1 \end{pmatrix} \times$$

$$\int_0^\infty \frac{u^2}{81+9u^2-2u^4+u^6} \left[\int_0^\infty Tr_{(L-Diff)} \left\{ Q_p^{1+} \exp\left[-\left(i\hat{L}_{(L-Diff)} + \frac{D_{diff}}{d^2} u^2 \hat{1} + i\omega_I \right) \tau \right] Q_q^1 \right\} d\tau \right] du$$

(12.25)

Let us first evaluate the 'internal' spectral density:

$$\widetilde{K}_{1,1}^{DD}(u,-\omega_I) = \sum_{p,q=-1}^{1} \begin{pmatrix} 2 & 1 & 1 \\ 1-p & p & -1 \end{pmatrix} \begin{pmatrix} 2 & 1 & 1 \\ 1-q & q & -1 \end{pmatrix} \times$$

$$\int_0^\infty Tr_{(L-Diff)} \left\{ Q_p^{1+} \exp\left[-\left(i\hat{L}_{(L-Diff)} + \frac{D_{diff}}{d^2} u^2 \hat{1} + i\omega_I \hat{1} \right) \tau \right] Q_q^1 \right\} d\tau$$

(12.26)

which results from the average over all lattice degrees of freedom, except the translational motion, however includes the diffusion term, $\exp\left(-\frac{D_{Diff}}{d^2} u^2 \tau\right)$. It is very important to stress clearly that in Eq.12.25 and Eq.12.26 we do not average over the translation degrees of freedom, because this has been already done by Eq.12.3 resulting in the diffusion term. This issue has been discussed in detail in Section XII.1 devoted to the analytical description of the nuclear spin relaxation, under the assumption that the electron spin subsystem fulfils the conditions of the Redfield relaxation theory. Even though now we do not keep this assumption any more, we deal obviously with the same Liouville operators as in the previous section, except that we may not introduce now the relaxation superoperator for the electron spin. Therefore the sum of the Liouville operators in the parentheses in Eq.12.26 is nothing else than the operator $\hat{M}(u)$ of Eq.12.8. To calculate the spectral density, $\widetilde{K}_{1,1}^{DD}(u,-\omega_I)$ we follow the strategy described in Chapter X. First, we have to set up an appropriate basis including all relevant degrees of freedom of the lattice. In the next step we shall evaluate within this basis, by employing the Wigner - Eckart theorem [15, 16], matrix elements of the Liouville operators constituting the operator $\hat{M}(u)$. At this moment we should realize consequences of the fact that it is always possible to separate the translational correlation function $C_{Diff}(t)$ from the other degrees of freedom of the lattice. As it has been already discussed, it is allowed to proceed in this way because of the assumption that the translational motion is uncorrelated with the molecular tumbling and does not affect the electron spin dynamics. The problem whether the electron spin dynamics can be described in terms of well defined relaxation rates or not, does not matter here. It implies that the translation degrees of freedom are excluded from the averaging in Eq.12.26 ($Tr_{(L-Diff)}$) and therefore the infinite Liouville basis $\{O_i\}$ appropriate for the present case does not need to contain components associated with the translational motion. Thus, the basis set can be formed in full analogy to the case considered in Chapter X as a direct product of orthonormal basis operators: $|ABC)$ for the distortional motion modelled as pseudorotational diffusion, $|LKM)$ for the molecular tumbling, and $|\Sigma\sigma)$ for the electron spin system: $\{O_i\} = \{|ABC)\} \otimes \{|LKM)\} \otimes \{|\Sigma\sigma)\}$; in other words we can use exactly the same Liouville basis as previously. This has a great advantage. It simplifies considerably the calculations by reducing the effective dimension of the matrix representation of the operator $\hat{M}(u)$. Despite the computational advantages, we profit mainly from the fact that we do not need to evaluate once again in an extended basis matrix elements of the Liouville operators $\hat{L}^S_{ZFS}, \hat{L}^T_{ZFS}, \hat{L}_Z(S)$ and $\hat{L}_R$ and $\hat{L}_D$. They are just given by the expressions of Eq.10.11-10.16. To set up the matrix $\left[\hat{M}(u)\right]$ we need to evaluate only the remaining matrix elements corresponding to the operator $\frac{D_{Diff}}{d^2}u^2\hat{1}$. The evaluations are particularly simple and one gets immediately:

$$(A'B'C'|(L'K'M'|(\Sigma'\sigma'|\frac{D_{Diff}}{d^2}u^2\hat{1}|\Sigma\sigma)LKM)ABC) =$$

$$\delta_{A'A}\delta_{B'B}\delta_{C'C}\delta_{L'L}\delta_{K'K}\delta_{M'M}\delta_{\Sigma'\Sigma}\delta_{A'A}\frac{D_{Diff}}{d^2}u^2$$

(12.27)

The reader should be aware that usually when one includes additional degrees of freedom characterising a spin system under interest it is necessary to extend the Liouville basis by including explicitly vectors associated with these degrees of

Chapter XII - Relaxation processes caused by translational diffusion of interacting spins

freedom. The situation which takes place here should be treated just as a fortuitous case.

When the matrix $\left[\hat{M}(u)\right]$ is set up, the spectral density $\widetilde{K}_{1,1}^{DD}(u,-\omega_I)$ can be obtained in analogy to Eq.10.17, as:

$$\widetilde{K}_{1,1}^{DD}(u,-\omega_I) = \left[c_1^{(Diff)*}\right] \times \left[\hat{M}(u)\right]^{-1} \times \left[c_1^{(Diff)}\right] \tag{12.28}$$

One can easily understand from Eq.12.26 that the vectors $\left[c_1^{(Diff)}\right]$ (the index (Diff) refers to the translational diffusion) project out the sum of the operators $\sum_{p=-1}^{1}\begin{pmatrix}2 & 1 & 1\\ 1-p & p & -1\end{pmatrix}Q_p^1$. Therefore, they contain only three non-zero elements associated with the basis vectors: $|ABC\rangle|LKM\rangle|\Sigma\sigma) = |000\rangle|000\rangle|1p)$, where $p = -1, 0, 1$. The corresponding coefficients are equal to: $\begin{pmatrix}2 & 1 & 1\\ 2 & -1 & -1\end{pmatrix}$, $\begin{pmatrix}2 & 1 & 1\\ 1 & 0 & -1\end{pmatrix}$ and $\begin{pmatrix}2 & 1 & 1\\ 0 & 1 & -1\end{pmatrix}$, respectively. At this stage some comments can be appropriate. One could easily conclude that the values of the projection coefficients are the same as in the case considered in Chapter X, when the $I-S$ dipole-dipole coupling has been modulated by the rotational motion. However, one should notice at the same time that the coefficients are now associated with different vectors $|ABC\rangle|LKM\rangle|\Sigma\sigma)$ of the Liouville space. The difference concerns the rotational states $|LMK\rangle$. The fact that the rotational motion does not modulate directly the dipole-dipole interaction implies that the Wigner matrices $D_{0,1-p}^2(\Omega_{ML})$ depending on the rotational variables do not appear in the expression for the tensor operator $T_1^{1(DD)}$ (Eq.12.24). Since, formally one can say that the rotational contribution to the operator $T_1^{1(DD)}$ is represented by the Wigner matrix $D_{KM}^L(\Omega_{ML}) = D_{0,0}^0(\Omega_{ML}) \equiv 1$, the non-zero projection coefficients are now linked to the vectors corresponding to the rotational state $|LMK) = |000)$, i.e. $|ABC\rangle|LMK\rangle|\Sigma\sigma) = |000\rangle|000\rangle|1p)$.

333

In the case considered in Chapter X when the *I-S* dipole-dipole coupling is affected by the rotational motion encoded in the matrices $D^2_{0,1-p}(\Omega_{ML})$ (which are associated with the *S* spin tensor components S^1_p) the projection coefficients correspond to the states $|ABC\rangle|LMK\rangle|\Sigma\sigma\rangle = |000\rangle|20(1-p)\rangle|1p\rangle$. The different origin of the modulations of the dipole-dipole interaction necessitates also some modifications in the structure of the matrix $\left[\hat{M}(u)\right]$ compared to the matrix $\left[\hat{M}\right]$, which we set up in Chapter X. The relationship between the rotational and electron spin quantum numbers: $\sigma = 1 - K$ (one can formulate it looking at the form of the vectors corresponding to the non-zero elements of the projection coefficients: $|ABC\rangle|LMK\rangle|\Sigma\sigma\rangle = |000\rangle|20(1-\sigma)\rangle|1\sigma\rangle$) can be treated as a selection rule; it allows one to exclude from the Liouville basis (which is infinite anyway) vectors, which do not fulfil this relation. I did not comment on this in Chapter X, because I did not want to provide at that stage too many computational details. This makes the numerical calculations presented in Chapter X less time consuming, however does not have any other consequences.

a)

$\|000\rangle\|000\rangle\|11\rangle$			
$\|000\rangle\|000\rangle\|21\rangle$			
$\|000\rangle\|1-11\rangle\|10\rangle$			
....			
$\|000\rangle\|202\rangle\|1-1\rangle$		R_{1I} no translation diffusion	
$\|000\rangle\|201\rangle\|10\rangle$			
$\|000\rangle\|200\rangle\|11\rangle$			
....			

b)

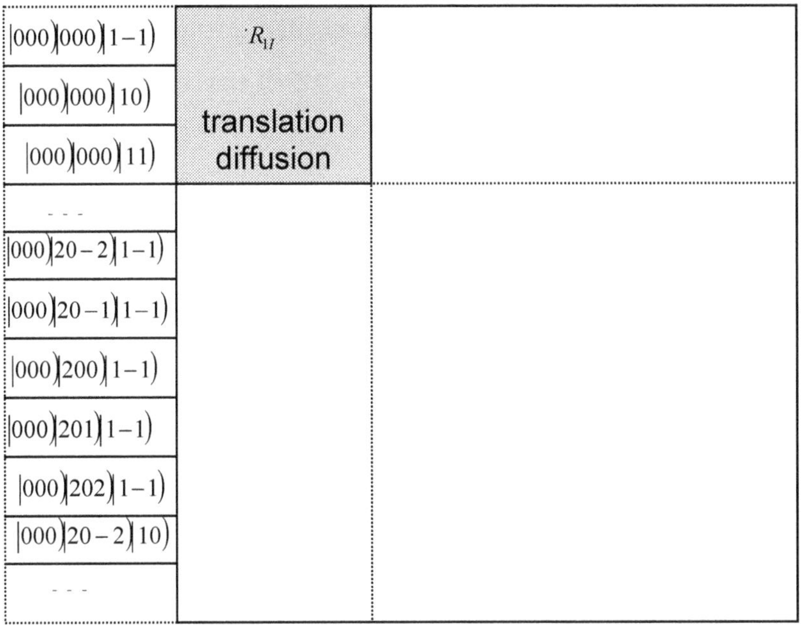

$\|000\rangle\|000\rangle\|1-1\rangle$	R_{1I}	
$\|000\rangle\|000\rangle\|10\rangle$	translation diffusion	
$\|000\rangle\|000\rangle\|11\rangle$		

$\|000\rangle\|20-2\rangle\|1-1\rangle$		
$\|000\rangle\|20-1\rangle\|1-1\rangle$		
$\|000\rangle\|200\rangle\|1-1\rangle$		
$\|000\rangle\|201\rangle\|1-1\rangle$		
$\|000\rangle\|202\rangle\|1-1\rangle$		
$\|000\rangle\|20-2\rangle\|10\rangle$		

Fig.12.4

The structures of the matrices $\left[\hat{M}\right]$ (a) and $\left[\hat{M}(u)\right]$ (b). In the first case the relationship between the rotational and electron spin quantum numbers: $\sigma = 1 - K$ reduces the basis to the vectors $|ABC\rangle|LMK\rangle|\Sigma\sigma\rangle = |000\rangle|20(1-\sigma)\rangle|1\sigma\rangle$ (therefore, for example, the vectors $|000\rangle|000\rangle|10\rangle$ and $|000\rangle|000\rangle|1-1\rangle$ are excluded). When the translational diffusion is present this selection rule does not apply.

In the present case there is no relationship between the rotational and spin quantum numbers and therefore the structure of the matrix $\left[\hat{M}(u)\right]$ cannot be simplified. It is illustrative to compare directly the structures of the matrices $\left[\hat{M}\right]$ and $\left[\hat{M}(u)\right]$ appropriate for the cases when the $I - S$ dipole-dipole coupling is modulated by the rotational and translational motions, respectively.

The matrices are presented in a graphical form in Fig.12.4a,b. In both the cases only 3×3 fragments of the inverted matrices are needed for computing the spectral densities $K_{1,1}^{DD}(\omega_I)$ and $\widetilde{K}_{1,1}^{DD}(u,-\omega_I)$, however as it has been explained above they correspond to different basis vectors $|O_i\rangle$.

So far I have explained how to evaluate the spectral density $\widetilde{K}_{1,1}^{DD}(u,-\omega_I)$ for an arbitrary value of the u variable. To obtain the dipolar spin relaxation rate $R_{1I}(\omega_I)$ we have to integrate these spectral density over u:

$$R_{1I}^{DD}(\omega_I) = 48(a_{DD}^T)^2 S(S+1)\frac{1}{d^3} N_S \int_0^\infty \frac{u^2}{81+9u^2-2u^4+u^6} \widetilde{K}_{11}^{DD}(u,-\omega_I) du \qquad (12.29)$$

Since we have to evaluate the spectral density $\widetilde{K}_{1,1}^{DD}(u,-\omega_I)$ many times the present calculations are much more time consuming that in the case of rotational modulations of the $I-S$ dipole-dipole coupling. The result of Eq.12.29 provides a very general recipe for the calculation of the dipolar spin relaxation in the presence of translational motion. It is a counterpart of the description formulated in Chapter X. Employing the general treatment of the nuclear spin relaxation modulated by translational motion, we can discuss the efficiency of the resulting nuclear spin relaxation outside of the validity regimes of the analytical Redfield description of the electron spin relaxation.

Fig.12.5 shows some examples of relaxation profiles predicted by the general theory, when the electron spin cannot be treated within the Redfield relaxation theory.

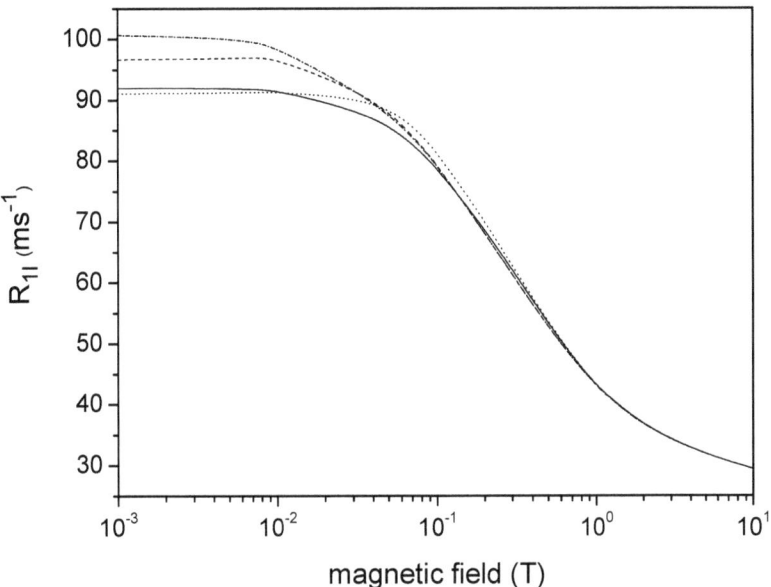

Fig.12.5

Proton spin-lattice relaxation profiles calculated by applying the general treatment when the electron spin S=1 does not fulfil the conditions of the Redfield relaxation theory. The calculations have been performed assuming that the relative diffusion coefficient corresponds to the water diffusion coefficient at ambient temperature $D_{diff} = 2.3 \cdot 10^{-9} m^2 /s$; the distance of closest approach has been set to $d = 300 pm$. The static as well as the transient ZFS interactions do not include the rhombic components.

Solid line: $D_S = 0.05 cm^{-1}$, $D_T = 0.02 cm^{-1}$, $\tau_D = 10 ps$, $\tau_R = 100 ps$. Because of the relatively fast molecular tumbling the Redfield relaxation theory cannot be applied to the electron spin in the intermediate range of magnetic fields (see Chapter IX). Dashed line shows the proton spin-lattice relaxation rate obtained for the same D_S, D_T and τ_D, but for $\tau_R = 1 \mu s$. In this case the electron spin can be described within the Redfield theory; this curve has been presented in Fig.12.3 (solid line). Comparing the two curves one can see the influence of the molecular tumbling.

Dashed line: $D_S = 0.02 cm^{-1}$, $D_T = 0.05 cm^{-1}$, $\tau_D = 10 ps$, $\tau_R = 1\mu s$. *One cannot apply the perturbation relaxation theory to the electron spin (at low field) since the transient ZFS dominates over the static part.*
Dashed – dotted line: $D_S = 0.02 cm^{-1}$, $D_T = 0.05 cm^{-1}$, $\tau_D = 10 ps$, $\tau_R = 100 ps$. *There are two reasons that the perturbation treatment breaks down: the transient ZFS has larger amplitude than the static one and in addition the molecular systems rotates.*

XII.3. Effects of intermolecular forces

In this section we shall take a further step in the theoretical development and remove the assumption that the translational motion is described by the force free diffusion model. The force-free diffusion mechanism corresponds to the case of non-interacting molecules, and is, in fact, a highly idealized picture of a rather limited applicability. Generally, intermolecular interactions affecting the translational motion and depending on the actual distance r between the molecules carrying the I and S spins can be described by a radial distribution function $g(r)$, In our case, the distribution function reflects the density of the molecules possessing the I spin versus the distance from the molecule with the S spin. The function $g(r)$ is related to a potential of mean force $U(r)$ between the interacting molecules as [5, 17, 18]:

$$\ln[g(r)] = -\frac{U(r)}{k_B T} \qquad (12.30)$$

where k_B is Boltzmann constant and T the absolute temperature of the system. Thus, an uniform radial distribution function for $r \geq d$ (scaled by a density of the spins) means that intermolecular interactions are negligible, $U(r) = 0$ At this moment we deal here with interaction potentials of spherical symmetry. They are independent of the orientation of the interspin vector, *i.e.* they depend only on the intermolecular distance r. The key to investigate effects of intermolecular forces on the translational diffusion, and in consequence on the dipolar spin relaxation, is an appropriate form of the correlation function $C_{Diff}(t)$ depending on the interaction potential. There are two ways of dealing with this problem. One can attempt to provide an analytical (semi-analytical) solution of the diffusion equation including an interaction potential [5, 187, 18], and evaluate on this background the correlation function. Unfortunately, such a treatment leads to an effective, 'relatively closed' form of the correlation function only in few cases of the potential functions (for example when $U(r) \propto r^2$ or $U(r) \propto r^{-1}$). There is one more argument against analytical approaches to this problem. Usually, when we intend to take into account effects of intermolecular interactions on the relaxation processes we have at our disposal a distribution function, obtained as a numerical result of computations based on an effective interaction potential being a superposition of many terms. Therefore the most suitable way of proceeding is to evaluate the correlation function numerically, as it has been done in [5, 19, 20], by considering instead of a continuous diffusion a succession of finite jumps between discrete distances r_i. Since we need to describe the effective translational motion, we can consider the motion of the molecule carrying the I spin with respect to a reference frame (a coordinate system) associated with the molecule possessing the S spin. According to [6, 18, 20, 21] the function $C_{Diff}(\tau)$ can be expressed as:

$$C_{Diff}(t) = \sum_{i,j,k=0}^{N_r} \frac{\Delta r_i}{r_i^2} (T^{-1})_{i,j} T_{j,k} R_k \exp(-\gamma_j \tau) \qquad (12.31)$$

The rate coefficients γ_j are eigenvalues of the matrix $C = -W + B$, where the B matrix has a diagonal form, with elements $B_{i,i} = 6 D_{Diff}/r_i^2$, while the W matrix contains transition probabilities between discrete values of the distance r_i; its elements include forces, $F(r_i)$, derived from the potential $V(r)$: $F(r) = -\nabla(V(r))$. The symbol R denotes the column vector with elements determined by the radial distribution function: $R_k = g(r_k)/r_k^2$. The number of discrete distances r_i is given by N_r. The correlation function is multiexponential; the amplitude $a_j = \sum_{i,k=0}^{N_r} \left(\frac{\Delta r_i}{r_i^2}\right)(T^{-1})_{i,j} T_{j,k} R_k$ describes the contribution of the exponential decays with the rate constant γ_j to the entire function. This formulation can be incorporated quite straightforwardly to the analytical as well as general descriptions of the dipolar spin relaxation presented in Sections XII.1 and XII.2, respectively, by replacing the force free model of the translational diffusion. Comparing the two correlation functions one can easily conclude that in fact we have replaced the "continuous form" of the correlation function $C_{Diff}(\tau)$ appropriate for free diffusion by its 'discrete' counterpart appropriate for discrete jump between the positions r_{j-1} and r_j driven by an effective force depending on the actual position of the molecule. The rate constants γ_j are determined by the free diffusion coefficient D_{Diff} and the shape of the radial distribution function in the distance range between r_{j-1} and r_j. So the function $\xi(u) = \frac{72}{5}\frac{1}{d^3}\frac{u^2}{81+9u^2-2u^4+u^6}$ corresponds to the set of amplitudes $\frac{4\pi}{5}a_j$, the exponential decay $\exp\left(-\frac{D_{diff}}{d^2}u^2\tau\right)$ to $\exp(-\gamma_j\tau)$, while the integration over u corresponds to the summation over j. Consequently we can write down directly the form of Eq.12.5 including the effects of intermolecular forces:

$$K_{1,1}^{DD}(-\omega_I) = 30\left(\frac{\mu_0\hbar\gamma_I\gamma_S}{4\pi}\right)^2\frac{4\pi}{5}\frac{1}{2S+1}N_S\sum_{p=-1}^{1}\begin{pmatrix}2 & 1 & 1\\1-p & p & 1\end{pmatrix}^2 \times$$
$$\sum_{j=1}^{N_r}a_j\int_0^\infty Tr_{(L-Diff)}\left\{S_p^{1(L)*}\left[\exp\left(-\left(i\hat{\hat{L}}_{(L-Diff)}+\gamma_j\hat{\hat{1}}\right)\tau\right)S_p^{1(L)}\right]\right\}\exp(-i\omega_I\tau)d\tau \quad (12.32)$$

Using this formulation as a starting point, we can derive expressions describing the nuclear spin relaxation depending on the time scale of the molecular tumbling and the applied magnetic field in the same manner as it has been done for the force free model of translational motion. In particular, for slowly rotating molecular systems (assuming obviously that the electron spin dynamics can be described within the Redfield relaxation theory) one gets as a counterpart of Eq.12.13:

$$K_{11}^{DD}(-\omega_I) = 30\left(\frac{\mu_0\hbar\gamma_I\gamma_S}{4\pi}\right)^2\frac{4\pi}{5}\frac{1}{2S+1}N_S\sum_{m,k=-1}^{1}\sum_{j=1}^{N_j}a_j\langle g_{mk}(\Omega_{P_SL})s_{mk}^{Diff}(\Omega_{P_SL},j,\omega_I)\rangle \quad (12.33)$$

The spectral densities, $s_{mk}(\Omega_{P_SL},u,\omega_I)$, reflect the contribution to the modulations of the $I-S$ dipole-dipole coupling resulting from the electron spin dynamics of molecules oriented under the angle Ω_{P_SL} with respect to the laboratory frame, when the interspin distance ranges between r_{j-1} and r_j. They are defined in full analogy to Eq.12.14 as:

$$s_{mk}^{Diff}(\Omega_{P_SL}, j, \omega_I) = \int_0^\infty Tr_S \left\{ S_m^{1+(P_S)} \left[-\left(i\hat{L}_S(\Omega_{P_SL}) + \gamma_j \hat{1} + i\omega_I \hat{1}\right) \right] S_k^{1(P_S)} \right\} dt =$$

$$\left[S_m^1(\Omega_{P_SL}) \right]^+ \left[\hat{M}(\Omega_{P_SL}, j) \right]^{-1} \left[S_k^1(\Omega_{P_SL}) \right] \quad (12.34)$$

where $\hat{M}(\Omega_{P_SL}, j) = -i\hat{L}_Z(\Omega_{P_SL}) - i\hat{L}_{ZFS}^S + \hat{R}^T(\Omega_{P_SL}) + (i\omega_I + \gamma_j)\hat{1}$. Finally, it implies that the nuclear spin relaxation rate is given as:

$$R_{1I} = \left(a_{DD}^T\right)^2 \frac{8\pi}{5} \frac{1}{2S+1} N_S \times$$

$$\text{Re}\left\{ s_{-1-1}^{Diff}(\omega_I) + s_{00}^{Diff}(\omega_I) + s_{11}^{Diff}(\omega_I) + 2s_{-11}^{Diff}(\omega_I) + 2s_{01}^{Diff}(\omega_I) + 2s_{-10}^{Diff}(\omega_I) \right\} \quad (12.35)$$

where, now the spectral densities $s_{mk}^T(\omega_I)$ result from the summation over j of the averaged quantities, after the averaging over molecular orientations of the quantities $\langle g_{mk}(\Omega_{P_SL}) s_{mk}^{Diff}(\Omega_{P_SL}, j, \omega_I) \rangle$, taken with the appropriate amplitudes a_j:

$$s_{mk}^{Diff}(\omega_I) = \sum_{i=j}^{N_r} a_j \langle g_{mk}(\Omega_{P_SL}) s_{mk}^{Diff}(\Omega_{P_SL}, u, \omega_I) \rangle \quad (12.36)$$

An important matter is to extend the general description of the nuclear spin relaxation allowing for intermolecular interactions. As long as the interaction potential depends only on the distance between the interacting molecules, it can be done quite straightforwardly.

Comparing Eq.12.26 and Eq.12.32 one can write a closed form expression for the nuclear spin relaxation rate, evaluated in the already described, general manner:

$$R_{1,1}^{DD}(\omega_I) = \frac{10}{3} \left(a_{DD}^T\right)^2 \frac{4\pi}{5} S(S+1) N_S \sum_{j=1}^{N_r} a_j \tilde{K}_{1,1}^{DD}(j, -\omega_I) \quad (12.37)$$

where the spectral density $\tilde{K}_{1,1}^{DD}(j, -\omega_I)$ is defined in full analogy to Eq.12.28, as:

$$\tilde{K}_{1,1}^{DD}(j, -\omega_I) = \left[c_1^{(Diff)*}\right] \times \left[\hat{M}(j)\right]^{-1} \times \left[c_1^{(Diff)}\right] \quad (12.38)$$

Finishing this section, I wish to point out that when the intermolecular interactions depend only on the relative distance between the diffusing molecules the dipole-dipole axis changes its orientation independently of the molecular tumbling; there is no correlation between the motions. Isotropic interactions do not bind the $I-S$ dipole – dipole axis to any reference frame, which is fixed in the molecule carrying the electron spin. Therefore, the physical picture of the lattice dynamics determining the I spin relaxation considered in Chapter VI and in this section remains still essentially different. In fact, to built a bridge between the two limiting cases, when the dipole-dipole coupling is affected, beside the electron spin relaxation, by the molecular tumbling or by the force free translational motion, respectively, one needs to consider radial as well as angular distribution of the interacting spins determined by an anisotropic potential $U(\vec{r}) \equiv U(r,\Omega)$.

References:

1. A. Abragam, The principles of nuclear magnetism, *Oxford University Press, Oxford* (1961)
2. R. Kimmich, NMR - Tomography, Diffusometry, Relaxometry, *Springer – Verlag Berlin* (1997)
3. H.C. Torrey, Nuclear spin relaxation by translational diffusion, *Phys. Rev.* **92** (1953) 962-969
4. J.F. Harmon, B.H. Muller, Nuclear spin relaxation by translational diffusion in liquid ethane, *Phys. Rev.* **182** (1969) 400-410
5. L-P. Hwang, J.H. Freed, Dynamic effects of pair correlation functions on spin relaxation by translational diffusion in liquids, *J. Chem. Phys.* **63** (1975) 4017-4025
6. J.H. Freed, Dynamic effects of pair correlation functions on spin relaxation by translational diffusion in liquids: II. Finite jumps and independent T_1 processes, *J. Chem. Phys.* **68** (1978) 4034-4037
7. Y. Ayant, E. Belorizky, J, Alizon, J. Gallice, Calculations of the spectral densities for relaxation resulting from random translational motion by magnetic dipolar coupling in liquids, *J. Phys. (Paris)* **36** (1975) 991-1004
8. Y. Ayant, E. Belorizky, P.H. Fries, J. Rosset, Effect of intermolecular dipolar magnetic interactions on the nuclear spin relaxation of polyatomic molecules in liquids, *J. Phys. (Paris)* **38** (1977) 325-337
9. D. Kruk, T. Nilsson, J. Kowalewski, Outer-sphere nuclear spin relaxation in paramagnetic systems: a low field theory, *Mol. Phys.* **99** (2001)1435-1445
10. D. Kruk, T. Nilsson, J. Kowalewski, Nuclear spin relaxation in paramagnetic systems with zero-field splitting and arbitrary electron spin, *Phys. Chem. Chem. Phys.* **3** (2001) 4907-4917
11. T. Bayburt, R. Sharp, Electron- and nuclear – spin relaxation in an integer spin system, *tris*-(acetylacetonato)Mn(III) in solution, *J. Chem. Phys.* **92** (1990) 5892-5899
12. S.M. Abernathy, R. Sharp, Spin dynamics calculations of electron and nuclear spin relaxation times in paramagnetic solutions, *J. Chem. Phys.* **106** (1997) 9032-9043

13. T. Bayburt, R. Sharp, Electron- and nuclear –spin relaxation in an integer spin system tris-(acetylacetotonato)Mn(III) in solution, *J. Chem. Phys.* **92** (1990) 1435-1445
14. J. Kowalewski, D. Kruk, G. Parigi, NMR relaxation in solution of paramagnetic complexes: Recent theoretical progress for S >= 1, *Advances in Inorganic Chemistry* **57** (2005) 41-104
15. D.A. Varshalovich, A.N. Moskalev, V.K. Khersonkii, Quantum theory of angular momentum, *Word Scientific Publishing, Singapore* (1988)
16. D. M. Brink, G.R. Satchler, Angular momentum, *Clarendon Press, Oxford* (1979)
17. D. Kruk, J. Kowalewski, Nuclear spin relaxation in ligands outside of the first coordination sphere in a gadolinium (III) complex: effects of intermolecular forces, *J.Chem. Phys.* **117** (2002)1194-1200
18. M. Doi, S. F. Edwards, The theory of polymer dynamics, *Clarendon Press, Oxford* (1986)
19. J. Pedersen, J. Freed, Theory of chemically induced dynamic electron polarization. I, *J. Chem. Phys.* **58** (1971) 2746-2762
20. J. Pedersen, J. Freed, Theory of chemically induced dynamic electron polarization. II, *J. Chem. Phys.* **59** (1973) 2869-2885

CHAPTER XIII

Quantum vibrations as an origin of electron spin relaxation

As pointed out in the previous chapters, the hierarchy of events is very important for the description of relaxation processes of the 'model' $I-S$ spin systems considered in this book. Let us take up the case of an electron spin – nuclear spin system, which exhibits some kind of vibrational motion. The vibrarional dynamics is completely independent of any spin variables, both the electron and the nuclear ones. The electron spin senses the vibrations via the ZFS interaction and therefore the vibrational relaxation is an origin of the electron spin relaxation. Therefore it is of primary importance to formulate an appropriate model of the ZFS coupling, linking the electron spin variables to the vibrational degrees of freedom. Different models of the ZFS interaction predict different electron spin dynamics resulting from the same vibrational motion. In other words, setting up the ZFS Hamiltonian (Liouvillian) we just 'decide' how the electron spin senses the molecular vibrations. Next, the electron spin dynamics is sensed via the electron spin - nuclear spin dipole-dipole coupling by the nuclear spin. Thus the electron spin (more generally, the S spin) plays a role of a bridge between the vibrational subsystem and the I spin dynamics. Therefore, a complete treatment of the entire system consists of three subsequent stages. The first step is to describe the vibrational dynamics. Next, one has to consider a more complex system which includes the vibrational as well as the electron spin degrees of freedom, coupled by the ZFS interaction. When the electron spin system is within the validity regimes of the Redfield relaxation theory, one ends up with a set of explicitly defined electron spin relaxation rates depending on the vibrational dynamics. Hence one is in a position to discuss the I spin relaxation resulting from the S spin dynamics. At this stage, we deal in principle with the entire system including the vibrations, the electron spin and the nuclear spin degrees of freedom. However, when the S spin relaxation is expressed in terms of spectral densities resulting from the vibrational motion, one does not need to invoke explicitly the vibrational variables to describe the I spin relaxation.

In fact, dealing with relaxation in multi-spin systems we follow always the outlined procedure. Depending on the motional conditions and the applied magnetic field the computational details are different; however the general idea remains unchanged.

We consider first the motion modulating this part of the ZFS (or of the quadrupolar interaction), which is responsible for the electron (quadrupolar) spin relaxation. Because of the rather simple model of the ZFS (quadrupolar) coupling used so far in this book, the considerations end up with the statement that the relevant motion can be treated classically and characterized by one time constant reflecting its timescale (τ_D, τ_R or τ_M depending on the physical origin of the motion; the time constants correspond to the distortional, rotational or exchange motion, respectively). A more advanced model assuming that the ZFS interaction is modulated by damped vibrations has been introduced in [1-5]. This model requires specifying frequencies of the relevant modes of the vibrations, their amplitudes and characteristic time constants. Nevertheless, the vibrational dynamics has been still treated classically within this model; in [2] the quantum nature of the vibrational motion has been anticipated.

Next, we aim for the perturbation description of the S spin dynamics resulting from the ZFS (quadrupolar) interaction connecting the S spin variables with the already discussed lattice dynamics. Having at last obtained the expressions for the electron (quadrupolar) spin relaxation we evaluate the I spin relaxation caused by the $I-S$ dipole – dipole coupling. The spin S subsystem is treated as a part of the quantum-mechanical composite lattice, however we do not ask at this stage about details of the vibrational (rotational, exchange) motion.

In this chapter I shall discuss a model of the ZFS interaction based on a quantum-mechanical treatment of the vibrational motion. There are two reasons for such considerations. The first one is that the model describes more realistically the physical mechanism of the vibrational dynamics and I believe that it can be applicable for several systems, not only paramagnetic complexes for which it was developed originally [6]. The second reason is that there are not so many relaxation theories in the literature treating the lattice quantum-mechanically [6 - 9]. In fact, this book deals largely with relaxation theories involving the concept of a composite lattice which includes classical as well as quantum – mechanical elements. The theories are mostly applied to 'model' $I - S$ spin systems and the spin S subsystem is treated as a quantum lattice for the spin I. It interesting and useful to find general analogies between this approach and a similar treatment applied to the spin S coupled to the vibrational, quantum degrees of freedom.

Basically we need to discuss two problems: the quantum – mechanical description of the vibrational dynamics and the model of the ZFS interaction leading to a specific form of the spectral densities which determine the electron spin relaxation. This is the subject of the next two sections.

XIII.1. Quantum description of vibrational dynamics

Let us begin the considerations from the vibrational subsystem. The total Hamiltonian H_V describing a vibrating molecule coupled to a bath (lattice) can be written, following the main concept of relaxation described in Section II.1, as $H_V = H_V^0 + H_{VB} + H_B$. I use here the terminology 'bath' instead of 'lattice'. We got used to think about a spin of interest and its lattice. Now, we do not deal with any spin system and therefore I prefer to use the different terminology to underline this fact.

The main part H_V^0 of the vibrational Hamiltonian determines the energy levels of the vibrating molecule. In what follows I assume that the molecular vibrations can be treated as harmonic oscillators. Therefore, the vibrational energy levels, E_μ, are determined by the expression [10,11]:

$$H_V^0|\mu\rangle = E_\mu|\mu\rangle = \left(\mu + \frac{1}{2}\right)\hbar\omega_V|\mu\rangle \qquad (13.1)$$

where ω_V is the characteristic frequency of the harmonic motion, while $|\mu\rangle$ denotes the vibrational eigenvectors. Taking into account typical vibrational frequencies one can safely restrict the description to a two-level quantum system, associated with the eigenvectors $|\mu = 0\rangle$ and $|\mu = 1\rangle$. One does not need to specify explicitly the bath Hamiltonian H_B. We shall use in the further description of the electron spin dynamics a set of effective vibrational relaxation times, not discussing in detail their physical origin. This should remind the reader the third (last) step of the procedure outlined above: we have used analogous arguments in the context of the I spin relaxation resulting from the S spin dynamics. The Hamiltonian H_{VB} describes the coupling of vibrational modes to the bath. It is related to an intermolecular potential V represented in terms of the normal modes. Generally one can express this Hamiltonian as the Taylor series expansion: $H_{VB} = \sum_i \left(\frac{\partial V}{\partial q_i}\right) q_i + \frac{1}{2}\sum_{i,j}\left(\frac{\partial^2 V}{\partial q_i \partial q_j}\right) q_i q_j + ...$, where the summation is performed over all normal modes q_i which are relevant for the considered molecular system. To resolve whether one can apply the perturbation theory to describe the vibrational dynamics one has to compare the timescales of the fluctuations of the intermolecular forces (represented by the first term of the Taylor expansion) and the vibrational relaxation resulting from the fluctuations. Characteristic correlation times for electrostatic forces are of a sub-picosecond order [12, 13], and that permits one to set up the Redfield equation for the vibrational subsystem, represented by the density operator $\rho_V(t)$:

$$\frac{d\rho_V(t)}{dt} = \left(i\hat{\hat{L}}_V^0 - \hat{\hat{R}}_V\right)\rho_V(t) \qquad (13.2)$$

where the Liouville operator $\hat{L}_V^0$ is generated by the Hamiltonian H_V^0. The relaxation superoperator $\hat{R}_V$ includes two vibrational relaxation times T_{1V} and T_{2V}. It is worthwhile to mention at this point that in [6] the vibrational relaxation times have been estimated experimentally from Raman linewidths. We need for the further consideration to set up a matrix representation of this equation. This requires special caution since the high temperature approximation [14] does not apply to the vibrational energy levels. Therefore, we have to take into account that the population probabilities of the vibrational states are different; they are determined by the Boltzmann distribution rule. Being aware of this fact, one can write the set of equations for the individual elements of the vibrational density operator $\rho_{\mu\mu'}^V$ in the form [6]:

$$\frac{d}{dt}\begin{bmatrix}\rho_{0,0}^V\\\rho_{1,1}^V\\\rho_{1,0}^V\\\rho_{0,1}^V\end{bmatrix}=\begin{bmatrix}-\frac{1}{T_{1V}}\left[\exp\left(-\frac{E_0\beta}{Z}\right)\right] & \frac{1}{T_{1V}}\left[\exp\left(-\frac{E_1\beta}{Z}\right)\right] & 0 & 0\\ \frac{1}{T_{1V}}\left[\exp\left(-\frac{E_0\beta}{Z}\right)\right] & -\frac{1}{T_{1V}}\left[\exp\left(-\frac{E_1\beta}{Z}\right)\right] & 0 & 0\\ 0 & 0 & i\omega_V-\frac{1}{T_{2V}} & 0\\ 0 & 0 & 0 & -i\omega_V-\frac{1}{T_{2V}}\end{bmatrix}\begin{bmatrix}\rho_{0,0}^V\\\rho_{1,1}^V\\\rho_{1,0}^V\\\rho_{0,1}^V\end{bmatrix} \quad (13.3)$$

As explained above the vibrational quantum number $\mu=0$ defines the fundamental state, $|\mu=0\rangle$, while $\mu=1$ describes the first, excited state $|\mu=1\rangle$. The partition function, Z, is defined as: $Z=\exp(-E_0\beta)+\exp(-E_1\beta)$ with $\beta=(k_BT)^{-1}$. Solving this set of equations, one gets explicit expressions for the time evolution of the particular elements of the density operator:

$$\rho_{1,0}^V(t)=\rho_{1,0}^V(0)\exp(i\omega_Vt)\exp\left(-\frac{t}{T_{2V}}\right) \quad (13.4a)$$

$$\rho_{0,1}^V(t)=\rho_{0,1}^V(0)\exp(-i\omega_Vt)\exp\left(-\frac{t}{T_{2V}}\right) \quad (13.4b)$$

$$\rho_{0,0}^V(t) = \frac{\exp(-E_0\beta)\rho_{0,0}^V(0) - \exp(-E_1\beta)\rho_{1,1}^V(0)}{Z}\exp\left(-\frac{t}{T_{1V}}\right) +$$
$$\frac{\exp(-E_0\beta)[\rho_{0,0}^V(0) + \rho_{1,1}^V(0)]}{Z} \quad (13.4c)$$

$$\rho_{1,1}^V(t) = \frac{-\exp(-E_0\beta)\rho_{0,0}^V(0) + \exp(-E_1\beta)\rho_{1,1}^V(0)}{Z}\exp\left(-\frac{t}{T_{1V}}\right) +$$
$$\frac{\exp(-E_1\beta)[\rho_{0,0}^V(0) + \rho_{1,1}^V(0)]}{Z} \quad (13.4d)$$

where the symbols $\rho_{\mu\mu'}^V(0)$ denote the initial vibrational density matrix elements. The vibrational dynamics, described by the above set of equations, is sensed by the electron spin through the ZFS interaction. We are concerned with this issue in the next section.

XIII.2. Electron spin spectral densities resulting from molecular vibrations

Following the notation used in this book, the ZFS Hamiltonian is expressed as $H_{ZFS} = \sum_{m=-2}^{2}(-1)^m \widetilde{V}_{-m}^2 T_m^2(S)$ (Eq.3.6). According to the model of the ZFS interaction proposed in [6] the spatial functions $\widetilde{V}_{-m}^2$ are related via a set of normal coordinates $q = [q_1,...,q_N]$ to the internal geometry of the considered molecule: $\widetilde{V}_m^2 = \sum_{k=-2}^{2} h_k(q)$. Since in the present approach we take into account the quantum nature of the vibrational motion, the normal coordinates should be treated as operators. The functions $\widetilde{V}_{-m}^2$ have been introduced to describe a classical lattice of the electron spin. To point out that now we deal with a quantum-mechanical lattice we should use different symbols to denote them (for example $T_{-m}^{2(ZFS)}$ in analogy to $T_{-m}^{1(DD)}$, see Section IV.4). It is very important to keep this in mind and be aware of the different physical nature of the lattice, but just for practical purposes I do not like to introduce new symbols.

Independently of the model proposed to describe the ZFS interaction, one can decompose it always into the static ZFS defined as the average: $H_{ZFS}^{S} = \langle H_{ZFS}(t) \rangle$, and the transient part describing the deviation of the total Hamiltonian from its average value: $H_{ZFS}^{T}(t) = H_{ZFS}(t) - H_{ZFS}^{S}$. To extract the two components of the ZFS tensor let us expand the functions $h_k(q)$ (and in consequence the functions $\tilde{V}_m^2(q)$) in the Taylor series [5, 6, 15, 16]:

$$h_k(q) = \sum_i h_k(q_i^0) + \sum_i \left\langle \frac{\partial h_k}{\partial q_i} \right\rangle_{q_i^0} q_i + \sum_{i,j} \left\langle \frac{\partial^2 h_k}{\partial q_i \partial q_j} \right\rangle_{q_i^0 q_j^0} q_i q_j + ... \quad (13.5)$$

The brackets $\langle \ \rangle$ denote here the derivatives of the ZFS tensor functions $h_k(q)$ over the normal coordinates at the points corresponding to the equilibrium values of the individual modes q_i^0. One can see from this expansion that the zero order term in the expansion corresponds to the static part of the ZFS interaction, while the transient ZFS is given by the higher order terms. It can be useful to write down the corresponding expressions explicitly: the static ZFS is determined by the equilibrium geometry of the molecule and therefore it is time independent: $H_{ZFS}^{S} = \sum_i h_k(q_i^0)$, while the transient part fluctuates in time due to changes of these normal coordinates caused by the vibrational dynamics:

$$H_{ZFS}^{T}(t) = \sum_i \left\langle \frac{\partial h_k}{\partial q_i} \right\rangle_{q_i^0} q_i(t) + \sum_{i,j} \left\langle \frac{\partial^2 h_k}{\partial q_i \partial q_j} \right\rangle_{q_i^0 q_j^0} q_i(t) q_j(t) + ... \ .$$ The coefficients $\left\langle \frac{\partial h_k}{\partial q_i} \right\rangle$ are related to the axial and rhombic parts of the transient ZFS tensor: $\left\langle \frac{\partial h_0}{\partial q_i} \right\rangle = \sqrt{\frac{2}{3}} \left\langle \frac{\partial D_T}{\partial q_i} \right\rangle$, while $\left\langle \frac{\partial h_2}{\partial q_i} \right\rangle = \left\langle \frac{\partial h_{-2}}{\partial q_i} \right\rangle = \left\langle \frac{\partial E_T}{\partial q_i} \right\rangle$.

Having seen from the ZFS Hamiltonian how the electron spin variables are coupled to the normal coordinates, let us work out the electron spin relaxation resulting from this coupling. Before we start the evaluations, I wish to explain the planned strategy. The vibrational density operator ρ_V, discussed in the previous section, as well as the normal coordinates operator q can be expressed in terms of the vibrational eigenvectors $|\mu\rangle$, since they form a complete, orthonormal basis.

Therefore, the time dependence of the vibrational density operator is transferred through this common representation to the normal coordinates included in the ZFS Hamiltonian. This is, in fact, the idea of the electron spin relaxation caused by quantum vibrations.

Since in this chapter I intend to discuss alternative models of the ZFS and the quantum nature of the fluctuations of this interaction, let me just assume that the electron spin fulfills the conditions of the Redfield relaxation theory. The case when the electron spin is beyond the validity regimes of the perturbation theory has been discussed in detail in [6] within the framework of the general treatment presented in Chapter X. Matters are greatly simplified if we assume, in addition, that the molecular tumbling is slow. Due to this assumption, we do not need to consider any effects of the rotational motion on the ZFS interaction, so we can focus attention on the quantum vibrations.

The Redfield equation of motion of the electron spin density operator $\rho_S(t)$ takes the form:

$$\frac{d\rho_S(t)}{dt} = \left(i\hat{L}_S^0 - \hat{R}_S \right) \rho_S(t) \tag{13.6}$$

where the operator $\hat{L}_S^0$ contains the interactions determining the energy level structure of the electron spin, *i.e.* the Zeeman- and the static ZFS couplings: $\hat{L}_S^0 = \hat{L}_{Zeeman} + \hat{L}_{ZFS}^S$ as explained in Chapter VI. The electron spin relaxation superoperator $\hat{R}_S$ results from the transient ZFS modulated by the vibrational motion:

$$\hat{R}_S = i\hat{L}_{ZFS}^T + i\hat{L}_V^0 + \hat{R}_V \tag{13.7}$$

It is of great importance to notice that the above expression contains the operator $i\hat{L}_V^0$, responsible for the discrete vibrational energy levels. We shall see soon consequences of this fact for the electron spin relaxation. The individual electron spin relaxation rates can be obtained in terms of appropriate electron spin spectral densities from the expressions of Eq.2.14, which have been invoked many times in this book. Thus, the description of the electron spin dynamics simplifies to the problem of a proper formulation of the spectral densities $J_m^S(\omega)$ within the framework of the quantum vibration model. We start from the ordinary definition of the spectral density:

$$J_m^S(\omega) \equiv J^S(\omega) \sum_i \text{Re} \left\{ \int_0^\infty \langle \tilde{V}_m^{2*}(\tau)\tilde{V}_m^2(0)\rangle_i \exp(-i\omega\tau)d\tau \right\} \quad (13.8)$$

where we sum up over all normal modes $\hat{q}_i$. Ignoring in the Taylor series expansion of the transient ZFS (Eq.13.5) terms of the order higher than two, the correlation function $\langle \tilde{V}_m^{2*}(\tau)\tilde{V}_m^2(0)\rangle$ yields the form:

$$\langle \tilde{V}_m^{2*}(\tau)\tilde{V}_m^2(0)\rangle = \left[\sum_{k\in(0,\pm 2)} \left\langle \frac{\partial h_k}{\partial q}\right\rangle\right]^2 \langle q^+(\tau)q(0)\rangle + \left[\sum_{k\in(0,\pm 2)} \left\langle \frac{\partial^2 h_k}{\partial q^2}\right\rangle\right] \langle (q^2)^+(\tau)q^2(0)\rangle \quad (13.9)$$

The first term corresponds to the linear coupling of the electron spin to the quantum bath, while the second term comes from the quadratic coupling (the second order term in the Taylor expansion). The above equation concerns a single vibrational mode. Since there is no dynamic coupling between the normal coordinates, the generalization of the formulation to several modes q_i is straightforward. Therefore, for simplicity reasons, I shall continue the derivations in terms of a single mode. To proceed further we need to link the normal coordinates q to the vibrational density operator ρ_V. The way how to achieve this has been sketched above. We should expand the operators q, q^2 and ρ_V in the Liouville basis $\{|\mu\rangle\langle\mu'|\}$ constructed from the vibrational eigenstates $\{|\mu\rangle\}$. Let us begin from the first two operators. They can be expressed as: $O = \sum_{\mu,\mu'} c^O_{\mu,\mu'} |\mu\rangle\langle\mu'| = \sum_{\mu,\mu'} \langle\mu|O|\mu'\rangle |\mu\rangle\langle\mu'|$ ($O = q, q^2$). Making use of the properties of a harmonic oscillator [11]:

$$\langle\mu|q|\mu'\rangle = \left(\frac{\hbar}{2m\omega_V}\right)^{1/2} \left[\delta_{\mu,\mu'+1}(\mu'+1)^{1/2} + \delta_{\mu,\mu'-1}(\mu)^{1/2}\right] \quad (13.10a)$$

and

$$\langle\mu|q^2|\mu'\rangle = \delta_{\mu,\mu'} \frac{E_\mu}{m\omega_V^2} \quad (13.10b)$$

one obtains the following representations of the operators q and q^2:

$$q = q(\tau) = \left(\frac{\hbar}{2m\omega_V}\right)^{1/2} [\rho^V_{1,0}(\tau) + \rho^V_{0,1}(\upsilon)] =$$

$$\left(\frac{\hbar}{2m\omega_V}\right)^{1/2} [|1\rangle\langle 0|\exp(i\omega_V\tau) + |0\rangle\langle 1|\exp(-i\omega_V\tau)]\exp\left(-\frac{\tau}{T_{2V}}\right) \quad (13.11)$$

$$q^2 = q^2(\tau) = \left(\frac{E_0}{m\omega_V^2}\right)^{1/2} \rho^V_{0,0}(\tau) + \left(\frac{E_1}{m\omega_V^2}\right)^{1/2} \rho^V_{1,1}(\tau) =$$

$$\frac{|0\rangle\langle 0| + |1\rangle\langle 1|}{Z}\left[\left(\frac{E_0}{m\omega_V^2}\right)^{1/2}\exp(-E_0\beta) + \left(\frac{E_1}{m\omega_V^2}\right)^{1/2}\exp(-E_1\beta)\right] + \quad (13.12)$$

$$\frac{1}{Z}\left[\left(\frac{E_1}{m\omega_V^2}\right)^{1/2} - \left(\frac{E_0}{m\omega_V^2}\right)^{1/2}\right][\exp(-E_1\beta)|1\rangle\langle 1| - \exp(-E_0\beta)|0\rangle\langle 0|]\exp\left(-\frac{\tau}{T_{1V}}\right)$$

The time dependencies of the $\rho^V_{\mu,\mu'}$ come from the vibrational dynamics and have been evaluated in Section XIII.1; Eq.13.4a-d. I have changed the notation in the above expressions replacing the initial elements of the density operator $\rho^V_{\mu,\mu'}(0)$ by the 'bracket' terms: $|\mu\rangle\langle\mu'|$. This formalism leads to more straightforward evaluations of the appropriate correlation functions as we shall see below. The quantity m is called the reduced mass of the harmonic oscillator. The explicit time dependencies of the normal coordinates $q(\tau)$ as well as their square $q^2(\tau)$ include the vibrational frequency ω_V and the vibrational relaxation times T_{1V} and T_{2V}.

The correlation functions $\langle O^+(\tau)O(0)\rangle$ ($O = q, q^2$) result from the averaging over the vibrational degrees of freedom, according to the formula [8]:

$$\langle O^+(\tau)O(0)\rangle = \sum_{\mu,\mu'=0} \frac{\exp(-E_\mu\beta)}{Z} \langle \mu|O(\tau)|\mu'\rangle\langle\mu'|O(0)|\mu\rangle \quad (13.13)$$

This expression can be treated as a quantum-mechanical counterpart of the classical correlation function defined in Chapter I. The classical quantities $A^*(\tau)$ and $A(0)$ correspond to the operator matrix elements: $\langle\mu|O(\tau)|\mu'\rangle$ and $\langle\mu'|O(0)|\mu\rangle$, respectively. The Bolzmann distribution function describing the occupation probability of the energy level E_μ, $\dfrac{\exp(-E_\mu \beta)}{Z}$, has the meaning of the classical equilibrium distribution function $P_{eq}(x)$. The integration over the classical degrees of freedom, represented in the classical limit by a continuum of states, is replaced in the quantum approach by the summation over the discrete states. At this stage we can evaluate explicitly the correlation functions $\langle q^+(\tau)q(0)\rangle$ and $\langle q^{2+}(\tau)q^2(0)\rangle$ by substituting into Eq.13.13 the time dependencies of the corresponding operators, given by Eq.13.11 and Eq.13.12. The electron spin spectral densities $J^{S(1)}(\omega)$ and $J^{S(2)}(\omega)$ resulting from the linear as well as quadratic couplings to the lattice, respectively, (the first and the second order terms in the ZFS series expansion) are obtained as Fourier transforms of these correlation functions. They have closed analytical forms [6]:

$$J^{S(1)}(\omega) = \sum_i J_i^{S(1)}(\omega) =$$
$$\sum_i \left(\sum_{k \in (0, \pm 2)} \frac{\partial h_k}{\partial q_i}\right)^2 \frac{\hbar}{4 m_i \omega_{V,i}} \left[\frac{T_{2V,i}}{1+(\omega_{V,i}+\omega)^2 T_{2V,i}^2} + \frac{T_{2V,i}}{1+(\omega_{V,i}-\omega)^2 T_{2V,i}^2}\right] \quad (13.14)$$

$$J^{S(2)}(\omega) = \sum_i J_i^{S(2)}(\omega) =$$
$$\sum_i \left(\sum_{k \in (0, \pm 2)} \frac{\partial^2 h_k}{\partial q_i^2}\right)^2 \frac{5\hbar^2}{2 m_i^2 \omega_{V,i}^4} \frac{\exp(-\omega_V \beta)}{[1+\exp(-\omega_V \beta)]^2} \frac{T_{1V,i}}{1+\omega^2 T_{1V,i}^2} \quad (13.15)$$

From these expressions one can easily extract the forms of the corresponding correlation functions (Fourier transform of an exponential function gives a Lorentzian function). For completeness of the derivations, the final formulations of the spectral densities contain the summation over all relevant normal modes. Correlation functions between the quantities q and q^2 vanish due to the properties of the harmonic oscillator eigenfunctions, Eq.13.10a,b.

The considerations presented in this chapter concern mainly the electron spin relaxation mechanism caused by vibrational motion, however they can be treated in a more general sense, as an example of relaxation theories dealing with a quantum mechanical-lattice. Therefore some more comments on this subject can be of interest. According to the present theory, we need to consider two types of vibrational relaxation processes: those affecting the populations of the vibrational levels, T_{1V}, and those related to the decay of the vibrational coherences, T_{2V}. If the lattice is characterized by many quantum states, one has to consider more time constants describing the life-times of the energy levels, as well as the decays of the individual coherences, as we have seen in the previous chapters. Both relaxation processes of the vibrational subsystem, described by the time constants T_{1V} and T_{2V}, are responsible for the electron spin relaxation. However, the T_{2V} process is associated with the linear term in the ZFS interaction, while the population vibrational relaxation T_{1V} is sensed by the electron spin through the second order terms. This fact has profound consequences for the electron spin relaxation. It implies that the spectral densities resulting from the linear coupling contain (because of its association to the T_{2V} process) the energy difference ω_V between the quantum levels of the lattice. Since the vibrational frequency fulfills usually the condition $\omega_V \gg \omega_S$ (where ω_S is the electron spin Larmor frequency) up to high magnetic fields of the order of 10T, one can expect, in principle, electron spin relaxation rates independent of the magnetic field. One can see this clearly from

a)

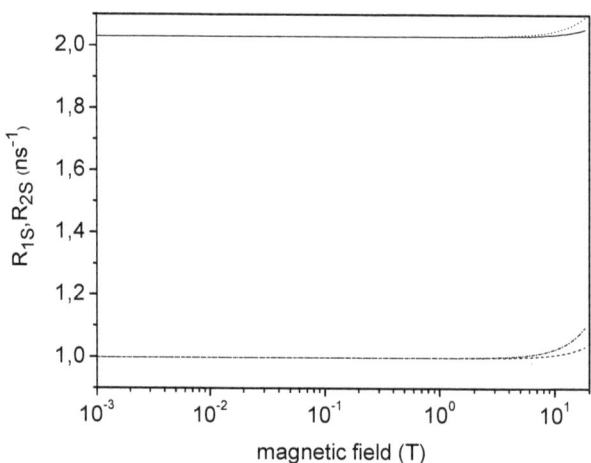

b)

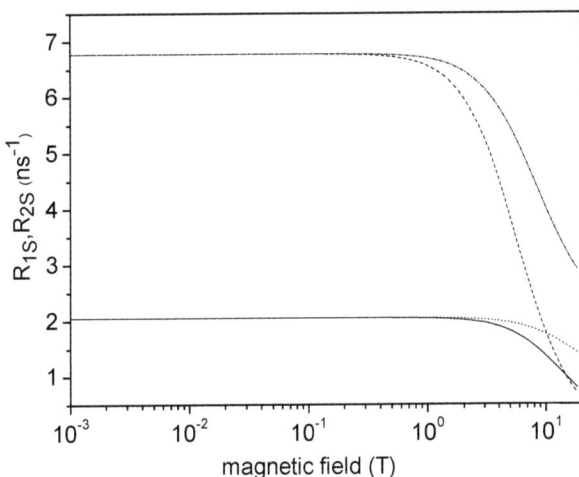

Fig.13.1
Electron spin relaxation rates (S = 1) caused by the first order term of the ZFS interaction (a) and the second order term (b)

Solid line: R_{1S} - $\omega_V = 350 cm^{-1}, T_{2V} = 0.2 ps, \left(\sum_{k \in (0, \pm 2)} \frac{\partial h_k}{\partial q} \right)^2 a^2 = 5$; ($\frac{\hbar}{2m\omega_V} = a^2$,

where a is the amplitude of the vibration [10])
Dotted line: R_{2S} - *the same parameters as for solid line*

Dashed line: R_{1S} - $\omega_V = 200 cm^{-1}, T_{2V} = 0.5 ps, \left(\sum_{k \in (0, \pm 2)} \frac{\partial h_k}{\partial q} \right)^2 a^2 = 5$

Dashed – dotted line: R_{2S} - *the same parameters as for dashed line*
It has been assumed that the ratio between the second and first order responses of the ZFS tensor: $\left(\sum_{k \in (0, \pm 2)} \frac{\partial^2 h_k}{\partial q^2} \right)^2 a^2 \Big/ \left(\sum_{k \in (0, \pm 2)} \frac{\partial h_k}{\partial q} \right)^2 = 0.01$, $T_{1V} = T_{2V}$ *and T=300K.*

Eq.13.14 written for $J^{S(1)}(\omega_S)$. The second order spectral densities $J^{S(2)}(\omega)$ are not affected by the high vibrational frequency, because they are associated with the relaxation of the vibrational system to the bath. Thus, one could expect that they provide a more efficient contribution to the electron spin relaxation. However, formulating such statement one should be aware that the amplitudes of the $J^{S(2)}(\omega)$ functions are determined by the second order responses of the ZFS tensor with respect to the changes of the normal coordinates, which can be significantly smaller than the first order derivatives.

Fig.13.1 shows examples of the electron spin relaxation profiles predicted by the present model, depending on the ratio between the vibrational relaxation times T_{1V}/T_{2V} and between the second and first order derivatives of the ZFS tensor over the normal coordinates. It helps to estimate the role of the second order term in the Taylor expansion of the transient ZFS interaction.

I am aware that the present approach is based on a series of parameters, which are quite difficult to estimate. This is the price to pay for a more advanced and realistic description of the electron spin relaxation. Here, molecular dynamics (MD) simulations are very useful. Analysis of the electron as well as nuclear spin relaxation for hexaaquonickel (II) within the present model, strongly supported by MD calculations, has been presented in [15, 16].

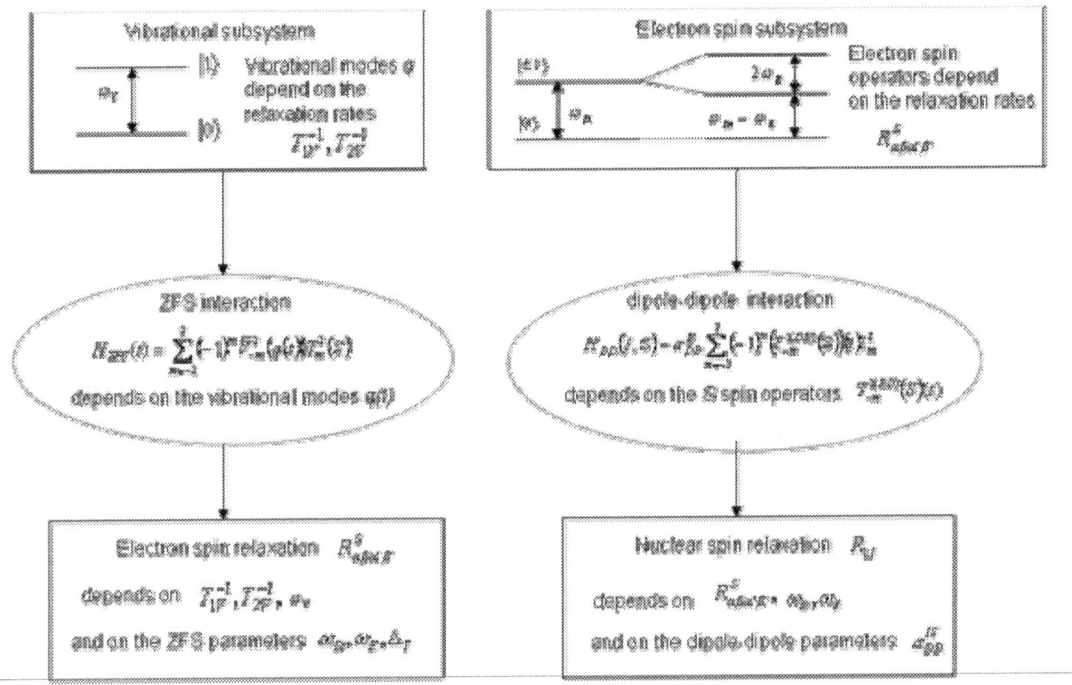

Fig.13.2

Graphical view of the hierarchy of relaxation processes in the electron spin – vibrations and nuclear spin – electron spin systems. The formal, perturbation treatment of the two systems is analogous.

Finishing this chapter I would like to turn attention once again to the analogies between the description of the I spin relaxation caused by the dipole-dipole coupling with the spin S (the spin S subsystem plays the role of a quantum lattice for the spin I), and the 'lower level considerations', namely the description of S spin relaxation resulting from the ZFS coupling to the vibrational subsystem. Fig.13.2 presents the main steps of the basically equivalent treatments applied to the $I-S$ and $S-V$ (V denotes the vibrational dynamic) systems. I hope that this figure will turn out to be useful to understand the essence of the approach to relaxation processes caused by a quantum lattice, which is, in principle, the main subject of this book.

References:

1. P.-O. Westlund, Nuclear paramagnetic spin relaxation theory: Paramagnetic spin probes in homogenous and microheterogenous solutions, in Dynamics of solutions and fluid mixtures by NMR, J. J. Delpuech (Ed.), *Wiley, Chichester* (1995) 173-229
2. P.-O. Westlund, N. Benetis, H. Wennerström, Paramagnetic proton nuclear magnetic relaxation in the Ni^{2+} hexa-aquo complex: A theoretical study, *Mol. Phys.* **61** (1987) 177-194
3. P.-O. Westlund, T.P. Larsson, O. Teleman, Paramagnetic enhanced proton spin-lattice relaxation in the Ni^{2+} hexa-aquo complex: A theoretical and molecular dynamics simulation study of the Bloembergen-Morgan decomposition approach, *Mol. Phys.* **78** (1993) 1365-1384
4. P.-O. Westlund, T.P. Larsson, Proton-enhanced relaxation in low-symmetry paramagnetic complexes (S=1): Beyond the Solomon-Bloembergen and Morgan theory. 1. The Smoluchowski distortion model of the ZFS interaction, *Acta Chem. Scand.* **45** (1991) 11-18
5. D. Kruk, J. Kowalewski, Vibrational motions and nuclear spin relaxation in paramagnetic complexes: Hexaaquonickel (II) as an example, *J. Chem. Phys.* **116** (2002) 4079-4086
6. D. Kruk, J. Kowalewski, P.-O. Westlund, Nuclear and electron spin relaxation in paramagnetic complexes in solution: Effects of the quantum nature of molecular vibrations, *J. Chem. Phys.* **121 (5)** (2004) 2215-2227
7. S. A. Al'tshuler, K.A. Valiev, On the theory of longitudinal relaxation of paramagnetic salt solutions, *Sov. Phys. JETP* **8 (4)** (1959) 661-668
8. A. Abragam, The principles of nuclear magnetism, *Oxford University Press, Oxford* (1961)
9. M. Goldman, Formal theory of spin – lattice relaxation, *J. Magn. Reson.* **149** (2001) 160-187
10. R. Shankar, Principles of Quantum Mechanics, *Plenum, New York* (1980)
11. P.W. Atkins, R.S. Friedman, Molecular Quantum Mechanics, *Oxford University Press, Oxford* (1997)

12. P.-O. Westlund, R.M. Lynden-Bell, A study of vibrational dephasing of the A1 modes of CH3CN in a computer-simulation of the liquid – phase, *Mol. Phys.* **60** (1987) 1189-1209
13. P.-O. Westlund, R.M. Lynden-Bell, The effects of pressure and temperature on vibrational dephasing in a simulations of liquid CH_3CN, *Mol. Phys.* **61** (1987) 1541-1547
14. C.P. Slichter, Principles of magnetic resonance, *Springer-Verlag, Berlin* (1990)
15. M. Odelius, C. Ribbing, J. Kowalewski, Molecular dynamics simulation of the zero-field splitting fluctuations in aqueous Ni (II), *J. Chem. Phys.* **103** (1995) 1800-1811
16. M. Odelius, C. Ribbing, J. Kowalewski, Spin dynamics under the Hamiltonian varying with time in discrete steps: Molecular dynamics-based simulation of electron and nuclear spin relaxation in aqueous nickel (II), *J. Chem. Phys.* **104** (1996) 3181-3186

APPENDIX

Some useful expressions...

Let me summarize here the interaction Hamiltonians and their transformations:

Quadrupole Hamiltonian:
General form:
$$H_Q(S) = \frac{1}{2}\sqrt{\frac{3}{2}} \frac{a_Q}{S(2S-1)} \sum_{m=-2}^{2}(-1)^m A^2_{-m} T^2_m(S) = \sum_{m=-2}^{2}(-1)^m \tilde{A}^2_{-m} T^2_m(S)$$

$a_Q = e^2 qQ/\hbar$ eQ - nuclear quadrupolar moment

The second rank spin tensor operators $T^2_m(S)$ are defined as follows:

$$T^2_0(S) = \frac{1}{\sqrt{6}}\left[3S_z^2 - S(S+1)\right]$$

$$T^2_{\pm 1}(S) = \mp\frac{1}{2}\left[S_z S_\pm + S_\pm S_z\right]$$

$$T^2_{\pm 2}(S) = \frac{1}{2} S_\pm S_\pm$$

In the principal axis system of the electric field gradient tensor (P) the quantities A^2_{-m} and $\tilde{A}^2_{-m}$ have the form:

$$A^{(P)}_0 = 1, \quad A^{(P)}_{\pm 1} = 0, \quad A^{(P)}_{\pm 2} = \eta/\sqrt{6}$$

$$\tilde{A}_0^{(P)} = \frac{1}{2}\sqrt{\frac{3}{2}}\frac{a_Q}{S(2S-1)}, \quad \tilde{A}_{\pm 1}^{(P)} = 0, \quad \tilde{A}_{\pm 2}^{(P)} = \frac{a_Q}{4S(2S-1)}\eta$$

η - asymmetry parameter

Zero field splitting (ZFS) Hamiltonian:

General form:

$$H_{ZFS}(S) = \sqrt{\frac{2}{3}}D\sum_{m=-2}^{2}(-1)^m V_{-m}^2 T_m^2(S) = \sum_{m=-2}^{2}(-1)^m \tilde{V}_{-m}^2 T_m^2(S)$$

In the principal axis system of the ZFS tensor (P) the quantities V_{-m}^2 and $\tilde{V}_{-m}^2$ have the form:

$$V_0^{2(P)} = 1, \quad V_{\pm 1}^{2(P)} = 0, \quad V_{\pm 2}^{2(P)} = \sqrt{\frac{3}{2}}\frac{E}{D}$$

$$\tilde{V}_0^{2(P)} = \sqrt{\frac{2}{3}}D, \quad \tilde{V}_{\pm 1}^{2(P)} = 0, \quad \tilde{V}_{\pm 2}^{2(P)} = E$$

D, E - axial and rhombic components of the ZFS, respectively

Dipole – dipole Hamiltonian:

General form:

$$H_{DD}(I,S) = a_{DD}^{IS}\sum_{m=-2}^{2}(-1)^m F_{-m}^2 T_m^2(I,S) = \sum_{m=-2}^{2}(-1)^m \tilde{F}_{-m}^2 T_m^2(I,S)$$

$a_{DD}^{IS} = \sqrt{6}\frac{\mu_0}{4\pi}\frac{\gamma_I \gamma_S \hbar^2}{r_{IS}^3}$ r_{IS} inter-spin distance, γ_I and γ_S gyromagnetic factors for spins I and S, respectively

The second rank spin tensor operators $T_m^2(I,S)$ are defined as follows:

$$T_0^2(I,S) = \frac{1}{\sqrt{6}}\left[2I_z S_z - \frac{1}{2}(I_+ S_- + I_- S_+)\right]$$

$$T_{\pm 1}^2(I,S) = \mp\frac{1}{2}[I_z S_\pm + I_\pm S_z]$$

$$T_{\pm 2}^2(I,S) = \frac{1}{2}I_\pm S_\pm$$

In the dipole-dipole frame the functions F^2_{-m} have the form:

$F^{2(DD)}_0 = 1$, $F^{2(DD)}_{\pm 1,\pm 2} = 0$

$\widetilde{F}^{2(DD)}_0 = 1$, $\widetilde{F}^{2(DD)}_{\pm 1,\pm 2} = 0$

Zeeman Hamiltonian:
in the laboratory frame (L)

$$H^{(L)}_Z(P) = \gamma_P B_0 P_z \qquad P = I, S$$

γ_P - gyromagnetic factor for spin P

Transformation rules:
The relationship between tensor functions B^l_{-m} ($B^l_{-m} = A^l_{-m}, \widetilde{A}^l_{-m}, V^l_{-m}, \widetilde{V}^l_{-m}, F^l_{-m}, \widetilde{F}^l_{-m}$) expressed in reference frames (P_1) and (P_2) is:

$$B^{l(P_2)}_{-m} = \sum_{k=-l}^{l} B^{l(P_1)}_k D^l_{k,-m}(\Omega_{P_1 P_2})$$

the Wigner rotation matrices $D^l_{k,-m}(\Omega_{P_1 P_2})$ with the Euler angles $\Omega_{P_1 P_2}$ describe the relative orientation of the frames

Redfield relaxation formula:

$$\frac{d\rho_{\alpha\alpha'}(t)}{dt} = -i\omega_{\alpha\alpha'}\rho_{\alpha\alpha'}(t) + \sum_{\substack{\beta\beta' \\ \omega_{\alpha\alpha'}=\omega_{\beta\beta'}}} \Gamma_{\alpha\alpha'\beta\beta'} \rho_{\beta\beta'}(t)$$

$\Gamma_{\alpha\alpha'\beta\beta'} = R_{\alpha\alpha'\beta\beta'} - iL_{\alpha\alpha'\beta\beta'}$

$= \mathfrak{J}_{\alpha\beta\alpha'\beta'}(\omega_{\alpha\beta}) + \mathfrak{J}_{\alpha\beta\alpha'\beta'}(\omega_{\beta'\alpha'}) - \delta_{\alpha'\beta'}\sum_\gamma \mathfrak{J}_{\alpha\gamma\beta\gamma}(\omega_{\gamma\beta}) - \delta_{\alpha\beta}\sum_\gamma \mathfrak{J}_{\beta'\gamma\alpha'\gamma}(\omega_{\beta'\gamma})$

$$\Im_{\alpha\gamma\beta\gamma}(\omega) = \int_0^\infty \langle\langle\alpha|H_{IL}(t)|\gamma\rangle\langle\beta|H_{IL}(t-\tau)|\gamma\rangle\rangle \exp(-i\omega\tau)d\tau$$

$|\alpha\rangle, |\beta\rangle, |\gamma\rangle$ - eigenstates of the spin system

Some useful values...

- ✓ typical amplitudes of ZFS, Δ_{ZFS}, for gadolinium paramagnetic complexes range between $0.02 - 0.05 cm^{-1}$ ($600 - 1500 MHz$)
- ✓ typical distortional correlation times, τ_D, for gadolinium paramagnetic complexes are 10-50ps
- ✓ the product $\Delta_{ZFS}\tau_D$ are:

    $\Delta_{ZFS}\tau_D \cong 0.04$ ($\Delta_{ZFS} = 0.02 cm^{-1}$, $\tau_D = 10 ps$), $\Delta_{ZFS}\tau_D \cong 0.5$ ($\Delta_{ZFS} = 0.05 cm^{-1}$, $\tau_D = 50 ps$); Δ_{ZFS} is expressed in angular frequency units
- ✓ rotational correlation time for small gadolinium complexes are rather not shorter than 100ps and, in consequence, the Redfied condition must be treated with caution: $\Delta_{ZFS}\tau_R \cong 1$ ($\Delta_{ZFS} = 0.05 cm^{-1}$, $\tau_R = 100 ps$); Δ_{ZFS} is expressed in angular frequency units
- ✓ quadrupole couplings in some selected compounds take the value: LaF$_3$ (^{139}La) $a_Q \cong 15 MHz$ ($0.0005 cm^{-1}$), paranitrotoluene (^{14}N) $a_Q \cong 1.35 MHz$, trinitrotoluene (^{14}N) $a_Q \cong 1.0 MHz$ ($0.000033 cm^{-1}$), urea (^{14}N) $a_Q \cong 3.45 MHz$, urotropine (^{14}N) $a_Q \cong 4.5 MHz$
- ✓ exchange motion (jump diffusion) in some selected crystal lattices is of the order: fluorine spin dynamics in BaF$_2$: $\tau_F \approx 10^{-5} - 10^{-6} s$ (650-700K), in LaF$_3$: $\tau_F \approx 5*10^{-9} - 10^{-9} s$ (750-850K)
- ✓ static ZFS of $1 cm^{-1}$ corresponds to electron spin Zeeman splitting at c.a. 1T

List of selected symbols

$\vec{B}_0$	external magnetic field
ω_I, ω_S	Larmor frequencies for spins I, S, respectively
γ_I, γ_S	gyromagnetic factors of spins I, S, respectively
T_1	spin-lattice (longitudinal) relaxation time
T_2	spin-spin (transversal) relaxation time
R_{1I}, R_{1S}	spin-lattice relaxation rate for spins I, S, respectively
R_{2I}, R_{2S}	spin-spin relaxation rate for spins I, S, respectively
L	lattice
H	total Hamiltonian
H_I	spin Hamiltonian
H_L	lattice Hamiltonian
H_{IL}	Hamiltonian of the spin-lattice coupling
ω_{IL}	amplitude of H_{IL}, expressed in angular frequency units
O	operator
$|O) \equiv \hat{\hat{O}}$	Liouville space representation of the operator O
$\hat{\hat{L}}$	Liouville operator
$\hat{\hat{L}}_I$	Liouville operator for spin I

List of selected symbols

$\hat{\hat{L}}_L$	lattice Liouvillian
$\hat{\hat{L}}_{IL}$	spin-lattice Liouvillian
$\hat{\hat{R}}$	relaxation operator (superoperator)
$\hat{\hat{\Gamma}}$	dynamic operator (superoperator)
$\rho'(t)$	density operator $\rho(t)$ in the interaction representation
$H'_{IL}(t)$	interaction representation of the spin-lattice Hamiltonian H_{IL}
$\rho(t)$	density operator
$\rho_L(t)$	lattice density operator
$\rho_I(t)$	density operator of spin I
ρ_L^{eq}	equilibrium density operator for the lattice
$R_{\alpha\alpha'\beta\beta'}$	relaxation matrix elements (real part of $\Gamma_{\alpha\alpha'\beta\beta'}$)
$L_{\alpha\alpha'\beta\beta'}$	dynamic frequency shift elements (imaginary part of $\Gamma_{\alpha\alpha'\beta\beta'}$)
I_m^l	spin tensor operators
$F_{-m}^l(t)$	classical lattice functions corresponding to I_m^l
$C_{mm'}^l(\tau)$	lattice correlation function
$J_{mm'}^l(\omega)$	spectral density (Fourier transform of $C_{mm'}^l(\tau)$)
T_{-m}^l	quantum-mechanical lattice functions
$G_{m,m'}(\tau)$	quantum-mechanical correlation function
$K_{m,m'}(\omega)$	quantum-mechanical spectral density (Fourier transform of $G_{m,m'}(\tau)$)
$|\Sigma,\sigma)$	orthonormal, irreducible tensor operators
I_σ^Σ	standard irreducible spin operators
T_m^2	components of the second rank spin tensor operator
$H_{ZFS}(S)$	zero field splitting (ZFS) Hamiltonian of spin S
$V_{-m}^2, \tilde{V}_{-m}^2$	spatial functions of the ZFS Hamiltonian
D, E	axial and rhombic components of the ZFS tensor, respectively

$H_{ZFS}^{S(P_S)}(S)$	permanent (static) ZFS Hamiltonian expressed in its principal axis system (P_S)
$\tilde{V}_{-m}^{2S(P_S)}$	spatial functions of the permanent ZFS Hamiltonian expressed in the (P_S) frame
$H_{ZFS}^{S(L)}(S)$	permanent (static) ZFS Hamiltonian expressed in the laboratory frame (L)
$\tilde{V}_{-m}^{2S(L)}$	spatial functions of the permanent ZFS Hamiltonian expressed in the (L) frame
D_S, E_S	axial and rhombic components of the permanent (static) ZFS, respectively
$\Omega_{P_S L}$	Euler angles describing the relative orientation of the (P_S) and (L) frames
$H_{ZFS}^{T(P_T)}(S)$	transient ZFS Hamiltonian expressed in its principal axis system (P_T)
$H_{ZFS}^{T(P_S)}(S)$	transient ZFS Hamiltonian expressed in the principal axis system of the permanent ZFS tensor (P_S)
$\tilde{V}_{-m}^{2T(P_S)}$	spatial functions of the transient ZFS Hamiltonian expressed in the (P_S) frame
$H_{ZFS}^{T(L)}(S)$	transient ZFS Hamiltonian expressed in the laboratory frame
$\tilde{V}_{-m}^{T(L)}$	spatial functions of the transient ZFS Hamiltonian expressed in the laboratory frame
D_T, E_T	axial and rhombic components of the transient ZFS, respectively
$\Omega_{P_T P_S}$	Euler angles describing the relative orientation of the (P_S) and (P_T) frames
$\Omega_{P_T L}$	Euler angles describing the relative orientation of the (P_T) and (L) frames
Δ_S, Δ_T	amplitude of the static and transient ZFS, respectively

List of selected symbols

$H_Q(S)$	quadrupolar Hamiltonian of spin S
a_Q	quadrupolar coupling constant
η	asymmetry parameter
$A^2_{-m}, \tilde{A}^2_{-m}$	spatial functions of the quadrupolar Hamiltonian
$H_Q^{(P)}(S)$	quadrupolar Hamiltonian expressed in its principal axis system (P)
$H_Q^{(L)}(S)$	quadrupolar Hamiltonian expressed in the laboratory frame (L)
$\tilde{A}^{2(L)}_{-m}(t)$	$\tilde{A}^2_{-m}$ expressed in the laboratory frame
Ω_{PL}	Euler angles describing the relative orientation of the (P) and (L) frames
$H_Q^S(S)$	static (averaged) quadrupolar Hamiltonian
$H_Q^T(S)$	transient quadrupolar Hamiltonian
$\Delta_{S(Q)}, \Delta_{T(Q)}$	amplitude of the averaged (static) and transient quadrupolar couplings, respectively
$H_{DD}(I,S)$	dipole–dipole Hamiltonian for spins I and S
a_{DD}^{IS}	dipole-dipole constant for spins I and S
r_{IS}	distance between spins I and S
$\tilde{F}^2_{-m}$	spatial functions of the dipole-dipole Hamiltonian
Ω_{DDL}	Euler angles describing the orientation of the dipole-dipole axis in the laboratory frame
Ω_{IS}^L	Euler angles describing the orientation of the dipole-dipole axis in the laboratory frame in the presence of translational motion
$H_Z^{(L)}$	Zeeman Hamiltonian expressed in the laboratory frame
H_0	main, unperturbed Hamiltonian
$H_1(t)$	perturbing Hamiltonian
$T^{1(DD)(L)}_{-n}$	tensor operators of the composite lattice
$K^{DD}_{1,1}(-\omega_I)$	dipole–dipole spectral density

$\hat{\hat{L}}_{ZFS}^{S}(S)$	permanent (static) ZFS Liouvillian
$\hat{\hat{L}}_{ZFS}^{T}(S)$	transient ZFS Liouvillian
$\hat{\hat{L}}_{Z}$	Zeeman Liouville operator
$\hat{\hat{L}}_{R}$	Liouville operator for rotational motion
$\hat{\hat{L}}_{D}$	Liouville operator for distortional motion
$\hat{\hat{L}}_{Diff}$	Liouville operator for translational diffusion
ω_D, ω_E	D_S, E_S in angular frequency units
ω_{ZFS}^{S}	amplitude of the permanent ZFS Hamiltonian in angular frequency units
ω_{Q}^{S}	amplitude of the averaged quadrupolar Hamiltonian in angular frequency units
ω_{ZFS}^{T}	amplitude of the transient ZFS Hamiltonian in angular frequency units
ω_{Q}^{T}	amplitude of the transient quadrupolar Hamiltonian in angular frequency units
ω_{DD}^{IS}	amplitude of the I-S dipole-dipole Hamiltonian in angular frequency units
τ_c	correlation time
τ_R	rotational correlation time for second order Wigner matrices
τ_D	distortional correlation time for second order Wigner matrices
τ_Q	quadrupole distortional correlation time for second order Wigner matrices
τ_{IS}	dipole-dipole correlation time due to exchange motion of spins I and S
H^{in}	Hamiltonian determining the unitial state of a spin system
$J_m^S(\omega)$	spectral density of spin S

List of selected symbols

$J_m^{S(S)}(\omega)$	static (permanent) ZFS spectral density of spin S
$J_m^{S(T)}(\omega)$	transient ZFS spectral density of spin S
$L(\omega)$	lineshape function
$D_{K,M}^L$	Wigner rotation matrices representing reorientational motion
$D_{B,C}^A$	Wigner rotation matrices representing distortional motion
$\hat{D}_{tr}$	translational diffusion tensor
D_{Diff}	translational diffusion coefficient
$C_{Diff}(\tau)$	correlation function for translational diffusion
d	distance of closest approach of spins I and S
$\hat{D}_{rot}$	rotational diffusion tensor
$U(\vec{r})$	interaction potential
$V(\vec{r})$	interaction potential in thermal energy units
H_V	Hamiltonian describing a vibrating molecule coupled to a bath
T_{1V}, T_{2V}	vibrational relaxation times
q_i	normal mode
q_i^0	equilibrium value of the normal mode q_i
$\rho_V(t)$	vibrational density operator
ω_V	characteristic frequency of harmonic motion
$h_k(q)$	ZFS tensor functions related to normal coordinates

Index

A

amplitude: spin-lattice interaction	28
amplitude: zero field splitting	64, 67, 117, 247, 266, 276
amplitude: quadrupole interaction	115, 192, 195, 266
angular momentum	8
angular momentum operator(s)	41, 46, 85
approximation: secular	32, 33, 37, 77, 92, 117, 124, 141, 183, 192, 254, 255, 257, 286, 297
approximation: high temperature	91, 214, 349
asymmetry parameter	55, 131, 194, 220, 221

B

BaF_2	189, 190
Bloch equations	12, 42
BPP formula	76

C

Coherences	33, 40, 42, 46, 76, 78, 82, 111, 141, 148, 150. 162, 165, 166, 170, 177, 182, 193, 212, 297, 304, 306, 310, 356

condition: Redfield condition	24, 26, 28, 31, 97, 117, 163, 182, 192, 229, 235, 236, 242, 248, 251, 255, 256, 257, 259, 261, 290
correlation function	8, 14, 15, 36, 39, 40, 45, 46, 66, 67, 68, 75, 83, 84, 87, 90, 91, 92, 93, 95, 103, 106, 148, 159, 187, 188, 189 190, 199, 236, 240, 241, 242, 251, 252, 253, 258, 299, 314, 315, 317, 318, 319, 320, 323, 327, 329, 331, 338, 339, 340, 353, 354, 355
correlation function: quantum-mechanical	45, 83, 84, 106, 159, 318
correlation function: rotational	91, 106, 251, 328
correlation function: translational	329, 331
correlation time	22, 24, 28, 66, 68, 75, 96, 111, 115, 140, 141, 152, 163, 266, 276, 328
correlation time: distortional	111, 115, 140, 141, 163, 276
correlation time: rotational	66, 68, 75, 96, 136, 152, 266, 276, 328
coupling: spin-lattice coupling	11, 23, 26, 27, 28, 29, 35, 39, 42, 44, 53, 58, 73, 84, 88, 281

D

density matrix(ces)	13, 24, 25, 32, 33, 34, 41, 46, 76, 77, 78, 92, 212, 215, 249, 294, 299, 304, 305, 306, 350
density operator(s)	8, 13, 17, 18, 19, 20, 27, 37, 41, 47, 90, 116, 211, 215, 222, 227, 249, 306, 308, 348, 351
density operator: equilibrium	37, 90, 94
density operator: initial	24, 212, 215, 216, 222, 227, 306, 308
diffusion coefficient	15, 319, 337, 340
diffusion operator	88, 273
diffusion: rotational	64, 80, 90, 258, 273, 314

diffusion tensors	14, 15
diffusion: translational	11, 14, 282, 314, 315, 316, 317, 318, 319, 325, 326, 327, 329, 330, 333, 335, 338, 340
distortion(s)	11, 64
distortional degrees of freedom	269
distortional states	269, 273, 283
dipole-dipole interaction (coupling)	9, 10, 11, 55, 72
dipole-dipole Hamiltonian	57, 58, 59
dipole-dipole axis	61, 62, 63, 68
distance of closest approach	318, 319, 337
distribution function: Boltzmann	354
distribution function: radial	338, 39, 340
dynamic frequency shift	35
dynamic(s): Redfield dynamic matrix	33, 35
dynamic(s): vibrational	346, 347, 348, 350, 351, 354, 359

E

energy level structure	22, 32, 43, 75, 85, 96, 100, 105, 114, 124, 128, 131, 138, 151, 158, 164, 168, 171, 183, 188, 193, 200, 235, 246, 256, 267, 292, 296, 310, 328, 352
energy level crossing(s)	183, 220, 310
equation: diffusion equation	14, 64, 66, 273, 338
equation: Redfield relaxation equation	29, 33, 38, 41, 42, 44
equation: Liouville -von Neumann equation	13, 17, 18, 22, 23, 26, 28, 38, 47, 210, 211, 225, 249, 259, 264
equilibrium state	10, 11, 79
equilibrium: thermal	10, 38, 210
evolution: time evolution	17, 22, 25, 26, 27, 32, 33, 75, 82, 112, 116, 126, 141, 143, 197, 210, 211, 212, 114, 216, 222, 225, 294, 303, 304, 308, 310, 349

Index

exchange motion	11, 181, 191, 202, 235, 238, 240, 243, 290, 346
expansion coefficients	13, 41, 148, 182, 253, 268, 274, 282, 322

F

fluorine relaxation	189, 190, 193, 194, 198, 200, 201
forces: intermolecular	14, 338, 340, 348
Fourier transform	15, 35, 36, 40, 87, 91, 103, 175, 266, 355
formula: Redfield formula	81, 100, 102, 131, 294, 297, 298

G

general theory (approach, treatment)	18, 235, 259, 265, 275, 283, 284, 285, 286, 329, 330, 336, 337, 352
gyromagnetic factor	10, 58, 73, 215, 227

H

Hamiltonian: main	23, 31, 43, 74, 75, 77, 100, 110, 116, 117, 118, 124, 127, 128, 129, 130, 131, 134, 138, 139, 151, 165, 166, 173, 176, 192, 244, 246, 247, 248, 249, 250, 251, 255, 258, 260, 267, 292, 293, 296, 304, 307, 324
Hamiltonian: unperturbed	73, 124

I

interaction(s): time independent	210, 211, 293
interaction representation	23, 24, 31, 47, 116, 249

375

J
jump diffusion   153, 187, 188, 189, 190, 191, 200, 204, 235, 238, 204, 235, 238, 240, 243, 260, 290, 292, 298, 314, 315, 318

K
Kramers degeneracy (doublets)   138, 160, 161, 163, 165, 168, 171

L
LaF$_3$ 197, 198, 200, 201, 204, 228, 229

Larmor frequency(ies)   10, 12, 73, 77, 86, 160, 162, 166, 171, 281, 356

lattice   10, 11, 12, 17, 18, 19, 22, 23, 25, 27, 31, 38, 53, 73, 79, 83, 84, 86, 87, 91, 93, 94, 96, 97, 101, 159, 249, 255, 347350

lattice: classical   38, 53, 73, 350

lattice: crystal   153, 186, 189, 191, 201, 243, 292, 318

lattice: composite   79, 83, 84, 86, 87, 91, 93, 94, 96, 97, 101, 159, 249, 255, 347

lattice tensor component(s)   54

limit: Redfield limit   24, 118, 123, 182, 205, 266

lineshape   204, 265, 266, 268, 274, 275, 276, 283, 300, 301

Liouville space formalisms   18, 20

M
magnetic spin quantum number   9, 76, 170, 267

moderately slow rotation   258

multipole representation (treatment)   17, 40, 42, 43, 44, 45, 83, 96, 97, 105, 264

Index

multipoles: state multipoles	41, 44, 45
multiquantum order	33

N

normal modes	40, 270, 348, 353, 355

O

observable(s)	13, 40, 143, 144, 148, 211, 212, 215, 218, 222, 299, 303, 304, 305, 306, 307, 309, 310
operator formalism	19, 22, 24, 36, 37, 39, 40, 45, 46, 51, 53, 79, 93, 159, 205, 283
operator: Liouville operator	29, 36, 37, 40, 45, 53, 72, 79, 88, 106, 159, 175, 178, 187, 238, 243, 252, 266, 268, 270, 272, 273, 315, 323, 331, 332, 348
orthogonality properties	39, 102, 142, 146

P

paramagnetic: relaxation enhancement	183, 186
perturbation expansion	24
perturbing Hamiltonian (interaction)	22, 23, 26, 61, 73, 75, 81, 100, 110, 116, 124, 134, 135, 139, 142, 145, 146, 236, 244, 246, 247, 259, 260, 292, 297, 321
polarization transfer	201, 202, 212, 213, 218, 220, 221, 222, 225, 227, 228, 229, 230, 292, 320
potential of mean force	338
profiles: relaxation (dispersion) profiles	137, 178, 179, 180, 189, 190, 201, 205, 229, 284, 285, 325, 336, 37, 358
projection vector(s)	161, 166, 167, 168, 177, 268, 274, 281

pseudorotational model   64, 66, 67, 136, 146, 236

Q

quadrupolar coupling(interaction)   10, 44, 53, 57, 60
quadrupolar interaction: averaged (static)   60, 108
quadrupolar interaction: transient   66, 109
quadrupolar Hamiltonian   54, 55, 56, 60
quadrupolar coupling constant   54, 220, 221, 227

R

relaxation: electron spin   81, 83, 85, 91, 98, 99, 100, 104, 105, 106, 117, 118, 123, 125, 128, 131, 134, 138, 142, 144, 148, 152, 158, 161, 168, 172, 175, 181, 183, 236, 239, 242, 245, 248, 251, 253, 256, 258, 266, 276, 280, 291, 322, 326, 336, 342, 245, 351, 355, 357

relaxation: enhancement   51, 183, 186, 228
relaxation: multiexponential   23, 137, 149, 195
relaxation: nuclear spin   97, 104, 105, 118, 138, 158, 159, 163, 165, 167, 170, 172, 174, 177, 182, 230, 239, 241, 244, 248, 251, 253, 287, 322, 324, 331, 336, 340, 341, 358

relaxation: Redfield theory   24, 40, 116, 123, 125, 181, 186, 205, 241, 156, 323, 331, 336, 337, 340, 346, 352

relaxation: quadrupolar spin   118, 122, 123, 129, 144, 188, 191, 192, 193, 194, 200, 202, 204, 205, 209, 29, 239, 243, 246, 291

relaxation: spin-lattice   10, 12, 40, 76, 77, 78, 80, 86, 95, 98, 99, 103, 105, 112, 114, 126, 140, 143, 148, 151, 153, 159, 164, 168, 172, 176, 179, 180, 186, 199, 242, 268, 280, 281, 290, 300, 304, 309, 337

relaxation: spin-spin (transversal)	10, 12, 22, 43, 80, 83, 95, 99, 103, 112, 115, 127, 140, 141, 144, 146, 150, 164, 168, 169, 177, 310
relaxation: SBM theory	83, 84, 95, 104, 129, 159, 178, 179, 189, 322
relaxation: time	10, 11, 13, 26, 46, 48, 71, 77, 80, 83, 103, 112, 119, 120, 152, 155, 156, 163, 185, 192, 234, 263, 283, 313, 343, 348, 354, 358
relaxation: WBR theory	17, 31, 74, 97, 98, 105, 110
rotational states	73, 283, 333

S

slowly rotating systems	118, 122, 123, 128, 137, 153, 158, 159, 162, 179, 180, 181, 244, 278, 322, 325, 327, 340
space: Hilbert space	19, 20, 21, 45, 46, 269
space: Liouville space	18, 19, 20, 21, 22, 23, 24, 33, 37, 38, 42, 46, 92, 112, 113, 176, 253, 269, 293, 333
spectrum, spectra: ESR	265, 266, 267, 268, 274, 275, 276, 278, 279, 280, 281, 286
spectrum, spectra: quadrupolar	265, 286, 296
spectral density (densities)	15, 43, 44, 46, 66, 75, 80, 82, 83, 85, 86, 87, 88, 89, 91, 93, 94, 95, 96, 97, 98, 103, 105, 106, 107, 111, 112, 113, 115, 134, 139, 142, 146, 150, 160, 161, 168, 170, 172, 182, 187, 189, 195, 199, 200, 236, 237, 238, 241, 248, 252, 256, 261, 265, 286, 290, 294, 315, 317, 320, 324, 336, 340, 341, 343, 346, 350, 352, 355, 357
spectral density: quantum-mechanical	44, 85, 97, 160, 187, 241, 315, 329
spin operators: standard, irreducible	41
spin tensor components	44, 46, 84, 167, 177, 250, 253, 295, 327, 334
spin tensor operator(s): second rank	41, 74, 129, 135, 269

superoperator: relaxation superoperator	20, 40, 91, 106, 174, 191, 242, 252, 266, 269, 275, 331, 348, 352
superoperator: dynamic superoperator	37, 38
symbols: 3j, 6j	41, 43, 85, 95, 106, 252, 272

T

Taylor expansion	215, 348, 351, 153, 358
tensor operators: orthonormal, irreducible	269
theory: perturbation	17, 22, 24, 28, 72, 137, 145, 182, 192, 196, 210, 234, 244, 246, 264, 265, 284, 286, 292, 293, 320, 348, 352
theory: slow motion	285

U

unified treatment	122

V

vibrational energy levels	347, 349, 352
vibrational relaxation time(s)	348, 354, 358
vibrational states	349
vibrations: quantum	66, 345, 351, 352
vibrations: molecular	345, 347, 350
vibrations: lattice vibrations	11, 188, 204, 240, 259, 298

W

Wigner rotation matrix, matrices	55, 66, 74, 75, 85, 103, 105, 128, 129, 142, 146, 173, 223, 241, 270, 272, 328, 330
Wigner-Eckart theorem	265, 270, 271

Z

zero field splitting	9, 57, 60
zero field splitting: permanent, static	62, 63, 64, 67, 85, 89, 97, 100, 101, 105, 114, 116, 122, 124, 128, 137, 144, `147, 151, 158, 167, 173, 174, 178, 182, 230, 236, 237, 240, 244, 252, 260, 266, 272, 274, 276, 279, 281, 315, 321, 328, 350, 352
zero field splitting: transient	61, 62, 63, 64, 67, 81, 86, 87, 89, 91, 96, 106, 115, 117, 123, 134, 136, 139, 141, 145, 146, 149, 153, 236, 242, 247, 249, 251, 255, 260, 266, 270, 276, 284, 294, 318, 321, 323, 337, 351, 352, 353, 358
zero field splitting: rhombic, rhombicity	57, 63, 114, 115, 138, 144, 145, 148, 151, 160, 164, 167, 170, 177, 337, 351
Zeeman Hamiltonian	45, 58, 59, 88, 128, 129, 165, 166, 244

www.ingramcontent.com/pod-product-compliance
Ingram Content Group UK Ltd.
Pitfield, Milton Keynes, MK11 3LW, UK
UKHW051248180426
11947UKWH00020B/1602